Emerging Sustainable and Renewable Composites

This edited volume presents a comprehensive discussion of emerging sustainable and renewable composites from tropical fibres and provides an in-depth analysis of their prospective applications as replacements for conventional petroleum-based packaging and the challenges regarding this.

Readers will gain a comprehensive understanding of the development and characterization of sustainable and renewable composites from fibres such as sugar palm, kenaf, sisal, curau, and coir. They will also learn about new potential materials from such fibres and their potential use in various nano-electronics applications. Each chapter provides recent insight from some of the field's most prominent industry and academic professionals. Chapter contributors present valuable case studies and describe related environmental issues, environmental advantages, and challenges. Topics include biodegradability, tensile and other physical properties, and applications. Consequently, readers can apply this knowledge to the further development of sustainable and renewable composites toward their global use in place of petroleum-based materials and in new electronics products.

This book is an invaluable and accessible guide for researchers and postgraduate students of composites engineering and nanotechnology who wish to learn more about composites from tropical fibres and their applications. The practical information will benefit those who wish to advance research in this field and promote the adoption of these materials in areas including packaging and nanoelectronics.

A. Atiqah is a Senior Lecturer/Research Fellow at the Institute of Microengineering and Nanoelectronics, National University of Malaysia. She earned a BSc and MSc in materials engineering at the International Islamic University Malaysia in 2011 and 2014, respectively, and a PhD in biocomposites technology and design at University of Putra Malaysia in 2018. She has been recognized as a Top 2% Scientist Worldwide in her field by Stanford University.

S. M. Sapuan is Professor (Grade A) of Composite Materials and Head of Advanced Engineering Materials and Composite Research Centre, University of Putra Malaysia. He earned a BEng in mechanical engineering at the University of Newcastle, Australia; an MSc in engineering design at Loughborough University, UK; and a PhD in materials engineering at De

Montfort University, UK. He is a Fellow of the World Academy of Sciences, Society of Automotive Engineers International, and Academy of Sciences Malaysia. He has received the SAE Subir Chowdhury Medal of Quality Leadership, International Society of Bionic Engineering Outstanding Contribution Award, the World Academy of Sciences Award in Engineering Sciences, and William Johnson International Gold Medal.

R. A. Ilyas is a Senior Lecturer at the University of Technology Malaysia and holds Fellowships with the Institute of Advanced Materials, Sweden and the International Society for Development and Sustainability, Japan. He is a member of the Royal Society of Chemistry (UK), the Institute of Chemical Engineers (UK), and the Chair of Science Outreach for Young Scientists Network – Academy of Sciences Malaysia. He has received several awards, including the MVP PhD Gold Medal Award, Kreso Glavac Special Award (Republic of Croatia; MTE2022), and Outstanding Reviewer recognition. He is listed as a Top 2% Scientist Worldwide for polymer and materials engineering by Stanford University.

Z. M. A. Ainun is a Senior Research Officer/Head of Pulp and Paper and Pollution Control Program at the Laboratory of Biopolymer and Derivatives, Institute of Tropical Forestry and Forest Products, University of Putra Malaysia, Serdang, Selangor, Malaysia. She earned a BS and an MS in wood, paper, and coating technology (Hons) at the Science University of Malaysia in 1998 and 2000, respectively, and a PhD in materials science at the National University of Malaysia in 2006.

K. Z. Hazrati is a Senior Lecturer at the Industrial Automation Control Section, German-Malaysian Institute. She earned a BS at the Technical University of Malaysia, Malacca in 2011, an MS at the University of Putra Malaysia in 2014, and a PhD at the University of Putra Malaysia in 2022.

Emerging Sustainable and Renewable Composites

From Packaging to Electronics

Edited by
A. Atiqah, S. M. Sapuan, R. A. Ilyas,
Z. M. A. Ainun, and K. Z. Hazrati

CRC Press
Taylor & Francis Group
Boca Raton London New York

CRC Press is an imprint of the
Taylor & Francis Group, an **informa** business

Designed cover image: Cover art by A. Atiqah et al., "Sustainable composites", © 2023

First edition published 2025
by CRC Press
2385 NW Executive Center Drive, Suite 320, Boca Raton FL 33431

and by CRC Press
4 Park Square, Milton Park, Abingdon, Oxon, OX14 4RN
CRC Press is an imprint of Taylor & Francis Group, LLC

ISBN: 978-1-032-52752-9 (hbk)
ISBN: 978-1-032-52753-6 (pbk)
ISBN: 978-1-003-40821-5 (ebk)

DOI: 10.1201/9781003408215

Typeset in Palatino
by Deanta Global Publishing Services, Chennai, India

Contents

Contributors

Nur Hanani Zainal Abedin
Faculty of Food Science and
    Technology
University of Putra Malaysia
Selangor, Malaysia

Sharmiza Adnan
Forest Research Institute Malaysia
Selangor, Malaysia

S. R. Ahmad
Mechanical Section
Malaysian Spanish Institute
University of Kuala Lumpur
Kedah, Malaysia

H. A. Aisyah
Department of Mechanical and
    Manufacturing Engineering
University of Putra Malaysia
Selangor, Malaysia

A. Atiqah
Institute of Microengineering and
    Nanoelectronics
National University of Malaysia
Selangor, Malaysia

M. N. M. Azlin
Clothing Innovation and Textile
    Research Association
Department of Textile Technology
MARA Technological University
Selangor, Malaysia
and
Negeri Sembilan Branch
Kuala Pilah, Malaysia

Nik Athirah Azzra
Institute of Microengineering and
    Nanoelectronics
National University of Malaysia
Selangor, Malaysia

M. K. Nur Fitrah
Green Biopolymer, Coatings and
    Packaging Cluster
School of Industrial Technology
Science University of Malaysia
Pulau Pinang, Malaysia

K. Z. Hafila
German-Malaysian Institute
Jalan Ilmiah, Taman University
Selangor, Malaysia

M. F. Hamid
School of Innovation, Science and
    Technology
TAJ International College
Perak, Malaysia

M. M. Harussani
Department of Transdisciplinary
    Science and Engineering
School of Environment and
    Society
Tokyo Institute of Technology
Tokyo, Japan

K. Z. Hazrati
German-Malaysian Institute
Jalan Ilmiah, Taman University
Selangor, Malaysia

M. M. Huzaifa
Bioresource Technology Division
School of Industrial Technology
Science University of Malaysia
Pulau Pinang, Malaysia

Madeha Jabbar
School of Engineering and
    Technology
Department of Textile
    Engineering
National Textile University
Faisalabad, Punjab, Pakistan

Mariam Jabbar
School of Engineering and
    Technology
Department of Textile
    Engineering
National Textile University
Faisalabad, Punjab, Pakistan

A. Jalar
Department of Applied Physics
    Faculty of Science and
    Technology
National University of Malaysia
Selangor, Malaysia

Latifah Jasmani
Paper and Fibre Testing
    Laboratory
Laboratory of Biopolymer and
    Derivatives
Institute of Tropical Forestry and
    Forest Products
University of Putra Malaysia
Selangor, Malaysia

A. Khalina
Laboratory of Biocomposite
    Technology
Institute of Tropical Forestry and
    Forest Products
University of Putra Malaysia
Selangor, Malaysia

Mohd Ashadie Kusno
Paper and Fibre Testing Laboratory
Laboratory of Biopolymer and
    Derivatives
Institute of Tropical Forestry and
    Forest Products
University of Putra Malaysia
Selangor, Malaysia

Z. Leman
Advanced Engineering Materials
    and Composites Research Centre
Department of Mechanical and
    Manufacturing Engineering
University of Putra Malaysia
Selangor, Malaysia

Muhd Ridzuan bin Mansor
Faculty of Mechanical Engineering
Technical University of Malaysia
Melaka, Malaysia

Ainun Zuriyati Mohamed
Paper and Fibre Testing
    Laboratory
Laboratory of Biopolymer and
    Derivatives
Institute of Tropical Forestry and
    Forest Products
University of Putra Malaysia
Selangor, Malaysia

N. M. Nurazzi
Green Biopolymer, Coatings and
    Packaging Cluster
School of Industrial Technology
Science University of Malaysia
Pulau Pinang, Malaysia

Noor Fadhilah Rahmat
Department of Physics
Centre for Defence Foundation
    Studies
National Defence University of
    Malaysia
Kuala Lumpur, Malaysia

Mohd Shaiful Sajab
Department of Chemical and
    Process Engineering
Faculty of Engineering and Built
    Environment
National University of Malaysia
Kuala Lumpur, Malaysia

S. S. M. Saleh
Faculty of Chemical Engineering
    Technology
University of Malaysia
Kuala Lumpur, Malaysia

S. M. Sapuan
Advanced Engineering
    Materials and Composites
    Research Centre
Department of Mechanical and
    Manufacturing Engineering
University of Putra Malaysia
Selangor, Malaysia

N. H. Sari
Department of Mechanical
    Engineering
Faculty of Engineering
University of Mataram
Mataram, Indonesia

Mohd Adrinata bin Shaharuzaman
Faculty of Mechanical
    Engineering
Technical University of Malaysia
Melaka, Malaysia

Khubab Shaker
Department of Materials
School of Engineering and
    Technology
National Textile University
Faisalabad, Punjab, Pakistan

S. F. K. Sherwani
Advanced Engineering
    Materials and Composites
    Research Centre
Department of Mechanical
    and Manufacturing
    Engineering
University of Putra Malaysia
Selangor, Malaysia

Amna Siddique
Department of Textile Technology
School of Engineering and
    Technology
National Textile University
Faisalabad, Punjab, Pakistan

M. S. M. Sidik
Mechanical Section
Malaysian Spanish Institute
University of Kuala Lumpur
Kedah, Malaysia

Zakiah Sobri
Paper and Fibre Testing Laboratory
Laboratory of Biopolymer and
    Derivatives
Institute of Tropical Forestry and
    Forest Products
University of Putra Malaysia
Selangor, Malaysia

S. Sujita
Department of Mechanical
    Engineering
Faculty of Engineering
University of Mataram
Mataram, Indonesia

Y. A. Sutaryono
Department of Mechanical
    Engineering
Faculty of Engineering
University of Mataram
Mataram, Indonesia

S. Suteja
Department of Mechanical
    Engineering
Faculty of Engineering
University of Mataram
Mataram, Indonesia

E. Syafri
Politeknik Pertanian Negeri
    Payakumbuh
Koto Tuo, Indonesia

J. Tarique
Advanced Engineering
    Materials and Composites
    Research Centre
Department of Mechanical and
    Manufacturing Engineering
University of Putra Malaysia
Selangor, Malaysia

Yiow Ru Vern
Faculty of Mechanical
    Engineering
Technical University of Malaysia
Melaka, Malaysia

Hanur Meku Yesuf
Key Laboratory of Textile Science
    and Technology
Ministry of Education
College of Textiles
Donghua University
Shanghai, China
and
Ethiopian Institute of Textile and
    Fashion Technology
Bair Dar University
Bahir Dar, Ethiopia

E. S. Zainudin
Advanced Engineering
    Materials and Composites
    Research Centre
Department of Mechanical and
    Manufacturing Engineering
and
Laboratory of Biocomposites
Institute of Tropical Forestry and
    Forest Products
and
Faculty of Engineering
University of Putra Malaysia
Selangor, Malaysia

Acknowledgements

A heartfelt thank-you goes to the members of my editor team, whose dedication and diligence have significantly contributed to the depth and breadth of this study. Your commitment to excellence and your collaborative spirit have been a driving force behind the success of this project.

I extend my appreciation to all of the chapter contributors who generously shared their time, knowledge, and experiences. Your willingness to contribute to this research has enriched its quality and relevance.

To my colleagues and peers, thank you for the stimulating discussions, shared insights, and camaraderie. Your diverse perspectives have added valuable dimensions to the research, making it a more comprehensive and nuanced contribution to the field.

My heartfelt thanks go to my family for their unwavering support, patience, and understanding during the demanding phases of research. Your encouragement has been my source of strength.

Finally, to my friends and loved ones, thank you for your patience, understanding, and encouragement. Your belief in the importance of this book has motivated me to persevere through the challenges.

This research is the result of a collective effort, and I am grateful for the collaborative spirit that has fuelled its completion. Each individual mentioned here has played a crucial role in shaping this endeavour, and I extend my heartfelt appreciation to all.

1

Analysing the Evolution of Sugar Palm Fiber (Arenga pinnata Wurmb. Merr) Polymer Hybrid Composites: A Review

S. F. K. Sherwani, S. M. Sapuan, E. S. Zainudin,
Z. Leman, A. Khalina, and J. Tarique

1.1 Introduction

As a result of pollution produced by non-biodegradable materials such as synthetic fibres, research on environmentally friendly material development has started. To solve this issue, researchers moved from synthetic fibres (non-biodegradable material) to natural fibres (biodegradable material). The major benefits of adopting natural fibres are their renewability, ease of availability, biodegradability, non-toxicity, low specific gravity, high toughness, and better strength [1]. Natural fibres have a low density and a good strength to weight ratio, making them suitable for use as lightweight composite and reinforcing materials. The mechanical characteristics of fibres are influenced by their microstructure and chemical composition, with the fibre cross-sectional area being the greatest variable determining fibre strength [2]. Natural fibres absorb water easily, attributed to the existence of hemicellulose, which gives them hydrophilic characteristics, making them less suitable in interactions with the hydrophobic matrix [3]. Higher cellulose concentration and crystallinity are likely to result in higher fibre strength characteristics, whereas the reverse is the case with lignin [4]. Aside from that, fibre anatomical features change across and within species, influencing density and mechanical characteristics [5]. Other elements that influence the size and quality of natural fibres include environmental circumstances, mode of transportation, storage duration and conditions, and fibre extraction [4, 6]. The review concludes that sugar palm fibre polymer hybrid composites have substantially increased their reach in industrial applications over the last 30 years. They demonstrate good mechanical properties after alkaline treatment of SPF, hybridization with glass fibre, and inclusion with various polymer matrices. In conclusion, the usage of SPF polymer hybrid composites can aid in the future growth of sugar palm as a new industrial crop, reduce reliance on petroleum products, and reduce the negative environmental impact of synthetic polymers

DOI: 10.1201/9781003408215-1

and fibres. However, a lot of research remains to be done on these composites, such as lowering manufacturing costs, increasing availability, and opening new markets such as motorcycle body framing, toys, household items, pharmaceuticals, and electronic packaging to make hybrid composites more common in replacing petroleum-based plastic.

1.2 Sugar Palm Tree and Sugar Palm Fiber

Arenga pinnata (Wurmb.) Merr is the botanical name of the sugar palm tree (SPT). It is a famous multifunctional tree that is primarily found in tropical areas. It is a member of the Palmae family, which includes 181 genera and around 2600 species [7]. SPTs may be found in abundance along rivers and bushes in the rural regions of Bruas, Kampung Gajah, and Parit, Perak; Raub, Pahang; Jasin, Melaka; and Kuala Pilah and Negeri Sembilan (Malaysia) [8]. The sugar palm tree provides one of the most significant natural fibres. Sugar palm (SP) appears to be a good source of natural fibre–reinforcing materials due to its low cost and availability in Malaysia. SP tree products have a higher tensile strength and a longer durability until degradation; the fibre does not require any extraction method and is not affected by high temperatures or moisture absorption when compared with natural fibres such as coir fibre [9–12]. Until 1991, different sections of the sugar palm tree were used to make various basic household products; the root was used for boards, tool handles, water pipes, and musical instruments like drums; fibres from the leaf were used for ropes, filters, and road construction; and mature leaflets were used for wrapping material, fruit baskets, decoration and many more products [13]. Sugar palm fibre (SPF) may be utilized in harsh conditions due to its high strength and resilience properties.

Numerous studies on the usage of SP-reinforced polymer composites have previously demonstrated that fibre surface treatments can improve the physical, mechanical, and other properties of the composites [9, 14–16]. Figure 1.1 shows sugar palm trees and sugar palm fibre. SPF is black in colour, with length up to 1.19 m, and measures 0.5 mm in average diameter; heat is absorbed by SPF up to 150 C [11]. SPF has a hollow (lumen) space and random nodes that separate the fibre into different cells. As the lumen size reduces, the thickness of the secondary cell wall improves; due to this, the strength and Young's modulus also increase [17].

Sugar palm fibre is mostly composed of cellulose, hemicellulose, and lignin, which originated from sugar palm trees. Table 1.1 shows the chemical composition of sugar palm fibres. Cellulose supports fibres with mechanical strength. Hemicellulose molecules are cellulose-bonded hydrogen and serve as a cementing matrix between growing cellulose microfibrils, forming the cellulose-hemicellulose network, which contributes to the main structural part of the fibre cells. Lignin acts as a binding agent, which enhances the cellulose-hemicellulose stiffness and provides strength to the plant [11]. Cellulose fibrils consist of two regions: crystalline and amorphous. Crystalline cellulose is a healthy cellulose chain structure produced by strong intramolecular hydrogen bonds, while amorphous regions formed by an impaired cellulose chain promote the absorption of dyes and resins.

Table 1.2 displays the properties of SPF. The SPF has many benefits that allow it to be used as a biocomposite reinforcing material, e.g. greater heat resistance, low water absorption relative to coir fibre, long life span, strong tensile strength, and better sea water resistance [18]. Most work on sugar palm fibre composite has evaluated mechanical strength, including flexural strength, impact strength, and tensile strength, as SPF can benefit from structural applications.

FIGURE 1.1
Sugar palm tree and sugar palm fibres[4].

TABLE 1.1

Chemical Composition of SPFs [19]

Sample	Cellulose (%)	Hemicellulose (%)	Holocellulose (%)	Lignin (%)	Ash (%)
SPF	43.88	7.24	51.12	33.24	1.01

TABLE 1.2

Properties of SPFs [12]

Properties	Density (g/m³)	Hardness	Water absorption (%)	Oil absorption (%)	Moisture content (%)
SPF	1.2–1.3	87.34–98.08	2.04–7.21	1.90–6.45	2.63–6.51

1.3 SPF-Reinforced Polymer Composites

1.3.1 SPF/Thermoplastic Composites

A greater amount of literature on SPF reinforced with various thermoplastics has been published. The main challenge in these studies was interfacial adhesion between SPF and the thermoplastic matrix. Ismaila et al. [15] developed sugar palm fibre–reinforced polypropylene (PP), and a comparison analysis was done between untreated and treated SPF, indicating that following alkaline and sodium bicarbonate treatment, the water absorption character of the fibre was reduced by 18%. Due to fibre pullout and an increase in holes in the untreated SPF/PP composite, there is poor adhesion between the fibre and matrix. After alkaline and sodium bicarbonate treatments, good adhesive properties were exhibited between SPF and the PP matrix [15]. Atiqah et al. [9] developed SPF-reinforced thermoplastic polyurethane (TPU) composites and found that 2% silane treatment of SPF enhanced the interfacial adhesion between matrix and fibre, resulting in better mechanical properties, particularly tensile and flexural strength. The study also predicted that 40% of SPF had the best optimum load value. The SPF/TPU composite has tensile strength of 17.22 MPa and flexural strength of 13.96 MPa at this percentage [20]. After analyzing SPF-reinforced high-impact polystyrene (HIPS), the researchers concluded that low matrix-fibre adhesion is shown by a decrease in tensile strength over 40%. The SPF/HIPS composite has excellent mechanical and thermal properties [21]. In another study, SPF-reinforced polypropylene was produced by injection moulding. The SPF was first treated with silane water, which improved the composite's tensile strength and hydrophilic character of the fibre [22].

1.3.2 SPF/Thermoset Composites

Thermoset polymers like epoxy, unsaturated polyester, and phenol formaldehyde were used as matrices with SPF. SPF reinforced with epoxy was developed for determining its storage modulus by the ageing process. This study showed that the composite with a lower ageing period has a high storage modulus, but this modulus decreases as the temperature rises [23]. By adding 10%, 15%, and 20% of the filament volume fraction, each filament arrangement in epoxy reduces the tensile and flexural strength values compared with pure epoxy resin (0% filament), but the tensile modulus and flexural modulus increase [24]. Sugar palm fibre–reinforced epoxy composite was developed by the hand lay-up process. The study compared the long fibre with chopped fibre–reinforced epoxy composite and concluded that the impact strength of long SPF is higher than that of chopped SPF [25].

1.3.3 Other Matrices in SPF-Reinforced Composites

As demand for green materials increases, biocomposites and fully biodegradable materials are increasingly being developed. Biocomposites are made by combining biodegradable polymers with fibres from lignocellulose. There are two types of biodegradable polymers. The first is known as agro-polymer, based on starch or protein, and the second is known as biodegradable thermoplastic, based on lactic acid. Biodegradable polymer based on a poly(lactic acid) (PLA) is derived from renewable resources. A number of studies on both types of biodegradable polymers combined with SPF have been completed. Sea water treatment and different fibre lengths of SPF in sugar palm fibre–reinforced unsaturated polyester composite were analysed to improve its mechanical properties. It was shown that for sea water–treated SPF, if the fibre length was 5 cm, then the tensile strength value was 18.33 MPa, and the flexural strength was 80.80 MPa [26]. SP yarn/Glass fibre-reinforced unsaturated polyester hybrid composites were developed. Alkaline treatment of this hybrid composite with 1% NaOH for 1 h improved its thermal properties [27]. Two biodegradable materials were used to develop a unique biocomposite, i.e. *ijuk* fibre and PLA as a matrix. After alkaline treatment of *ijuk* fibre with 0.25M NaOH for 30 mins, the mechanical properties of the composite were enhanced. But, this treatment also reduces both the diameter and the density of *ijuk* fibre [28]. In SPF-reinforced plasticized sugar palm starch (SPF/SPS) biocomposites, 10%, 20%, and 30% (weight%) of SPF were used with glycerol to act as a plasticizer for SPS. The moisture absorption of fibre in this composite was decreased, which improved the interfacial adhesion bonding between fibre and plasticizer. For this reason, this biocomposite is thermally stable and shows good mechanical properties [29]. Table 1.3 shows different compounding processes and properties for various SPF/polymer composites.

1.4 Natural Fibre–Reinforced Polymer Composites

Natural fibre reinforced polymer composites (NFRPCs) are composites made of natural fibres and synthetic or biopolymers. NFRPC materials are gaining popularity because of their eco-friendliness, lightweight nature, excellent life-cycle performance, biodegradability, low cost, and excellent mechanical characteristics [39]. NFRPCs are widely used in a variety of engineering applications, and this study area is constantly evolving. However, because of the innate properties of natural fibres, researchers face several obstacles to the development and implementation of NFPRCs [40]. These difficulties include fibre quality, thermal stability, water permeability, and incompatibility with

TABLE 1.3

Compounding Process, Properties of SPF/Polymer Composites

	SPF/Polymer Composite	Compounding Process	Properties/ Outcomes	References
1	SPF (S)/Ramie (R) fibre/epoxy composites	Hand lay-up/compression moulding machine	Among the five-layer SPF and SRSRS hybrid composites, RSRSR hybrid composite has higher *tensile* (52.66 MPa) and *flexural* (80.70 MPa) strength.	[30]
2	SPF/GF/thermoplastic polyurethane (TPU) hybrid composites	Melt-mixing compounding/hot pressing moulding	When compared with untreated SPF, the combination of silane and alkaline treatment produced considerably better results in terms of *density, thickness swelling, water absorption*, and thermal stability.	[31]
3	SP yarn/GF/unsaturated polyester hybrid composites	Yarning process/hot press machine	Dynamic mechanical analyser (DMA) and thermogravimetric analysis (TGA) and thermogravimetric analysis (TGA). The *storage modulus* (E'), *loss modulus* (E''), and *damping factor* (tan δ) were improved after alkaline treatment of SP yarn and hybridization with GF.	[27]
4	SPF/GF/epoxy hybrid composites	Hand lay-up technique	The SPF was benzoylated. The SPF/GF/epoxy hybrid composites' *flexural and compressive* characteristics were significantly enhanced.	[32]
5	SPF/GF/TPU hybrid composites	Melt compounding technique/compression moulding process	Due to the superior hybrid performance of the two fibres, the *tensile and impact* characteristics of the hybrid composites were enhanced with increasing SPF content (30%/10% SPF/GF) as compared with GF-reinforced composites (0%/40% SPF/GF). When a greater amount of GF was added at 40 wt.%, the *flexural* characteristics improved.	[20]
6	Corn husk/SPF/cornstarch hybrid composites	Conventional solution casting technique	The polymer matrix and reinforcement fibre had a strong interfacial contact and high biocompatibility, which resulted in increased *tensile strength* and *Young's modulus*. In general, the hybridization of CS/CH composites with SP fibre has contributed to the improvement of biocomposites for biomaterials applications, particularly for 6% SP fibre loading.	[33]
7	SPF/TPU composites	Extruder/hot press machines	The best tensile strength was 18.42 MPa with a microwave temperature of 70 C and a pretreatment with 6% NaOH.	[34]

(Continued)

TABLE 1.3 (CONTINUED)

Compounding Process, Properties of SPF/Polymer Composites

	SPF/Polymer Composite	Compounding Process	Properties/ Outcomes	References
8	Roselle RF/SPF/TPU hybrid composites	Brabender plastograph/hot press moulding	25%RF/75%SPF hybrid composites had the lowest *water absorption* (7.35%) and thickness swelling (7.15%) values. These values increase as the % content of SPF increases.	[35]
9	SPF/phenolic (PF) composites	Disk-shaped films/ Alpha Analyzer/ temperature controller	Dielectric relaxation spectroscopy to examine the effect of treatment on SPF composite. Interfacial bonding is stronger in alkaline-treated composites than in untreated and sea water–treated composites.	[36]
10	Cassava/SPF/cassava starch hybrid composites	Casting technique	The addition of SPF caused changes in the film characteristics of cassava starch, potentially compromising the film's performance.	[37]
11	Seaweed SW/SPF/ thermoplastic SP starch/ agar hybrid composites	Brabender plastograph hydraulic thermo-press	The study demonstrated that hybrid composites have enhanced tensile and flexural characteristics while having a reduced impact resistance. SW/SPF (50/50) hybrid composite had the highest tensile strength (17.74 MPa) and flexural strength (31.24 MPa). *Water absorption*, thickness swelling, and soil burial tests revealed that the hybrid composites were more resistant to water.	[38]
12	SPF/High Impact Polystyrene HIPS composites	Melt compounding technique/ compression moulding process	The findings showed that increasing the short SPF loading in the HIPS matrix enhanced the composites' tensile strength and modulus.	[21]

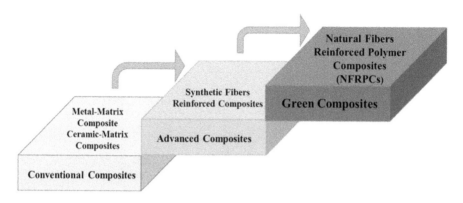

FIGURE 1.2
Composite material advancements. (From Bharanichandar, J., *Am. J. Environ. Sci.*, 9, 494–504, 2014.)

polymer matrices [41]. Figure 1.2 shows the composite material advancements. Previous research on NFRPCs has revealed attractive properties in terms of strong fundamental mechanical properties and low cost [1, 42].

1.5 Polymer Matrices

There are a variety of polymers available for use in NFRPCs, including thermoset, thermoplastic, and biopolymers [43]. There is no benefit over thermoset when using thermoplastic polymers as a matrix in natural fibre composite. Due to the linear or branched chain arrangement of molecules, thermoplastics can be reshaped after melting and solidification by applying heat and pressure. During the curing process, no chemical bonding takes place, which helps in recycling. Thermoplastics can be easily moulded or shaped into any product as well as having good impact resistance, and the environment also remains green when using these polymers [44]. Examples of thermoplastics are PLA, polypropylene (PP), polystyrene (PS), low-density polyethene (LDPE), high-density polyethene (HDPE), polyurethane (PU), etc. But, thermoset polymers are cheaper than thermoplastics, and thermoplastics will quickly melt even at low temperatures. These are drawbacks of using thermoplastic rather than thermoset polymers. Another disadvantage of using thermoplastics with natural fibres is the low degradation temperature of natural fibres, which is only 180 C. So, the working or processing temperature cannot exceed 180 C [14]. Table 1.4 shows different physical and mechanical properties of various thermoplastic and thermoset polymers used in natural fibre composite (NFC).

TABLE 1.4

Various Physical and Mechanical Properties of Polymers Used as Matrices in NFC [45–47]

Property	PLA	PP	PS	LDPE	HDPE	Polyester	Epoxy
Density (g/cm³)	1.25	0.89	1.04–1.06	0.910–0.925	0.94–0.96	1.2–1.5	1.1–1.4
Water absorption–24 hours (%)	–	0.01–0.02	0.03–0.10	<0.015	0.01–0.2	0.1–0.3 (@ 20 C)	0.1–0.4 (@ 20 C)
Tensile strength (MPa)	52–66	26–41.4	25–69	40–78	14.5–38	40–90	35–100
Elastic modulus (GPa)	2.4–3.02	0.95–1.77	4–5	0.055–0.38	4–5	2–4.5	3–6
Elongation (%)	10–100	15–700	1–2.5	90–800	2–130	2	1–6
Impact strength	–	21.4–267	1.1	>854	26.7–1068	–	–

1.5.1 Poly(Lactic Acid) (PLA)

PLA is obtained from renewable resources and is capable of decomposition by bacteria, thereby avoiding the pollution created by petrochemical-based plastic [48]. As PLA also easily decomposes, it gives H_2O, CO_2, and humus, the black material in the soil. PLA is a thermoplastic polymer that can be widely used for making plastic bags, big planting cups, paper coating, fibres, films, and packaging, and extensively as matrix material in composites. PLA is brittle in nature, but this can be improved by using a plasticizer.

Figure 1.3 shows the life cycle of PLA. PLA is made by ring opening polymerization of the cyclic lactide. Neat PLA resin can be processed using the traditional extrusion process [49]. PLA products can be thermally decomposed. Acetaldehyde is a primary thermal decomposition product of PLA, and carbon monoxide and hexanal may also exist during decomposition at normal room temperature. These can naturally be degraded after use without contaminating the environment. PLA can be hydrolysed in the body into lactic acid and acetic acid and metabolized by enzymes to CO_2 and H_2O. Siakeng et al. [40] reported that PLA is an environmentally attractive biopolymer with unique properties, such as good transparency and processability, a glossy look, and high stiffness; however, it does have certain drawbacks, such as brittleness and a high rate of crystallization.

Oksman et al. [50] used the extruded product of PLA obtained from compression moulding and compared its properties with polypropylene

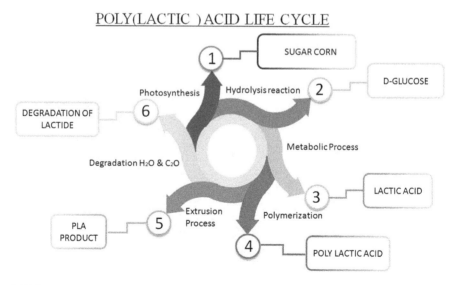

FIGURE 1.3
Life cycle of poly(lactic acid) (PLA) [49].

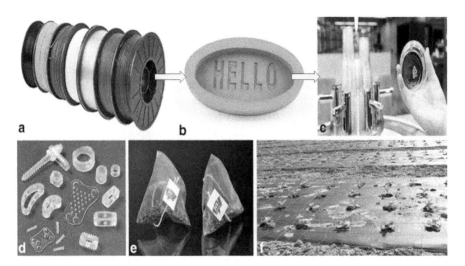

FIGURE 1.4
(a) PLA-3D Printing thread, (b) 3-D printed soap dish from coloured PLA, (c) biodegradable PLA cups used at restaurant, (d) PLA-bio absorbable implants, (e) tea bags made of PLA, (f) mulch film made of PLA-blend "bio-flex" [40].

flax fibre composites (PP/flax). The mechanical properties of PLA are very good; for example, its composite strength is almost 50% higher than the composite strength of PP/flax. This makes it useful in automotive components. Extrusion and compression moulding can be easily done using PLA. Figure 1.4 shows the applications of PLA.

1.6 Treatment of Sugar Palm Fibres

The main issue in polymer composite research is plant hydrophilicity. Several experiments have been conducted to investigate the extent to which interfacial bonding may be enhanced by plasticizing fibre or modifying the surface [23, 51, 52]. Because natural hydrophilic fibre and the hydrophobic polymer matrix are incompatible, the main problem encountered is a fibre–matrix adhesion. Chemically treating the surface of fibres improves this problem. There are different processes for the treatment of natural fibres, such as alkaline, silane, acetylation, benzoylation, and permanganate.

1.6.1 Alkaline Treatment

Alkaline treatment is the simplest and most effective treatment process among all these treatments. It is a low-cost, efficient surface modification method that improves mechanical properties and results in a rough fibre surface [14]. The rough surfaces produced following alkaline treatment improve fibre interlocking for matrix penetration, establishing a large region of contact between the matrix and the fibre [27]. Vigneshwaran et al. [53] addressed the water absorption process of natural fibre (NF) polymer composites, which is primarily governed by three diffusion mechanisms. The first process is water molecule diffusion caused by the existence of tiny gaps or holes in the polymer chain. The second kind of transport is capillary transfer across the spaces between the fibre and matrix interfaces. The diffusion of water via the tiny fractures in the composite surface is the third mechanism. Figure 1.5 depicts the water diffusion effect in NF and the reason for fibre–matrix de-bonding. Pineapple leaf and kenaf fibres were treated with a 6% alkaline solution, which helped remove contaminants, lignin, and hemicelluloses while increasing the tensile strength [9]. According to the previous study, the tensile strength of alkaline-treated fibres was much higher than that of untreated fibres [15]. Maleque et al. [54], reported that 6% alkaline treatment of kenaf fibre enhanced the impact and flexural properties of kenaf/glass composites.

Afzaluddin et al. [20] reported that a 6% alkaline treatment of kenaf mat improved the interfacial bonding between kenaf fibre and the unsaturated polyester matrix, which improved the flexural strength of the composite. Several studies have examined surface improvements of SPF after various treatments. Rashid et al. [51] reported that alkaline treatment of SPF enhanced adhesion to fibre by removing wax and contaminants from the surface. Figure 1.6 shows a schematic diagram of the untreated and alkaline-treated natural fibre. Atiqah et al. [10] indicated that a 6% alkaline

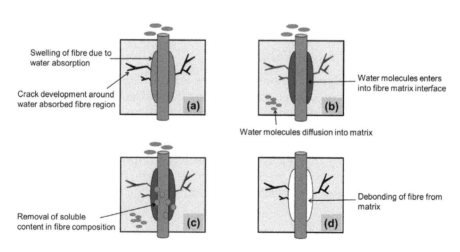

FIGURE 1.5
The effect of water absorption in natural fibre composites [53].

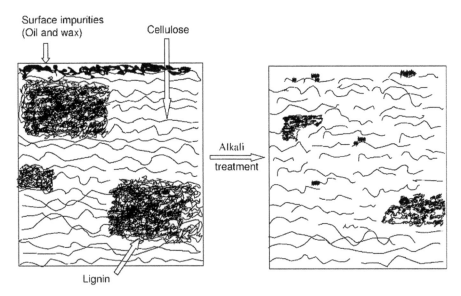

FIGURE 1.6
Schematic diagram of the untreated and alkaline-treated natural fibre [55].

treatment of SPF was used to remove impurities and hemicelluloses from the fibre for SPF/GF-reinforced TPU hybrid composites. The 6% alkaline treatment of SPF improved adhesion between the fibre and the matrix, which resulted in the mechanical properties of Roselle fibre/SPF hybrid composite being improved [14]. According to Atiqah et al. [20], 6% alkaline treatment of sugar palm fibre produced SPF/GF/TPU hybrid composites with tensile, impact, and flexural strength increased by 16%, 18%, and 39%, respectively. This increase in tensile strength was caused by the fibre's enhanced interfacial bonding. Among a number of advantages of alkaline SPF treatment, two advantages are essential to highlight: it reduces the water absorption characteristics of the fibre and improves the interfacial bonding between fibre and matrix [56]. The cellulose content in SPF also increases to 22.6% after alkaline treatment [19].

1.6.2 Benzoyl Chloride Treatment

Benzoyl chloride treatment is also effective in reducing the hydrophilic nature of natural fibres and improving fibre attachment to the matrix, which increases the strength of the composites. Siakeng et al. [40] and Kabir et al. [3] confirmed that after treating *Cannabis indica* fibre with a 5% benzoyl chloride solution, fibres were more resistant to water absorption. Kushwaha et al. [57] used benzoylation with bamboo fibre/polyester composites and found that tensile strength and modulus were increased by 71% and 181%, respectively. Benzoyl chloride treatment was also reported on Palmyra palm-leaf stalk fibre composites, increasing tensile strength and modulus by 60%. Fibre surface roughness is improved by eliminating as much starch, cellulose, hemicelluloses, and lignin as possible during the benzoyl chloride treatment [58]. Benzoyl chloride treatment improved the wettability between *Ipomoea pes-caprae* fibres/epoxy composites, resulting in stronger bonding, and as a result, increased the overall strength of the composites [59]. Benzoyl chloride reduces the hydrophilic nature of SPF and enhances the interaction with an epoxy resin matrix by enhancing fibre and matrix adhesion, improving strength, and lowering the overall composite's water absorption properties [32]. The tensile properties of polyvinyl chloride–epoxidized natural rubber–kenaf core powder composites were increased after benzoylation treatment using 10% sodium hydroxide agitated with 50 mL benzoyl chlorides in pretreatment kenaf core powder [60]. Izwan et al. (2020) studied SPF that was pretreated with 18% NaOH, then benzoylated with 10% NaOH, and determined that the highest tensile strength was observed after 15 minutes of soaking time [61].

1.7 Effect of Fiber Loading

Increased fibre loading will often result in a considerable improvement in composite stiffness, as mechanical strengths are improved by the addition of natural fibres. To date, several research works have been conducted to study the optimal natural fibre content inside composites in order to improve the mechanical performance of the composites. Shalwan and Yousif [62] report that there is no uniform value for NF content (weight) and fibre volume fraction at which greater tensile characteristics may be obtained. This is because each NF type may have a type-specific optimal loading for exhibiting excellent tensile strength. This may be related to the chemical composition of natural fibres as well as their intrinsic strength, interfacial adhesion, and physical properties [63]. According to Venkateshwaran et al. [64], when the volume fraction of fibre loading exceeds 50%, there is a tendency for the fibres to form aggregations. This leads to a decrease in the interfacial area and a debonding between the fibres and the matrix. Hanafee et al. [65] revealed that if the percentage of NF was higher than 50%, the mechanical characteristics of the composites deteriorate. This is because the fibres do not sufficiently absorb the matrix loading, which causes fibre–matrix debonding, enhanced sample brittleness, and therefore, failure of the specimen composite at the start of the flexural test. This will have an impact on the composites' overall

mechanical performance. The maximum mechanical properties of variation SP yarn reinforced unsaturated polyester composites were found at
30 and 40 weight percent, according to Nurazzi et al. [66].

Muthalagu et al. [67] researched the failure mechanism of composites.
The fracture mechanism occurs when a fibre is combined with a brittle
matrix, resulting in stress concentration at the crack junction, which leads
to crack propagation due to fibre debonding between the fibre and the
matrix. Figure 1.7 shows different failure modes occurring between the
fibre and the matrix. Because the primary function of fibre reinforcement is to improve toughening properties, shear failure can occur in the
interphase region between the fibre and the matrix if the fracture yields
around the matrix.

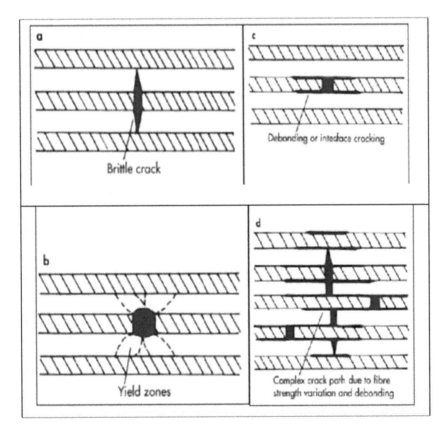

FIGURE 1.7
Failure propagation: (a) brittle cracking of matrix, (b) matrix shear yielding, (c) interfacial failure, and (d) fracture propagation including propagation of (a) and (c) [68].

1.8 Physical Properties

Several researchers have investigated the physical properties of natural fibre–reinforced polymer composites, particularly the effects of water absorption. The process through which molecules of a substance move through a layer or area, such as the surface of a solution, is known as the diffusion process. Diffusion can occur in the absence of a membrane or a gas–liquid barrier. The rate of diffusion of a substance across a unit area (such as a surface or membrane) is proportional to the concentration gradient, according to Fick's law [69]. Due to the fibre's high cellulose content, Girisha et al. [70] found that the water absorption percentage grows as the fibre volume fraction increases in sisal/coconut coir fibre–reinforced epoxy composites. At room temperature, these composites' water absorption trends were seen to follow Fickian behaviour, according to Akash et al. [71]; however, at higher temperatures, Fick's law did not apply to the water absorption characteristics.

With an increase in cellulose fibre content, sisal/coir fibre–reinforced composites absorb more water. Using distilled water, saltwater (5% NaCl solution), and sub-zero temperature (–25 C), Bera et al. [72] examined the water absorption behaviour of luffa fibre/epoxy composites and discovered that the water absorption pattern followed Fickian diffusion behaviour. Water transport is significantly impacted by the fibre–matrix interface's characteristics. If the contact is strong, water molecules find it challenging to diffuse through the composite structure. Zamri et al. [2] conducted water absorption tests for distilled water, sea water, and acidic water in three varied conditions. Glass/jute fibre–reinforced unsaturated polyester hybrid composites have been reported to exhibit non-Fickian behaviour at room temperature. According to Atiqah et al. [73], water absorption results in a buildup of moisture in the cell walls of the fibre, which promotes fibre swelling and a decrease in dimensional stability.

1.9 Flammability of Natural Fiber Composites

Natural fibre use is encouraged by environmental concerns, but when compared with glass and carbon fibre–reinforced composites, natural fibre–reinforced composites have the poorest flammability characteristics [71]. Flammability is a touchy subject in many industrial applications, especially in the transformation area, where compact spaces present a serious fire risk. The flammability of the used fire-retardant products and the fire-retarding final goods being processed can be assessed using flammability testing methods. Flammability tests are conducted in various sectors and academic laboratories on various scales (including small, medium, and large) [74].

The flammability of natural fibre–reinforced thermoplastic biocomposites has been investigated using horizontal and vertical burning experiments. The mass loss rate and flame propagation rate of treated composites were reportedly decreased in the UL 94V and UL 94HB examinations, but their flame or fire resistance was increased, according to Bharath et al. [75]. As the burning rate decreased, the thermal stability of treated composites increased. The lack of adhesion between the fibre particles and the polymer matrix allowed void spaces to form around the fibre particles. The enhanced flammability of the composite material as a result of treating coconut tree leaf sheath fibres suggests possible uses for the material in building and decorating. The experimental setup for UL-94 flammability horizontal testing is shown in Figure 1.8 [75]. Since natural fibres tend to burn well, adding cellulosic fibre makes materials more flammable. The pyrolysis of the cellulosic fibre results in the formation of a surface layer, making it a poor flame retardant with limited fire resistance. By acting as a fire supporter, this layer stops heat from reaching the pyrolyzed material [71]. UL-94 tests were also used to establish the flammability of treated sisal fibre (SF)–reinforced recycled polypropylene (RPP) composites.

When compared with RPP and RPP/SF composites that had not been treated, the results showed a burning rate reduction of up to 16% and 7.42%, respectively [76]. In comparison to untreated pineapple leaf fibre (PALF) and kenaf fibre (KF), treated PALF and KF phenolic composites had a 50% worse fire retardancy, as reported by Asim et al. [77]. According to Chen et al. [78], PLA composite had minimal flame retardancy in both the limiting

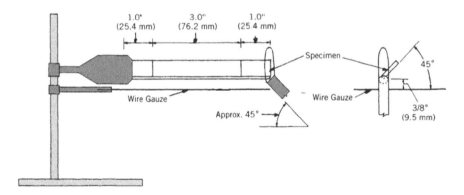

FIGURE 1.8
Experimental setup of UL-94 Flammability horizontal testing [75].

oxygen index (LOI) and UL-94 tests; this improved with the addition of chitosan (CS) alone. The flame retardant property of PLA composites improved after CS and ammonium polyphosphate loading. To ascertain the impact of compatibilizer and graphene nanoplatelets on PLA/polybutylene adipate co-terephthalate nanocomposites, LOI and UL-94 tests were conducted [79]. These composites were examined utilizing horizontal burning analysis, as claimed by Suriani et al. [80]. The specimen containing $Mg(OH)_2$ as a flame retardant and 35% SPF (Epoxy/PET/SPF-35) had the lowest burning rate at 13.25 mm/min.

1.10 Creep, Hardness, and Compression Properties

There have been few studies investigating the prolonged load-bearing properties of treated biocomposites. Durante et al. [81] studied the creep behaviour of PLA reinforced with woven hemp fabric. Wang et al. [82] investigated the creep character of flax fibre–reinforced composites after treatment of flax fibre.

There has been very limited research to determine the hardness and compression of bio-based composites. Idris et al. [83] determined the Brinell hardness value for banana peel waste/phenolic resin to replace asbestos. Ruzaidi et al. [84] studied the hardness of a palm slag filler (asbestos-free) brake pad using the Rockwell hardness test. Rashid et al. [51] investigated the Rockwell hardness characteristics of SPF/phenolic composites and indicated that as the SPF loading in the composites increases, the findings confirm a decrease in Rockwell hardness. Recently, Syafiqah et al. [32] reported that SPF/glass fibre–reinforced epoxy hybrid composite shows excellent flexural and compressive properties after treatment of SPF. According to Badyankal et al. [85], increasing the percentage of SF improves the compression and water absorption properties of banana and sisal hybrid fibre polymer composites. Few research works have examined the effect of durability on the structural properties of NFC load-bearing structural components. The effects of high cycle compression fatigue and moisture ingress on the structural characteristics of flax and jute natural fibre–epoxy composite structural columns evaluated in pure compression were investigated by Bambach et al. [86]. The compression stiffness of saturated columns was reduced by up to 44%, the buckling load was reduced by up to 48%, and the ultimate strength was reduced by up to 48%.

1.11 Sugar Palm Fibre Reinforced Composite Fabrication Processes

Compression moulding, resin transfer moulding, injection moulding, hot pressing, and vacuum infusion moulding are the main current fabrication techniques utilized to manufacture Sugar palm fibre reinforced composites SPFRCs [87]. Figure 1.9 shows three of the most common techniques: (a) compression moulding, (b) injection moulding, and (c) extrusion of SPFRCs.

Because these conventional manufacturing methods are largely designed/developed on the basis of synthetic fibre–reinforced composites, the low stability of NFs during processing always creates problems in processing SPFRCs. As a result, pretreatment (chemical/physical) of natural fibres becomes an essential element of the processing of NFRCs using conventional moulding techniques. Fibre pretreatment increases fibre–matrix interfacial interactions, which improves the processability and characteristics of SPFRCs. Table 1.5 summarizes some of the most recent composites that have been fabricated using various processing techniques.

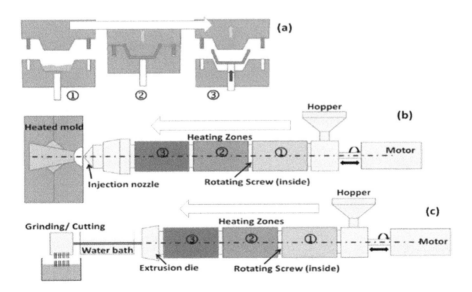

FIGURE 1.9
(a) Compression moulding, (b) injection moulding, and (c) extrusion of NFRCs [87].

TABLE 1.5

Sugar Palm Fiber Polymers Hybrid Composites That Have Been Fabricated Using
Various Processing Techniques

Natural Fibre/Other Fiber Reinforcements	Matrix	Fabrication	References
SPF/Ramie fibre	Epoxy	Compression moulding	[30]
SPF/GF	Thermoplastic polyurethane	Compression moulding	[31]
SP yarn/GF	Unsaturated polyester	Compression moulding	[88]
SPF/GF	Epoxy	Hand lay-up technique	[89]
SPF	TPU	Compression moulding	[34]
Roselle/SPF	Thermoplastic polyurethane	Compression moulding	[14]
Seaweed/SPF	Thermoplastic sugar palm starch/agar	Hydraulic thermo-press	[38]
SPF	High impact polystyrene	Compression moulding	[21]
Kenaf	Kevlar		[90]
Kenaf-derived cellulose	Poly(lactic acid)	Compression moulding	[91]
Kevlar/date palm	Epoxy	Hand lay-up technique	[67]
Kenaf	Magnesium hydroxide impregnated	Vacuum bag resin transfer moulding	[41]
Natural rubber/kenaf core powder	Poly(lactic acid)	Compression moulding	[92]
Pineapple leaf/kenaf	Phenolic	Compression moulding	[77]
Aloe vera	Poly(lactic acid)	Compression moulding	[93]
Ijuk fibre	Polypropylene	Injection moulding	[22]
Ipomoea pes-caprae	Epoxy	Hand lay-up technique	[59]
Ensete	Woven glass fibre fabric	Vacuum-assisted resin transfer moulding	[94]
Kenaf	Glass	Hand lay-up and hydraulic cold press	[95]
Sisal	Polyester	Resin transfer and compression moulding	[96]
Kenaf/glass	Epoxy	Dry filament winding and hand lay-up technique	[97]
Carbon/flax	Epoxy	Platen press process	[98]
Pineapple leaf fibre/chopstick	Poly(lactic acid)	Counter-rotating internal mixer	[99]

1.12 Applications of Sugar Palm Fiber–Reinforced Polymer Hybrid Composites

Sugar palm fibres may be used to weave hats and mats, as well as ropes, brooms, road construction, brushes, roof materials, cushions, and fish breeding shelters [100]. Also, the fibre stem core may be used to make sago flour and the root to make a tea to treat bladder stones, insect repellent, and posts for pepper, boards, tool handles, water pipes, and musical instruments such as drums [13]. According to Sapuan et al. [101], 12 sugar palm products, including SPF, sugar palm starch, roofing, rope, brooms, bottles, brushes, vinegar, berries, liquid sugar, fined sugar, and block sugar, have been successfully obtained, as shown in Figure 1.10. Recent advanced uses of sugar palm fibres include subterranean and underwater cables, the substitution of geotextile fibreglass reinforcement in road building for soil stabilization, and the use of polymer matrix composites as reinforcement in material engineering [7].

FIGURE 1.10
Twelve objects derived from the sugar palm tree [101].

Many researchers have investigated the possibilities of sugar palm fibres and biopolymers for industrial applications like automotive, packaging, bioenergy, and many others [11, 86, 87]. SPF is used in a variety of applications, including SPF-concrete composite for the building industry [102], sound-absorbing material for a ducting silencer capable of absorbing the noise produced by air conditioning [103], modified zinc roof with composites of sugar palm fibres for sound insulation and vibration reduction purposes [104], SPF/Sugar Palm Starch (SPS) for food packaging applications, and SPF/polystyrene for roof tiles [4, 105, 106]. The hand lay-up technique may be used to make pultruded SP composite rods for biocomposite drain covers and multifunctional tables out of SPF-reinforced composites [107]. Alaaeddin et al. [108] developed a new solar module with an innovative polyvinylidene fluoride–short sugar palm fibre (PVDF-SSPF) composite backsheet, as shown in Figure 1.11.

It has been found that combining GF with treated SPF improves the thermal properties of hybrid composites for structural and automotive applications [27]. SPF/GF/TPU hybrid composites can be utilized in applications that need high thermal resistance [109]. Ferdiansyah et al. [110] revealed that SPF may also be utilized for black SPF-reinforced concrete composite. Misri et al. [111] used a hand lay-up technique extensively in the production of SPF/GF-reinforced unsaturated polyester for a boat, as shown in Figure 1.12. The potential of SPF/GF-reinforced polyurethane thermoplastic hybrid composite to replace a steel-based anti-roll bar without compromising the component's stiffness is illustrated in Figure 1.12, which depicts an anti-roll bar for SPF/glass fibre–reinforced automotive applications [107]. Recently, In order to create char briquettes, Harussani et al. [112] applied sugar palm starch with slow pyrolysis at 450 °C to polypropylene (PP) powder made from COVID-19 isolation gown waste.

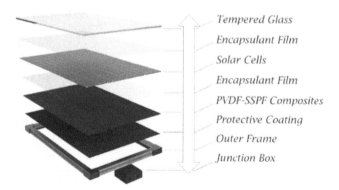

FIGURE 1.11
SPF biocomposite solar module [108].

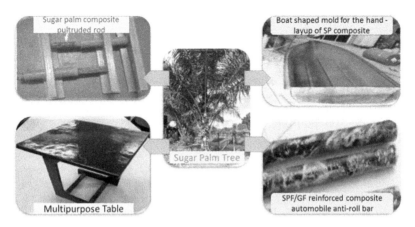

FIGURE 1.12
Various applications of sugar palm composites [107, 111].

1.13 Conclusions

The review concludes that sugar palm fibre polymer hybrid composites have greatly expanded their reach in industrial applications over the last 30 years. They exhibit good mechanical characteristics after alkaline treatment of SPF, hybridization with glass fibre, and incorporation into various polymer matrices. In summary, the use of SPF polymer hybrid composites can aid in the future development of sugar palm as a new industrial crop, the reduction of reliance on petroleum products, and the reduction of the negative environmental effects of synthetic polymers and fibres. However, there is still a lot of work to be done on these composites, such as reducing the carbon footprint after use, lowering manufacturing costs, increasing availability, and opening new markets such as motorcycle body framing, toys, household items, pharmaceuticals, and electronic packaging to make hybrid composites more common in replacing petroleum-based plastic. More work is also required to effectively organize these composites depending on their intended usefulness. This might entail looking at their optical, thermal, and technical properties. Composites may improve the sustainability and productivity of green production while reducing hazardous waste.

Acknowledgements

The authors gratefully acknowledge Universiti Putra Malaysia (UPM) for funding this research through Geran Putra Berimpak (GPB), UPM. RMC.800-3/3/1/GPB/2020/9694500. The current book chapter is extended from an International Conference on Sugar Palm and Allied Fibre Polymer Composites (SAPC) 2021 conference paper.

References

1. Singhaa AS, Thakura VK. Fabrication and study of lignocellulosic hibiscus sabdariffa fiber reinforced polymer composites. *BioResources* 2008;3(4):1173–86. https://doi.org/10.15376/biores.3.4.1173-1186.

2. Zamri MH, Akil HM, Bakar AA, Ishak ZAM, Cheng LW. Effect of water absorption on pultruded jute/glass fiber-reinforced unsaturated polyester hybrid composites. *J Compos Mater* 2012;46(1):51–61. https://doi.org/10.1177/0021998311410488.

3. Kabir MM, Wang H, Lau KT, Cardona F. Chemical treatments on plant-based natural fibre reinforced polymer composites: An overview. *Compos Part B Eng* 2012;43(7):2883–92. https://doi.org/10.1016/j.compositesb.2012.04.053.

4. Ilyas RA, Sapuan SM, Ishak MR, Zainudin ES. Sugar palm nanofibrillated cellulose (Arenga pinnata (Wurmb.) Merr): Effect of cycles on their yield, physic-chemical, morphological and thermal behavior. *Int J Biol Macromol* 2019;123:379–88. https://doi.org/10.1016/j.ijbiomac.2018.11.124.

5. Karimah A, Ridho MR, Munawar SS, Adi DS, Ismadi, DR, et al. A review on natural fibers for development of eco-friendly bio-composite: Characteristics, and utilizations. *J Mater Res Technol* 2021;13:2442–58. https://doi.org/10.1016/j.jmrt.2021.06.014.

6. Mukhtar I, Leman Z, Ishak MR, Zainudin ES. Sugar palm fibre and its composites: A review of recent developments. *BioResources* 2016;11(4):10756–82. https://doi.org/10.15376/biores.11.4.10756-10782.

7. Ishak MR, Sapuan SM, Leman Z, Rahman MZA, Anwar UMK, Siregar JP. Sugar palm (Arenga pinnata): Its fibres, polymers and composites. *Carbohydr Polym* 2013;91(2):699–710. https://doi.org/10.1016/j.carbpol.2012.07.073.

8. Sapuan SM, Ilyas RA. Characterization of sugar palm nanocellulose and its potential for reinforcement with a starch-based composite. *Sugar Palm Biofibers Biopolym Biocomposites* 2018:189–220. https://doi.org/10.1201/9780429443923-10.

9. Atiqah A, Jawaid M, Sapuan SM, Ishak MR. Effect of surface treatment on the mechanical properties of sugar palm/glass fiber-reinforced thermoplastic polyurethane hybrid composites. *BioResources* 2018;13(1):1174–88. https://doi.org/10.15376/biores.13.1.1174-1188.

10. Atiqah A, Jawaid M, Ishak MR, Sapuan SM. Effect of alkali and Silane treatments on mechanical and interfacial bonding strength of sugar palm fibers with thermoplastic polyurethane. *J Nat Fibers* 2018;15(2):251–61. https://doi.org/10.1080/15440478.2017.1325427.

11. Huzaifah MRM, Sapuan SM, Leman Z, Ishak MR, Maleque MA. A review of sugar palm (Arenga pinnata): Application, fibre characterisation and composites. *Multidiscip Model Mater Struct* 2017;13(4):678–98. https://doi.org/10.1108/MMMS-12-2016-0064.

12. Sapuan SM, Ilyas RA, Ishak MR, Leman Z, Huzaifah MRM, Ammar IM, et al. Development of sugar palm–based products: a community project. In *Sugar Palm Biofibers, Biopolym. Biocomposites*. 1st ed., Boca Raton: CRC Press/Taylor & Francis Group; 2018, p. 245–66. https://doi.org/10.1201/9780429443923-12.

13. Mogea J, Seibert B, Smits W. Multipurpose palms: The sugar palm (Arenga pinnata (Wurmb) Merr.). *Agrofor Syst* 1991;13(2):111–29. https://doi.org/10.1007/BF00140236.
14. Radzi AM, Sapuan SM, Jawaid M, Mansor MR. Effect of alkaline treatment on mechanical, physical and thermal properties of Roselle/sugar palm fiber reinforced thermoplastic polyurethane hybrid composites. *Fibers Polym* 2019;20(4):847–55. https://doi.org/10.1007/s12221-019-1061-8.
15. Mukhtar I, Leman Z, Zainudin ES, Ishak MR. Hybrid and nonhybrid laminate composites of sugar palm and glass fibre-reinforced polypropylene: Effect of alkali and sodium bicarbonate treatments. *Int J Polym Sci* 2019. https://doi.org/10.1155/2019/1230592.
16. Rashid B, Leman Z, Jawaid M, Ghazali MJ, Ishak MR. Physicochemical and thermal properties of lignocellulosic fiber from sugar palm fibers: Effect of treatment. *Cellulose* 2016;23(5):2905–16. https://doi.org/10.1007/s10570-016-1005-z.
17. Alves Fidelis ME, Pereira TVC, Gomes ODFM, De Andrade Silva F, Toledo Filho RD. The effect of fiber morphology on the tensile strength of natural fibers. *J Mater Res Technol* 2013;2(2):149–57. https://doi.org/10.1016/j.jmrt.2013.02.003.
18. Huzaifah M, Roslim M, Sapuan SM, Leman Z, Ishak MR. *A Review on Sugar Palm (Arenga Pinnata): Characterization of Sugar Palm Fibre 2018.*
19. Ilyas RA, Sapuan SM, Ishak MR, Zainudin ES. Development and characterization of sugar palm nanocrystalline cellulose reinforced sugar palm starch bionanocomposites. *Carbohydr Polym* 2018;202:186–202. https://doi.org/10.1016/j.carbpol.2018.09.002.
20. Afzaluddin A, Jawaid M, Salit MS, Ishak MR. Physical and mechanical properties of sugar palm/glass fiber reinforced thermoplastic polyurethane hybrid composites. *J Mater Res Technol* 2019;8(1):950–9. https://doi.org/10.1016/j.jmrt.2018.04.024.
21. Sapuan SM, Bachtiar D. Mechanical properties of sugar palm fibre reinforced high impact polystyrene composites. *Procedia Chem* 2012;4:101–6. https://doi.org/10.1016/j.proche.2012.06.015.
22. Zahari WZW, Badri RNRL, Ardyananta H, Kurniawan D, Nor FM. Mechanical properties and water absorption behavior of polypropylene / ijuk fiber composite by using Silane treatment. *Procedia Manuf* 2015;2:573–8. https://doi.org/10.1016/j.promfg.2015.07.099.
23. Khalid MFS, Abdullah AH. Storage modulus capacity of untreated aged Arenga pinnata fibre-reinforced epoxy composite. *Appl Mech Mater* 2013;393:171–6. https://doi.org/10.4028/www.scientific.net/AMM.393.171.
24. Siregar JP. Tensile and flexural properties of Arenga pinnata filament (IJUK Filament) reinforced epoxy composites tensile and flexural properties Of Arenga pinnata Filament (IJUK Filament) reinforced epoxy. COMPOSITES by Thesis Submitted to the School of Graduate 2015.
25. Leman Z, Sapuan SM, Saifol AM, Maleque MA, Ahmad MMHM. Moisture absorption behavior of sugar palm fiber reinforced epoxy composites. *Mater Des* 2008;29(8):1666–70. https://doi.org/10.1016/j.matdes.2007.11.004.

26. Maisara AMN, Ilyas RA, Sapuan SM, Huzaifah MRM, Mohd Nurazzi N, Saifulazry SOA. Effect of fibre length and sea water treatment on mechanical properties of sugar palm fibre reinforced unsaturated polyester composites. *Int J Recent Technol Eng* 2019;8(2S4):510–4. https://doi.org/10.35940/ijrte.B1100 .0782S419.

27. Mohd Nurazzi N, Khalina A, Sapuan SM, Ilyas RA, Ayu Rafiqah S, Hanafee ZM. Thermal properties of treated sugar palm yarn/glass fiber reinforced unsaturated polyester hybrid composites. *J Mater Res Technol* 2019;9:1606–18. https://doi.org/10.1016/j.jmrt.2019.11.086.

28. Chalid M, Prabowo I. The effects of alkalization to the mechanical properties of the ijuk fiber reinforced PLA biocomposites. *Int J Chem Mol Nucl Mater Metall Eng* 2015;9:342–6.

29. Sahari J, Sapuan SM, Zainudin ES, Maleque MA. Mechanical and thermal properties of environmentally friendly composites derived from sugar palm tree. *Mater Des* 2013;49:285–9. https://doi.org/10.1016/j.matdes.2013.01.048.

30. Siregar JP, Zalinawati M, Cionita T, Rejab MRM, Mawarnie I, Jaafar J, et al. Mechanical properties of hybrid sugar palm/ramie fibre reinforced epoxy composites. *Mater Today Proc* 2020;46:1729–34. https://doi.org/10.1016/j.matpr.2020.07.565.

31. Atiqah A, Jawaid M, Sapuan SM, Ishak MR, Ansari MNM, Ilyas RA. Physical and thermal properties of treated sugar palm/glass fibre reinforced thermoplastic polyurethane hybrid composites. *J Mater Res Technol* 2019;8(5):3726–32. https://doi.org/10.1016/j.jmrt.2019.06.032.

32. Safri SNA, Sultan MTH, Saba N, Jawaid M. Effect of benzoyl treatment on flexural and compressive properties of sugar palm/glass fibres/epoxy hybrid composites. *Polym Test* 2018;71:362–9. https://doi.org/10.1016/j.polymertesting .2018.09.017.

33. Ibrahim MIJ, Sapuan SM, Zainudin ES, Zuhri MYM. Preparation and characterization of cornhusk/sugar palm fiber reinforced Cornstarch-based hybrid composites. *J Mater Res Technol* 2020;9(1):200–11. https://doi.org/10.1016/j.jmrt .2019.10.045.

34. Mohammed AA, Bachtiar D, Rejab MRM, Siregar JP. Effect of microwave treatment on tensile properties of sugar palm fibre reinforced thermoplastic polyurethane composites. *Def Technol* 2018;14(4):287–90. https://doi.org/10.1016/j.dt .2018.05.008.

35. Radzi AM, Sapuan SM, Jawaid M, Mansor MR. Water absorption, thickness swelling and thermal properties of roselle/sugar palm fibre reinforced thermoplastic polyurethane hybrid composites. *J Mater Res Technol* 2019;8(5):3988–94. https://doi.org/10.1016/j.jmrt.2019.07.007.

36. Agrebi F, Ghorbel N, Rashid B, Kallel A, Jawaid M. Influence of treatments on the dielectric properties of sugar palm fiber reinforced phenolic composites. *J Mol Liq* 2018;263:342–8. https://doi.org/10.1016/j.molliq.2018.04.130.

37. Edhirej A, Sapuan SM, Jawaid M, Zahari NI. Cassava/sugar palm fiber reinforced cassava starch hybrid composites: Physical, thermal and structural properties. *Int J Biol Macromol* 2017;101:75–83. https://doi.org/10.1016/j.ijbiomac .2017.03.045.

38. Jumaidin R, Sapuan SM, Jawaid M, Ishak MR, Sahari J. Thermal, mechanical, and physical properties of seaweed/sugar palm fibre reinforced thermoplastic sugar palm Starch/Agar hybrid composites. *Int J Biol Macromol* 2017;97:606–15. https://doi.org/10.1016/j.ijbiomac.2017.01.079.

39. Khalid MY, Al Rashid A, Arif ZU, Ahmed W, Arshad H, Zaidi AA. Natural fiber reinforced composites: Sustainable materials for emerging applications. *Results Eng* 2021;11:100263. https://doi.org/10.1016/j.rineng.2021.100263.

40. Siakeng R, Jawaid M, Ariffin H, Sapuan SM, Asim M, Saba N. Natural fiber reinforced polylactic acid composites: A review. *Polym Compos* 2019;40(2):446–63. https://doi.org/10.1002/pc.24747.

41. Wu Y, Xia C, Cai L, Garcia AC, Shi SQ. Development of natural fiber-reinforced composite with comparable mechanical properties and reduced energy consumption and environmental impacts for replacing automotive glass-fiber sheet molding compound. *J Clean Prod* 2018;184:92–100. https://doi.org/10.1016/j.jclepro.2018.02.257.

42. Bharanichandar J. Natural fiber reinforced polymer composites for automobile accessories. *Am J Environ Sci* 2014;9:494–504. https://doi.org/10.3844/ajessp.2013.494.504.

43. Chapple S, Anandjiwala R. Flammability of natural fiber-reinforced composites and strategies for fire retardancy: A review. *J Thermoplast Compos Mater* 2010;23(6):871–93. https://doi.org/10.1177/0892705709356338.

44. Meon MS, Othman MF, Husain H, Remeli MF, Syawal MSM. Improving tensile properties of kenaf fibers treated with sodium hydroxide. *Procedia Eng* 2012;41:1587–92. https://doi.org/10.1016/j.proeng.2012.07.354.

45. Holbery J, Houston D. Natural-fiber-reinforced polymer composites in automotive applications. *JOM* 2006;58(11):80–6. https://doi.org/10.1007/s11837-006-0234-2.

46. Haleem A, Kumar V, Kumar L. Mathematical modelling & pressure drop analysis of fused deposition modelling feed wire. *Int J Eng Technol* 2017;9(4):2885–94. https://doi.org/10.21817/ijet/2017/v9i4/170904066.

47. Hinchcliffe SA, Hess KM, Srubar WV. Experimental and theoretical investigation of prestressed natural fiber-reinforced polylactic acid (PLA) composite materials. *Compos Part B Eng* 2016;95:346–54. https://doi.org/10.1016/j.compositesb.2016.03.089.

48. Huda MS, Drzal LT, Mohanty AK, Misra M. Effect of fiber surface-treatments on the properties of laminated biocomposites from poly(lactic acid) (PLA) and kenaf fibers. *Compos Sci Technol* 2008;68(2):424–32. https://doi.org/10.1016/j.compscitech.2007.06.022.

49. Bajpai PK, Singh I, Madaan J. Development and characterization of PLA-based green composites: A review. *J Thermoplast Compos Mater* 2014;27(1):52–81. https://doi.org/10.1177/0892705712439571.

50. Oksman K, Skrifvars M, Selin JF. Natural fibres as reinforcement in polylactic acid (PLA) composites. *Compos Sci Technol* 2003;63(9):1317–24. https://doi.org/10.1016/S0266-3538(03)00103-9.

51. Rashid B, Leman Z, Jawaid M, Ghazali MJ, Ishak MR, Abdelgnei MA. Dry sliding wear behavior of untreated and treated sugar palm fiber filled phenolic composites using factorial technique. *Wear* 2017;380–381:26–35. https://doi.org/10.1016/j.wear.2017.03.011.

52. Shaniba V, Sreejith MP, Aparna KB, Jinitha TV, Purushothaman E. Mechanical and thermal behavior of styrene butadiene rubber composites reinforced with silane-treated peanut shell powder. *Polym Bull* 2017;74(10):3977–94. https://doi.org/10.1007/s00289-017-1931-4.

53. Vigneshwaran S, Sundarakannan R, John KM, Joel Johnson RD, Prasath KA, Ajith S, et al. Recent advancement in the natural fiber polymer composites: A comprehensive review. *J Clean Prod* 2020;277:124109. https://doi.org/10.1016/j .jclepro.2020.124109.

54. Maleque MA, Atiqah A, Iqbal M. Flexural and impact properties of kenaf-glass hybrid composite. *Adv Mater Res* 2012;576:471–4. https://doi.org/10.4028/www .scientific.net/AMR.576.471.

55. Mohanty AK, Misra M, Dreal LT. Surface modifications of natural fibres and peformance of the resulting biocomposite. *Compos Interfaces* 2001;8(5):313–43.

56. Huzaifah M, Roslim M, Sapuan SM, Leman Z, Ishak MR. A review of sugar palm (Arenga pinnata): Application , fibre characterisation and composites Article information 2017. https://doi.org/10.1108/MMMS-12-2016-0064.

57. Kushwaha PK, Kumar R. Influence of chemical treatments on the mechanical and water absorption properties of bamboo fiber composites. *J Reinf Plast Compos* 2011;30(1):73–85. https://doi.org/10.1177/0731684410383064.

58. Thiruchitrambalam M, Shanmugam D. Influence of pre-treatments on the mechanical properties of palmyra palm leaf stalk fiber-polyester composites. *J Reinf Plast Compos* 2012;31(20):1400–14. https://doi.org/10.1177 /0731684412459248.

59. Vinod A, Vijay R, Manoharan S, Vinod A, Lenin Singaravelu D, Sanjay MR. Characterization of raw and benzoyl chloride treated Impomea pes-caprae fibers and its epoxy composites. *Mater Res Express* 2019;6(9). https://doi.org/10 .1088/2053-1591/ab2de2.

60. Abdul Majid R, Ismail H, Mat Taib R. Processing, tensile, and thermal studies of poly(vinyl chloride)/epoxidized natural rubber/kenaf core powder composites with benzoyl chloride treatment. *Polym Plast Technol Eng* 2018;57(15):1507–17. https://doi.org/10.1080/03602559.2016.1211687.

61. Mohd Izwan S, Sapuan SM, Zuhri MYM, Mohamed AR. Effects of benzoyl treatment on NaOH treated sugar palm fiber: Tensile, thermal, and morphological properties. *J Mater Res Technol* 2020;9(3):5805–14. https://doi.org/10.1016 /j.jmrt.2020.03.105.

62. Shalwan A, Yousif BF. In state of art: Mechanical and tribological behaviour of polymeric composites based on natural fibres. *Mater Des* 2013;48:14–24. https:// doi.org/10.1016/j.matdes.2012.07.014.

63. George G, Joseph K, Boudenne A, Thomas S. Recent advances in green composites. *Key Eng Mater* 2010;425:107–66. https://doi.org/10.4028/www.scientific.net /KEM.425.107.

64. Venkateshwaran N, Elayaperumal A, Sathiya GK. Prediction of tensile properties of hybrid-natural fiber composites. *Compos Part B Eng* 2012;43(2):793–6. https://doi.org/10.1016/j.compositesb.2011.08.023.

65. Zin MH, Abdan K, Mazlan N, Zainudin ES, Liew KE, Norizan MN. Automated spray up process for Pineapple Leaf Fibre hybrid biocomposites. *Compos Part B Eng* 2019;177:107306. https://doi.org/10.1016/j.compositesb .2019.107306.

66. Nurazzi NM, Khalina A, Sapuan SM, Rahmah M. Development of sugar palm yarn/glass fibre reinforced unsaturated polyester hybrid composites. *Mater Res Express* 2018. https://doi.org/10.1088/2053-1591/aabc27.

67. Muthalagu R, Murugesan J, Sathees Kumar S, Sridhar Babu B. Tensile attributes and material analysis of Kevlar and date palm fibers reinforced epoxy composites for automotive bumper applications. *Mater Today Proc* 2021;46:433–8. https://doi.org/10.1016/j.matpr.2020.09.777.

68. Ilyas RA, Sapuan SM, Ibrahim R, Abral H, Ishak MR, Zainudin ES, et al. Sugar palm (Arenga pinnata (Wurmb.) Merr) cellulosic fibre hierarchy: A comprehensive approach from macro to nano scale. *J Mater Res Technol* 2019;8(3):2753–66. https://doi.org/10.1016/j.jmrt.2019.04.011.

69. Davis PD, Parbrook GD, Kenny GNC. Diffusion and osmosis. *Basic Phys Meas Anaesth* 1995:89–102. https://doi.org/10.1016/b978-0-7506-1713-0.50012-3.

70. Girisha C, Srinivas GR. Sisal / coconut coir natural fibers – Epoxy composites : Water absorption and mechanical properties. 2012;2:166–70.

71. Akash GKG, Venkatesha Gupta NS, Sreenivas Rao KV. A study on flammability and moisture absorption behavior of sisal/coir fiber reinforced hybrid composites. *IOP Conf Ser Mater Sci Eng* 2017;191. https://doi.org/10.1088/1757-899X/191/1/012003.

72. Bera T, Mohanta N, Prakash V, Pradhan S, Acharya SK. Moisture absorption and thickness swelling behaviour of luffa fibre/epoxy composite. *J Reinf Plast Compos* 2019;38(19–20):923–37. https://doi.org/10.1177/0731684419856703.

73. Atiqah A, Jawaid M, Sapuan SM, Ishak MR. Effect of Surface Treatment on the Mechanical Properties of Sugar Palm/Glass Fiber-reinforced Thermoplastic Polyurethane Hybrid Composites 2017.

74. Saba N, Jawaid M, Paridah MT, Al-othman OY. A review on flammability of epoxy polymer, cellulosic and non-cellulosic fiber reinforced epoxy composites. *Polym Adv Technol* 2016;27(5):577–90. https://doi.org/10.1002/pat.3739.

75. Bharath KN, Basavarajappa S. Flammability characteristics of chemical treated woven natural fabric reinforced phenol formaldehyde composites. *Procedia Mater Sci* 2014;5:1880–6. https://doi.org/10.1016/j.mspro.2014.07.507.

76. Gupta AK, Biswal M, Mohanty S, Nayak SK. Mechanical, thermal degradation, and flammability studies on surface modified sisal fiber reinforced recycled polypropylene composites. *Adv Mech Eng* 2012. https://doi.org/10.1155/2012/418031.

77. Asim M, Jawaid M, Nasir M, Saba N. Effect of fiber loadings and treatment on dynamic mechanical, thermal and flammability properties of pineapple leaf fiber and kenaf phenolic composites. *J Renew Mater* 2018;6(4):383–93. https://doi.org/10.7569/JRM.2017.634162.

78. Chen C, Gu X, Jin X, Sun J, Zhang S. The effect of chitosan on the flammability and thermal stability of polylactic acid/ammonium polyphosphate biocomposites. *Carbohydr Polym* 2017;157:1586–93. https://doi.org/10.1016/j.carbpol.2016.11.035.

79. Shrivastava NK, Wooi OS, Hassan A, Inuwa IM. Mechanical and flammability properties of poly(lactic acid)/poly(butylene adipate-co-terephthalate) blends and nanocomposites: Effects of compatibilizer and graphene. *Malays J Fundam Appl Sci* 2018;14(4):425–31. https://doi.org/10.11113/mjfas.v14n4.1233.

80. Suriani MJ, Sapuan SM, Ruzaidi CM, Nair DS, Ilyas RA. Flammability, morphological and mechanical properties of sugar palm fiber/polyester yarn-reinforced epoxy hybrid biocomposites with magnesium hydroxide flame retardant filler. *Text Res J* 2021. https://doi.org/10.1177/00405175211008615.

81. Durante M, Formisano A, Boccarusso L, Langella A, Carrino L. Creep behaviour of polylactic acid reinforced by woven hemp fabric. *Compos Part B Eng* 2017;124:16–22. https://doi.org/10.1016/j.compositesb.2017.05.038.

82. Meng Q, Wang Z. Creep damage models and their applications for crack growth analysis in pipes: A review. *Eng Fract Mech* 2019;205:547–76. https://doi.org/10.1016/j.engfracmech.2015.09.055.

83. Idris UD, Aigbodion VS, Abubakar IJ, Nwoye CI. Eco-friendly asbestos free brake-pad: Using banana peels. *J King Saud Univ - Eng Sci* 2015;27(2):185–92. https://doi.org/10.1016/j.jksues.2013.06.006.

84. Ruzaidi CM, Kamarudin H, Shamsul JB, Abdullah Rafiza MMA. Mechanical properties and wear behavior of brake pads produced from palm slag. *Adv Mater Res* 2012;341–342:26–30. https://doi.org/10.4028/www.scientific.net/AMR.341-342.26.

85. Badyankal PV, Manjunatha TS, Vaggar GB, Praveen KC. Compression and water absorption behaviour of banana and sisal hybrid fiber polymer composites. *Mater Today Proc* 2019;35:383–6. https://doi.org/10.1016/j.matpr.2020.02.695.

86. Bambach MR. Durability of natural fibre epoxy composite structural columns: High cycle compression fatigue and moisture ingress. *Compos Part C Open Access* 2020;2:100013. https://doi.org/10.1016/j.jcomc.2020.100013.

87. Balla VK, Kate KH, Satyavolu J, Singh P, Tadimeti JGD. Additive manufacturing of natural fiber reinforced polymer composites: Processing and prospects. *Compos Part B Eng* 2019;174:106956. https://doi.org/10.1016/j.compositesb.2019.106956.

88. Norizan MN, Abdan K, Salit MS, Mohamed R. Physical, mechanical and thermal properties of sugar palm yarn fibre loading on reinforced unsaturated polyester composite. *J Phys Sci* 2017;28(3):115–36. https://doi.org/10.21315/jps2017.28.3.8.

89. Safri SNA, Sultan MTH, Shah AUM. Characterization of benzoyl treated sugar palm/glass fibre hybrid composites. *J Mater Res Technol* 2020;9(5):11563–73. https://doi.org/10.1016/j.jmrt.2020.08.057.

90. Bakar NH, Hyie KM, Jumahat A, Kalam A, Salleh Z. Effect of alkaline treatment on tensile and impact strength of kenaf/Kevlar hybrid composites. *Appl Mech Mater* 2015;763:3–8. https://doi.org/10.4028/www.scientific.net/amm.763.3.

91. Tawakkal ISMA, Talib RA, Abdan K, Ling CN. Mechanical and physical properties of kenaf-derived cellulose (KDC)-filled polylactic acid (PLA) composites. *BioResources* 2012;7(2):1643–55. https://doi.org/10.15376/biores.7.2.1643-1655.

92. Alias NF, Ismail H, Ishak KMK. The effect of kenaf loading on water absorption and impact properties of polylactic acid/ natural rubber/ kenaf core powder biocomposite. *Mater Today Proc* 2019;17:584–9. https://doi.org/10.1016/j.matpr.2019.06.338.

93. Ramesh P, Prasad BD, Narayana KL. Effect of MMT clay on mechanical, thermal and barrier properties of treated Aloevera Fiber/ PLA-hybrid biocomposites. *Silicon* 2020;12(7):1751–60. https://doi.org/10.1007/s12633-019-00275-6.

94. Negawo TA, Polat Y, Akgul Y, Kilic A, Jawaid M. Mechanical and dynamic mechanical thermal properties of Ensete fiber/woven glass fiber fabric hybrid composites. *Compos Struct* 2021;259:113221. https://doi.org/10.1016/j.compstruct.2020.113221.

95. Sharba MJ, Leman Z, Sultan MTH, Ishak MR, Azmah Hanim MA. Effects of kenaf fiber orientation on mechanical properties and fatigue life of glass/kenaf hybrid composites. *BioResources* 2016;11(1):1448–65. https://doi.org/10.15376/biores.11.1.1448-1465.

96. Sreekumar PA, Joseph K, Unnikrishnan G, Thomas S. A comparative study on mechanical properties of sisal-leaf fibre-reinforced polyester composites prepared by resin transfer and compression moulding techniques. *Compos Sci Technol* 2007;67(3–4):453–61. https://doi.org/10.1016/j.compscitech.2006.08.025.

97. Sapiai N, Jumahat A, Mahmud J. Flexural and tensile properties of kenaf /glass fibres hybrid composites. *J Teknol* 2015;3:115–20.

98. Assarar M, Zouari W, Sabhi H, Ayad R, Berthelot JM. Evaluation of the damping of hybrid carbon-flax reinforced composites. *Compos Struct* 2015;132:148–54. https://doi.org/10.1016/j.compstruct.2015.05.016.

99. Shih YF, Chang WC, Liu WC, Lee CC, Kuan CS, Yu YH. Pineapple leaf/recycled disposable chopstick hybrid fiber-reinforced biodegradable composites. *J Taiwan Inst Chem Eng* 2014;45(4):2039–46. https://doi.org/10.1016/j.jtice.2014.02.015.

100. Martini E, Roshetko JM, van Noordwijk M, Rahmanulloh A, Mulyoutami E, Joshi L, et al. Sugar palm (Arenga pinnata (Wurmb) Merr.) for livelihoods and biodiversity conservation in the orangutan habitat of batang Toru, North Sumatra, Indonesia: Mixed prospects for domestication. *Agrofor Syst* 2012;86(3):401–17. https://doi.org/10.1007/s10457-011-9441-0.

101. Sapuan SM, Ishak MR, Leman Z, Huzaifah MRM, Ilyas RA, Ammar IM, et al. 2017. Pokok Enau: Potensi dan Pembangunan Produk:1–160.

102. Wahyuni AS, Elhusna. The tensile behaviour of concrete with natural fiber from sugar palm tree. *Int Conf Eng Sci Res Dev* 2016:15–8.

103. Prabowo AE, Diharjo K, Ubaidillah, PI. Sound absorption performance of sugar palm trunk fibers. *E3S Web Conf* 2019;130:1–9. https://doi.org/10.1051/e3sconf/201913001003.

104. Imran M, Mohd Khairi,. Sound insulation and vibration reduction of modified zinc roof using natural fiber (Arenga Pinnata) 2015;Semantic Scholar Corpus ID: 113030658.

105. Ilyas RA, Sapuan SM, Ishak MR. Zainudin ESD and characterization of sugar palm nanocrystalline cellulose reinforced sugar palm starch bionanocomposites. Development and characterization of sugar palm nanocrystalline cellulose reinforced sugar palm starch bionanocomposites. *Carbohydr Polym* 2018;202:186–202. https://doi.org/10.1016/j.carbpol.2018.09.002.

106. Bachtiar D, Salit MS, Zainudin E, Abdan K, Dahlan KZHM. Effects of alkaline treatment and a compatibilizing agent on tensile properties of sugar palm fibrereinforced high impact polystyrene composites. *BioResources* 2011;6(4):4815–23. https://doi.org/10.15376/biores.6.4.4815-4823.

107. Sapuan SM. RA Ilyas and SFK Sherwani: Advanced applications of sugar palm reinforced polymer composites. INSTITUTE OF TROPICAL FORESTRY AND FOREST PRODUCTS INTROP, UPM MALAYSIA

108. Alaaeddin MH, Sapuan SM, Zuhri MYM, Zainudin ES. Development of photovoltaic module with fabricated and evaluated novel backsheet-based MDPI, Materials 2019, 12, 3007; doi:10.3390/ma12183007 2019.

109. Atiqah A, Jawaid M, Sapuan SM, Ishak MR. Dynamic mechanical properties of sugar palm/glass fiber reinforced thermoplastic polyurethane hybrid composites. *Polym Compos* 2019;40(4):1329–34. https://doi.org/10.1002/pc.24860.
110. Ferdiansyah T, Razak HA. Mechanical properties of black sugar palm fiber-reinforced concrete. *J Reinf Plast Compos* 2011;30(11):994–1004. https://doi.org/10.1177/0731684411411335.
111. Misri S, Leman Z, Sapuan SM, Ishak MR. Mechanical properties and fabrication of small boat using woven glass/sugar palm fibres reinforced unsaturated polyester hybrid composite. *IOP Conf Ser Mater Sci Eng* 2010;11:012015. https://doi.org/10.1088/1757-899x/11/1/012015.
112. Harussani MM, Sapuan SM, Rashid U, Khalina A. Development and characterization of polypropylene waste from personal protective equipment (Ppe)-derived char-filled sugar palm starch biocomposite briquettes. *Polymers (Basel)* 2021;13(11). https://doi.org/10.3390/polym13111707.

2

Environmental Degradation of Natural Fibre–Reinforced Composites

Amna Siddique, Khubab Shaker, and Hanur Meku Yesuf

2.1 Introduction

Degradation is the gradual loss of material properties as a result of exposure to environmental conditions [1]. Environmental degradation represents one of the biggest problems during the service life of a product. The
degradation may lead to material deformation (fracture, creep, etc.), corrosion, wear, fatigue, oxidation, etc. The factors affecting material degradation include temperature, light, electromagnetic radiation, mechanical
load, microorganisms, etc. [2]. The degradation type governed by the biological environment is termed biodegradation [3]. Natural fibre–reinforced
composites have received significant interest from researchers as the next
stage of composite products [4–6] This is due to the increased concern for
renewable green resources and environmental protection over the past few
decades [7]. Natural fibre–reinforced composite materials are more prone
to environmental degradation due to the high moisture absorption and low
thermal stability of the natural fibres [7]. Furthermore, polymer degradation
is also induced by oxidation, thermal activation, hydrolysis, or photolysis.
Therefore, the knowledge of degradation kinetics is essential for the safe use
of natural fibre-reinforced composites (NFRCs) [8]. The addition of a plasticizer may affect the degradation behaviour of the material. Pelegriny et al.
used a plasticizer (triacetin) and showed that it inhibits the degradation of
polylactic acid (PLA) for up to 45 days after exposure. However, it no longer
affects the degradation process after that period [9]. This chapter aims to
discuss various mechanisms involved in the environmental degradation of
natural fibre–reinforced composites, which are necessary to consider when
designing a composite material for specific purposes to ensure durability
and safe application.

2.2 Classification of Environmental Degradation

The environmental degradation of a polymer composite material can occur by chemical, physical, and biological processes or a combination of these. The degradation is the result of environmental factors such as temperature (thermal degradation), microorganisms (biodegradation), moisture (hydrolytic degradation), air (oxidative degradation), light and radiation (photodegradation), chemical agents (corrosion), and mechanical loads [8]. The environmental degradation of NFRCs can be categorized into chemical, physical, and biodegradation, as shown in Figure 2.1 [10].

These environmental factors lead to irreversible changes in the NFRC. There are two distinct phases identified for environmental degradation. The first stage is the deterioration of appearance, physical, morphological, or mechanical properties, while the second is attributed to the complete breakdown of material into its constituents (water, carbon dioxide, etc.). All the materials are affected by these factors; however, the rate of deterioration may vary depending on the nature of the material. The rate of deterioration of a material helps to distinguish between environmentally degradable and non-degradable materials.

Environmental degradation is the desired process at end-of-life, but its effects are undesirable during the service life of the material. A NFRC may be exposed to environmental factors for a number of years during its service life. The composite may perform effectively for a certain period, but after that, its performance becomes poor, and therefore, it no longer fulfils the designed properties [11]. Chemical ageing is due to the irreversible change of polymer chains. Irreversible changes are the cross-linking of chains or the splitting of chains. Oxidative, thermo-oxidative, and hydrolytic degradation are included in chemical ageing. The important parameters that greatly affect the composites are cross-linking and oxidation. By increasing the temperature and time, oxidative degradation is

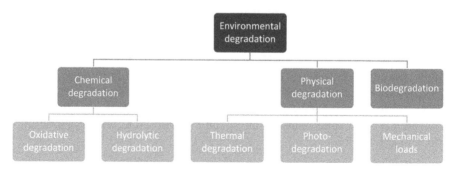

FIGURE 2.1
Schematic of factors involved in environmental degradation of NFRCs.

accelerated. Cross-linking is another factor that highly affects the physical properties of composites by changing the density of the polymers. Physical properties such as strength and stiffness change due to the cross-linking of molecules.

2.2.1 Oxidative Degradation

Oxidative degradation refers to material degradation caused by the action of oxygen. The oxidation reaction of materials is normally a radical chain reaction mechanism [12]. To determine the thermo-oxidative effect, a thermogravimetric analyser is used. This analysis is based on the weight loss of the material after oxidation. As the temperature increases, this effect also increases. The initial weight loss of the composites is due to the removal of moisture or volatile compounds from the sample. But after several hours, the weight will be stabilized. After stabilization, if again the sample has weight loss, then it is due to the thermo-oxidative effect.

Certain physical changes (colour, structure, crack density, etc.) are observed in the composites because of oxidative degradation. Polymers absorb water due to these physical changes, and the absorption depends upon the crystalline structure and orientations of the material. Change in mechanical properties occurs due to moisture uptake [13].

2.2.2 Hydrolytic Degradation

This is a type of environmental degradation in which vulnerable bonds in a polymer chain react with water molecules, causing them to break up into smaller chains. Hydrolytic degradation is normally due to the diffusion of water into the composites [14]. Crystallinity plays a significant role in limiting water diffusion, which is restricted in the crystalline regions as compared with the amorphous regions. In natural fibre–reinforced composites, generally, natural fibres are vulnerable to moisture and humidity due to the presence of substantial levels of hydrophilic chemical constituents such as cellulose and hemicellulose. Hence, water absorption plays a vital role in explaining the value and endurance of natural fibre–reinforced composites [15].

Moisture uptake may result in micro-cracks, which will lead to increased moisture absorption and hydrolytic degradation of the material [16–23]. The hydrolytic degradation of composite materials occurs mainly due to the accumulation of water within the fibre–matrix interface due to the slow development of superficial flaws.

2.2.3 Thermal Degradation

Generally, the degradation of composites is highly dependent on the temperature. Mostly, polymers change their properties with temperature. The thermal stability of natural fibres is very limited, which prohibits their use in some industrial practices and limits their usage for moderate–temperature applications [24]. Likewise, the probability of cellulose degradation and volatile material discharge increases due to the low thermal stability, which may have a negative effect on the properties of the composite. Thus, the processing temperatures are confined to around 200 C, but it is viable to employ higher temperatures for short times [25].

In structural applications of composites, thermal properties are also notable for things like material behaviour, load capacity at a definite temperature, temperature spread from one end to the other, and dimensional stability at high temperatures [25].

Thermogravimetric analysis (TGA) is performed to examine the thermal degradation of composites. The resultant features can be applied during processing to deal with the degradation of processed material. According to Yao et al., the decomposition temperature range of numerous natural fibres overlaps with the processing temperatures of certain thermoplastics. In the case of natural fibre–reinforced composites, the thermal stability is affected in two ways: first, direct thermal contraction or expansion; second, change in the quantity and pace of moisture absorption, which results in swelling of natural fibres. Due to thermal expansion, residual stresses are produced, which degrade the composite. As a result, micro-cracks, delamination, and debonding of fibres may occur. Micro-cracks can be produced in the matrix due to the thermal cycling, and degradation of mechanical properties of the composites occurs. The increased adhesion between the natural fibre and the matrix results from the oxidation process that happens during the production of composites, which is caused by thermal decay of these materials [26].

2.2.4 Photodegradation

This effect occurs at elevated temperature, because cross-linking of chains or chain breakage occurs at elevated temperature, due to which the composite degrades. UV rays affect the surface of the composite and degrade the matrix in the composite from the surface. There will be a change in the optical properties, such as a change in colour. UV rays of 330 nm wavelength easily break many bonds of the composites present in the polymer matrix [15].

2.2.5 Biodegradation

Biodegradation presents a substantial change in the chemical structure under environmental conditions. These changes result in a loss of mechanical as well as physical properties, as measured by their life cycle assessment. These changes are the result of the effect of microorganisms like fungi, algae, and bacteria [3]. Generally, the biodegradability of natural fibres is cited as a positive attribute that can be employed to justify the application of these fibres. But for the sake of various outdoor applications of these composites, they should be weatherproof. To ensure durable and long-lasting application of these composites, it is necessary to control their natural degradation processes. One way to slow down or stop natural decomposition is the modification of cell wall composition. Hence, unfavourable features of cellulosic fibres, such as chemical degradation, biodegradability, flammability, and dimensional instability, can be slowed or avoided. By impeding the closely packed hydroxyl (–OH) groups on the fibre surface, chemical treatments can decrease the amount of water absorption by the natural fibres, which in turn, reduces biological deterioration and swelling [15].

2.3 Degradation-Induced Defects in Composite Materials

Environmental degradation of composites observed on a macro scale is irreversible. There are various degradation mechanisms, such as breakage of fibre, cracking of matrix, degradation of interface, delamination, and plastic deformation of composite, which have a direct effect on mechanical properties like strength and stiffness [20]. The behaviour of the material becomes plastic if high stress is applied and plastic strain is produced in the material. The mechanical degradation of materials occurs after chemical and physical ageing. The main problem of thermosets is thermo-oxidative degradation, which produces micro-cracks in the matrix at the surface and edges. When these micro-cracks are formed, they provide sites for the acceleration of macro-cracks due to mechanical loading. Once macro-cracks are present, the chemical degradation rate increases due to reactive sites on the surface and interior of the matrix, and crack growth increases [27].

Among various degradation environments, the most significant degradation observed in the case of NFRCs include:

- Thermo-oxidation
- Hygrothermal degradation
- Physical and chemical degradation
- Biodegradation

Figure 2.2 shows the defects produced in a composite due to hygrothermal ageing, which ultimately reduces the material strength and stiffness [27].

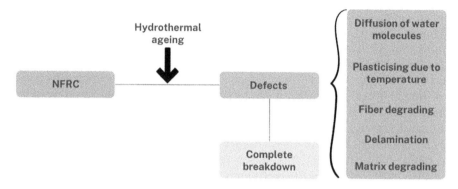

FIGURE 2.2
Defects produced in composites due to hydrothermal ageing.

2.4 Chemical Degradation Mechanisms and Processes of NFRCs

2.4.1 Hydrolytic Degradation

Natural fibres have a primary wall and three secondary walls, consisting of cellulose microfibrils cemented in a lignin matrix. The mechanical performance of a natural fibre depends on the thickness of the middle layer in the secondary wall. However, water molecules can diffuse through the cell wall from a high-concentration to a low-concentration area. As discussed earlier, the main limitation in the application of NFRCs is their susceptibility to moisture [7]. The higher moisture uptake of the composites decreases their mechanical properties due to the different moisture uptake of the fibre and the matrix. This difference in the moisture uptake of the fibre and the matrix results in differential swelling with respect to one another. It produces stresses at the fibre–matrix interface, which decrease the fibre–matrix bonding and decrease the mechanical properties of composites [28, 29]. Due to fibre swelling, micro-cracks form, and ultimately, fibre–matrix interface debonding occurs, as shown in Figure 2.3.

Natural fibres are rich in cellulose, pectin, lignin, and hemicelluloses, all containing hydroxyl groups that are strongly polar and hydrophilic, whereas matrix polymers show substantial hydrophobicity. Consequently, there are many challenges of compatibility between the fibre and matrix that cause the interface region between natural fibres and matrices to deteriorate. Water absorption occurs at the outer layers of composites and steadily decreases into the bulk of the matrix [30].

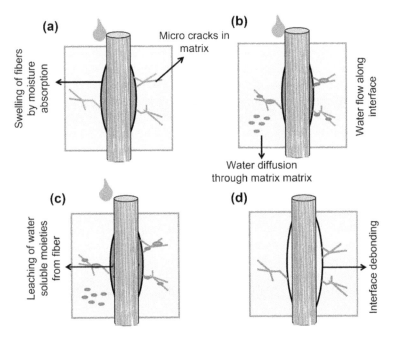

FIGURE 2.3
Effect of moisture absorption on fibre–matrix interface. Concept taken from [15].

2.4.1.1 The Effect of Moisture on Jute Fibre–Reinforced Composites

Jiang et al. presented the effect of hygrothermal ageing on jute/PLA composites using X-ray tomography. The composites aged for 28–56 days showed a small decrease in tensile modulus (>25% after 28 days) and tensile strength (>12% in 0–7 days), and matrix cracking was almost invisible. After 28–56 days of passage, the moisture uptake of composite samples increased significantly due to the formation of macro-cracks. The increase in porosity was due to matrix hydrolysis followed by embrittlement and internal cracking, which was mainly attributed to the loss of mechanical properties [31].

2.4.1.2 The Effect of Moisture on Flax Fibre–Reinforced Composites

Natural fibre–reinforced composites in humid environments have the main challenge of moisture absorption and loss of their durability. So, the mechanical properties of flax fibre and glass fibre composites were compared in a study [32]. Flax fibre has 12% more water absorption than glass fibre composites, which strongly reduces the mechanical properties of fibres.

The glass fibre composites were prepared with eight layers in epoxy resins and cured at 130 C. The flax fibre composites, with 11 plies of unidirectional fibres, were immersed in epoxy resin by the hand layup method. Both composites were immersed in water for some time, and their water absorption was checked by gravimetric analysis. Flax fibre composites had 13.5% water absorption, and glass fibre composites had 1.05%. As the immersion time for the glass fibre and flax fibre increased, failure stress decreased. For the glass fibre, failure stress was 9% after the first 10 days of ageing in water. For the flax fibres, the stress at failure was reduced by 13% on the first day. Both these composites were aged for 10 to 30 days. The failure stress decreased by 25% as compared with the unaged composite. Flax fibre strength decreased by 20% after 20 days of ageing [32].

After the immersion of composites in water, several tests were performed at different intervals of time to check ageing properties. It was concluded that water immersion degrades the modulus and the tensile strength of flax fibre in contrast to glass fibre. The failure stress is decreased for both composites by increasing the immersion time. The degradation of the mechanical properties of both composites was due to the poor matrix interface bonding as affected by water ageing [32].

Cellulosic fibres have high elastic modulus and specific strength. When natural cellulosic fibres like kenaf, flax, and jute are exposed to high humidity, their moisture absorption increases due to the presence of –OH groups [33]. One study reported a decrease in the mechanical properties of flax fibre by the absorption of moisture. A small decrease in stiffness and a large decrease in strength occurred after 38 days of ageing in water. This decrease was due to the plasticizing effect of water on the matrix. Several failure modes are observed in flax fibre failure, including axial splitting and pulling out of microfibrils. This was mainly attributed to matrix degradation.

2.4.1.3 The Effect of Moisture on Kraft Fibre–Reinforced Composites

The effect of water on the physical and mechanical properties of the kraft fibre–reinforced polypropylene composite was discussed. The penetration of water molecules into the composites leads to micro-cracks due to swelling and decreases the mechanical properties. Water diffuses into the micro-gaps between the molecular chains and then moves, due to capillary action, into the fibre–matrix interface. Due to the motion of water, the interface becomes weak, and debonding of composites starts. The water is then stored in these gaps and decreases the properties. This occurs due to the environment or service life of the composite [34].

Two types of samples were formed, one with the coupling agent MAPP and the other without MAPP. The tensile strength and modulus were increased by the addition of a coupling agent. The coupling agent strengthens the interface between the matrix and reinforcement and hence, increases the physical and mechanical properties of composites. By adding the coupling agent, the tensile strength increased by 64%, and the modulus increased by 220% [34]. By decreasing the length of the fibre, tensile strength and modulus decreased, while failure stress increased due to reduced reinforcing efficiency. Beating the fibres for 5.5 min increased the modulus from 41 to 45 MPa in beaten fibre composites. The diffusion of water is greater without than with the coupling composite, and it increases as the ageing temperature increases, which damages the fibre–matrix bonding and degrades the composite.

The material properties change when the composite is exposed to moisture in the environment. The degradation of the material starts, which results in debonding of the fibre–matrix interface and matrix cracks, and the properties of the material decrease. It is noted that the effect of moisture becomes more significant at high temperatures than at low temperatures [35, 36].

Hygrothermal ageing affects lignocellulosic composites by moisture diffusion, fibre swelling followed by crack formation, leaching of hydrophilic substances, and finally, fibre/matrix debonding upon drying [29]. However, the moisture absorption may be reduced by certain chemical treatments such as alkalization, acetylation, salinization, etc. [37]. The moisture absorption of flax fibre composites was reduced by up to 45.80% by chemical treatment, while the fibre/matrix interfacial bonding was improved.

Moisture absorption of natural fibre–reinforced composites is influenced by many factors, including fibre type, crystallinity, degree of fibre cross-linking, fibre volume fraction, and the response between water molecules and polymer. The temperature also affects the moisture uptake of the composite [38]. By increasing the temperature, the moisture uptake is increased due to thermal stresses. These thermal stresses may initiate the degradation of the composite and therefore, change the mechanical properties of the composite. The decrease in properties at higher temperatures is greater as compared with low temperatures [39].

2.4.2 Ageing of Composites Treated with NaOH

2.4.2.1 Jute Fibre Composites

In one study, unidirectional jute roving–reinforced polyester composites were prepared from jute roving of an average 2.5 mm diameter and unsaturated polyester that was wetted by using a wetting agent. The jute roving was treated with 10% NaOH, and after drying, it was used as reinforcement [40].

Natural fibres are mostly polar, but the polymer matrices are non-polar, so they have a weak interface and hence, weak mechanical properties. So, the jute fibre was treated with NaOH, which increases the surface energy of the fibres and increases the interface properties of the reinforcement and the matrix. NaOH removes the impurities from the surface and makes shorter fibres. So, after NaOH treatment, fibres provide more sites for the polymer impregnation. There were two samples: one was treated with NaOH, and the other was without NaOH treatment. The NaOH-treated sample presented good mechanical properties; however, after 8 days of ageing in water, it presented poor mechanical properties [40].

2.4.2.2 Sisal Fibre–Reinforced Composites

Abdullah et al. studied the environmental degradation of sisal fibre–reinforced composites. Benzene ethanol was used for the extraction of sisal fibres. The extracted sisal fibres were treated with NaOH for 1.5 hours, and then, benzylation was carried out in a jar containing benzyl chloride at a temperature of 115 C. Finally, the benzylated sisal fibres were obtained by drying the mixture in an oven. The self-reinforced sisal composite was formed by the hot press moulding technique, and the composites were kept in water for a specific time. The results revealed approximately a 15% loss in impact strength, flexural modulus, and flexural strength. This minor loss of composite properties was attributed to the hydroxyl groups having been substituted by benzyl groups in the sisal fibre [33].

2.4.3 Thermo-Oxidative Degradation of NFRCs

The degradation induced by the combined action of heat and oxygen is termed thermo-oxidative degradation. The thermal and thermally activated processes affect the performance of composites after they have been in service for some time. When degradation starts, various processes occur in the matrix:

- Oxygen diffusion into the matrix
- The evolution of degraded products formed due to the diffusion of the oxygen into the matrix
- The reaction of oxygen with the matrix
- The reaction within the matrix due to the thermal degradation

The products formed during oxidation move into and out of the product by mass transfer. This may wear out the surface of the composites [41].

Natural fibres present low temperature stability, which is one of their major drawbacks. For most natural fibres, their thermal stability temperature is up to 200 C for a short exposure time, while their decolouration and decomposition start above 100 C when exposed for a long time. Hence, it is problematic to use them with engineering thermoplastic matrices with melting temperatures >200 C. To overcome the low thermal stability issue of natural fibres, a commonly used method is to combine them with low–melting temperature thermoplastic resins (polypropylene or polyethylene) or low–curing temperature thermosetting resins [42].

Thermal stability is normally characterized by the glass transition temperature of the material. But, the glass transition temperature is not a good parameter to determine this type of effect. It measures the stiffness of the polymer. Some matrices show thermo-oxidative stability due to the presence of fluorine in their structure, especially polyimide-containing fluorinated groups. Various molecular features can increase the thermo-oxidative stability of polymers, including degree of cross-linking, presence of cyclic groups in the molecular chains, and isomeric differences. If these features are added to the molecular chains, then that matrix can behave like a thermo-oxidative stable matrix [43].

2.4.3.1 Effect of Thermo-Oxidation on Epoxy Resin

In NFRCs, thermo-oxidation mainly affects the properties of the matrix material; hence, the matrices used with natural fibres, such as epoxy resins, are studied for the thermo-oxidative effect. Those polymeric composites that are used in aircraft should withstand temperatures between 177 C and 232 C for a long time. The material should be consistent in its properties for the desired lifetime. Epoxy resin is used because it allows better evaluation for such applications [44].

The effect of different curing temperatures on epoxy resin was studied, and it was found that as the temperature of curing increases, more stable epoxy will be formed. It is found that the curing of epoxy too near to its T_g or above its T_g will not clarify the thermo-oxidative ageing effect properly because the behaviour will be nonlinear, so it does not give an actual prediction of material behaviour. So, the curing temperature should be below the T_g for good prediction of material properties [45].

It was noted that the thermo-oxidative degradation starts at the edges of the epoxy material, so a thin aluminium foil is used on the edges to minimize the effect of thermo-oxidation [44–46]. Compared with the interior of the matrix, the depth in the surface layer determined the loss of strength. Because if the surface layer depth is less, then the core of the material will be undamaged, and there will be less decrease in mechanical properties. There will be two factors that determine the loss in strength: the first is surface degradation depth, and the second is the reaction inside the core of the material [47].

2.4.3.2 Effect of Material Properties and Environment on Thermo-Oxidative Resistance of NFRCs

The toughness of the material also contributes to the stability of the composites against the thermo-oxidative effect. If the reinforcement is less affected by thermo-oxidation than the matrix, then the stability of this composite increases. Because the surface will be less degraded, and the material resists further degradation due to toughness, the rate of degradation decreases, and hence, the service life of the composite increases [43, 46].

A study of composites in a nitrogen environment was performed. The effect of ageing at 150 C in nitrogen and in normal air for 9 months was determined. There were several changes in the physical properties of composites; a 30–40% decrease in bending strength, a 40–60% decrease in strain, and an almost 20% increase in modulus were observed, along with a 1–2% loss in weight. So, it was concluded that ageing is more affected by ambient air but less affected by nitrogen [48].

2.5 Physical Degradation

The change in the mechanical and physical properties of composites with time is called physical ageing. If material is heated above its T_g, then it can instantly achieve thermodynamic equilibrium, but if it is heated below its T_g, then it takes time to achieve thermodynamic equilibrium. The material properties change during the time to equilibrium. This process of changing the properties of the material is called physical ageing [49].

Creep is one of the important characteristics to define the performance of composites. Several tests were performed on materials, and some conclusions were obtained by studying the creep behaviour of the material.

- A basic characteristic of the composite is its physical ageing, whether it is polymeric or organic.
- When materials have a glass transition temperature, their mechanical properties strongly depend upon the ageing time.
- All glassy materials have the same dependency on physical ageing irrespective of their chemical structures.
- The physical ageing phenomenon is thermo-reversible, i.e. when the material is heated above T_g, then it will be absent from the material, and if the temperature is below T_g, then physical ageing will be a property of the material [50].

First, the temperature of the material is increased to its T_g for some time, and then, the temperature is decreased to that limit on which the test must be performed. This process is called rejuvenation. The temperature of the material remains constant after the rejuvenation, and tests are performed for the creep behaviour.

2.5.1 Effect of Temperature and Moisture on Physical Ageing

This is related to the absorption of water into the composites. The moisture content increases the plastic behaviour of the material, and degradation starts. This effect is increased by the stress applied during service life. Substances having low molecular weights, like water, can diffuse into the matrix. The amount of liquid absorbed into the composites depends upon the chemical structure and the crystalline nature of the material. It also depends upon the impermeability of the fibres and the interface of the fibre–matrix composite. The properties of the material become poorer as the amount of water absorbed increases. The combined effect of moisture and temperature affects the material in various ways [51].

The absorption of water into the matrix decreases the glass transition temperature of the material, which directly affects the mechanical properties of the composites. The swelling of the material is also an important factor in the composites. By the absorption of water, the swelling of material occurs, which causes residual stresses in the composite. Due to the residual stresses, micro-cracks are produced in the material. Then, these micro-cracks increase the rate of water absorption. Due to this absorption, debonding of the fibre–matrix starts, and this causes macro-cracks in the composites [52].

There are several changes in the composites due to water absorption, such as matrix cracking, micro-crack generation, ply delamination, and blisters forming on the surface of the composites as a result of brief heating alone, even if no mechanical load is applied. The polymers that are used in the composite for the binding of fibres take up moisture from the environment and cause a change in the physical and mechanical properties of the composites [53]. Mixing the polymer with a miscible liquid decreases the glass transition temperature of the polymer. So when the moisture is absorbed by the polymer, the glass transition temperature decreases, which has a direct effect on the physical and mechanical properties of the composites [54, 55].

So, the ageing properties at constant temperature and moisture are different from those at variable temperature and moisture content.

2.5.2 Effect of UV Radiation

UV radiation, humidity, and temperature have a significant influence on the performance of NFRCs. When NFRCs are exposed to UV radiation, photodegradation starts due to the breakage of the constituents of natural fibres, such as cellulose, lignin, and hemicellulose. This reaction can affect the interfacial bonding between the matrix and the fibre. The weak interfacial bonding will ultimately decrease the effectiveness by which load transfer occurs in components of the composite, consequently reducing the mechanical performance of the composite [56].

The photochemical reactions occur when polymer composites are exposed to extreme weather conditions during application, such as composites used in the automobile industry [57]. The photochemical reaction is also known as photodegradation. In the photodegradation process, excessive thermal oxidation and UV penetration result in the strength loss of NFRCs [58]. To minimize the effect of environmental degradation on NFRCs used for outdoor applications, hybridization with highly moisture-resistant and photodegradation-resistant hydrophobic thermoplastic fibres should be considered.

2.5.3 Mechanical Loading

The cracks between the plies of a laminated composite and the micro-cracks in the composite are critical factors for mechanical degradation. The cracking of the composite starts with the cracking of the matrix, which may be the result of mechanical stress or fatigue loading. It may be due to hygrothermal or chemical effects on the composite matrix. When the thermal expansion occurs due to temperature changes, the chances of cracking increase between the plies of fibres or fabrics that are placed in transverse and longitudinal directions.

Due to these cracks, the mechanical properties of the composites are reduced. The properties of the matrix depend upon the fibre, polymer matrix, and fibre–matrix interface. But generally, the properties of these components change due to long-term ageing, mainly due to the changes in the mechanical properties of the polymer matrix. The polymers are characterized by their viscoelastic behaviour. To check the viscoelasticity of the material, the creep property is checked, which is a time-dependent deformation at constant load. Stiffness is not the only concern with polymeric composites; strength is also an important factor to be considered. The creep and the recovery behaviour of the composite highly depend upon the viscoelastic behaviour. During the modelling of long-term ageing, this factor should be considered. The long-term viscoelastic behaviour can be predicted by the time–temperature superposition. The long-term properties at low temperatures can correspond to the short-term properties at high temperatures [59].

Mechanical ageing depends on various factors, such as temperature, moisture, accelerated ageing, and degradation mechanisms [31]. The temperature effect that is normal for the polymer may become critical due to the moisture absorption of the composite matrix, because moisture absorption into the composite material decreases the T_g [60].

2.6 Test Methods and Parameters to Analyse Environmental Degradation

Several approaches are adopted by researchers to quantify the environmental degradation in NFRCs. Some common tests include physical appearance, weight loss measurement, strength/stiffness loss measurement, thermogravimetric analysis, etc.

2.6.1 Weight Loss Measurement

Weight loss can be correlated with the mechanical degradation of composites. As weight loss increases, mechanical properties decrease. A small change in weight loss can effect very large strength loss [43, 61, 62]. It shows greater degradation at the edges as compared with the interior of the matrix. The depth in the surface layer determines the loss of strength, because if the depth is less, then the core of the material will be undamaged, and there will be less decrease in mechanical properties. There will be two factors that determine the loss in strength: the first is surface degradation depth, and the second is the reaction inside the core of the material [45, 63, 64].

2.6.2 Thermogravimetric Analysis

For the thermal stability of materials, thermogravimetric analysis is performed. In this method, the results are calculated for the weight loss at different temperatures [61, 62, 65]. Later, it was considered that thermogravimetric analysis is not good for determining the ageing of composites. Tsotsis said that for a very small change in weight loss, a huge change in mechanical properties occurs. So, if there is a small error in the weight loss measurement, then there will be a large error in the mechanical properties of composites, and therefore, this method is not a good indicator for the long-term ageing of composites [60].

2.7 Conclusions and Future Perspectives

Environmental degradation of a polymer composite can occur due to various environmental factors, such as temperature (thermal degradation), microorganisms (biodegradation), moisture (hydrolytic degradation), air (oxidative degradation), light and radiation (photo-degradation), chemical agents (corrosion), and mechanical loads. The limiting factors in the use of natural fibre–reinforced composites in outdoor applications are water absorption and low thermal stability, since in wet conditions, NFRCs deteriorate, and their mechanical properties decrease. The low thermal stability of natural fibres makes it difficult to combine them with engineering thermoplastic resins. The hydrophilic nature of natural fibres influences the ageing behaviour of NFRCs in wet environments. To decrease their environmental degradation in outdoor applications, more in-depth studies on the mechanical behaviour of NFRCs exposed to aggressive environments for a long duration are required. Cellulose fibres contain several –OH groups and form hydrogen bonds by interaction with water molecules. Swelling by water uptake results in swelling of fibres, subsequently leading to micro-cracking, interface breakage, and damage to the mechanical properties. A good interfacial adhesion between the fibre and the matrix is a way to reduce water absorption at the composite interface. The predominant damage modes in NFRCs are matrix cracking, interfacial debonding, and delamination. In order to fully utilize the potential of natural fibres in NFRCs in various applications, strong interfacial adhesion between the fibre and the matrix should be guaranteed.

References

1. K. J. Kurzydłowski, M. Lewandowska, W. Święszkowski, and M. Lewandowska-Szumiel, "Degradation of engineering materials – Implications to regenerative medicine," *Macromol. Symp.*, vol. 253(1), pp. 1–9, 2007, doi: 10.1002/masy.200750701.
2. A. Findrik et al., "In vitro degradation of specimens produced from PLA / PHB by additive manufacturing in simulated conditions," *Polymers*, vol. 13, pp. 1542, 2021. https://doi.org/10.3390/polym13101542.
3. Y. Tokiwa, B. P. Calabia, C. U. Ugwu, and S. Aiba, "Biodegradability of plastics," *Int. J. Mol. Sci.*, vol. 10, no. 9, pp. 3722–3742, 2009, doi: 10.3390/ijms10093722.
4. A. Ali et al., "Hydrophobic treatment of natural fibers and their composites – A review," *J. Ind. Text.*, vol. 47, no. 8, 2016, doi: 10.1177/1528083716654468.
5. M. M. Rehman, M. Zeeshan, K. Shaker, and Y. Nawab, "Effect of micro-crystalline cellulose particles on mechanical properties of alkaline treated jute fabric reinforced green epoxy composite," *Cellulose*, vol. 26, no. 17, pp. 9057–9069, 2019, doi: 10.1007/s10570-019-02679-4.
6. A. Ali et al., "Impact of hydrophobic treatment of jute on moisture regain and mechanical properties of composite material," *J. Reinf. Plast. Compos.*, vol. 34, no. 24, pp. 2059–2068, Oct. 2015, doi: 10.1177/0731684415610007.
7. M. Usman et al., "Performance evaluation of jute / glass-fiber-reinforced different layering configurations," *Materials (Basel)*, vol. 15, pp. 3–4, 2022.
8. M. Brebu, "Environmental degradation of plastic composites with natural fillers-a review," *Polymers (Basel)*, vol. 12, no. 1, 2020, doi: 10.3390/polym12010165.
9. K. Pelegrini et al., "Degradation of PLA and PLA in composites with triacetin and buriti fiber after 600 days in a simulated marine environment," *J. Appl. Polym. Sci.*, vol. 133, no. 15, 2015, doi: 10.1002/app.43290.
10. A. F. Balogová et al., "In vitro degradation of specimens produced from pla/phb by additive manufacturing in simulated conditions," *Polymers (Basel)*, vol. 13, no. 10, pp. 1–17, 2021, doi: 10.3390/polym13101542.
11. K. Joseph and S. Thomas, "Effect of aging on the physical and mechanical properties of Sisal fiber reinforced polyethylene composites," *Compos Sci Technol.*, vol. 53, no. 94, pp. 99–110, 1995.
12. J. Izdebska, "Aging and degradation of printed materials," In *Printing on Polymers*, J. Izdebska and S. Thomas, Eds. William Andrew Publishing, 2016, pp. 353–370.
13. M. Brebu, "Environmental degradation of plastic composites with natural fillers—A review," *Polymers (Basel)*, vol. 12, no. 1, 2020, doi: 10.3390/polym12010166.
14. S. P. Lyu and D. Untereker, "Degradability of polymers for implantable biomedical devices," *Int. J. Mol. Sci.*, vol. 10, no. 9, pp. 4033–4065, 2009, doi: 10.3390/ijms10094033.

15. M. Mohammed et al., "Challenges and advancement in water absorption of natural fiber-reinforced polymer composites," *Polym. Test.*, vol. 124, p. 108083, 2023, doi: 10.1016/j.polymertesting.2023.108083.

16. J. P. Sargent, "Durability studies for aerospace applications using peel and wedge tests," *Int. J. Adhes. Adhes.*, vol. 25, no. 3, pp. 247–256, 2005, doi: 10.1016/j.ijadhadh.2004.07.005.

17. B. F. Boukhoulda, E. Adda-Bedia, and K. Madani, "The effect of fiber orientation angle in composite materials on moisture absorption and material degradation after hygrothermal ageing," *Compos. Struct.*, vol. 74, no. 4, pp. 406–418, 2006, doi: 10.1016/j.compstruct.2005.04.032.

18. J. M. Hutchinson, "Physical aging of polymers," *Prog. Polym. Sci.*, vol. 20, no. 4, pp. 703–760, Jan. 1995, doi: 10.1016/0079-6700(94)00001-I.

19. M. C. Lafarie-Frenot, "Damage mechanisms induced by cyclic ply-stresses in carbon–epoxy laminates: Environmental effects," *Int. J. Fatigue*, vol. 28, no. 10, pp. 1202–1216, 2006, doi: 10.1016/j.ijfatigue.2006.02.014.

20. D. Lévêque, A. Schieffer, A. Mavel, and J. F. Maire, "Analysis of how thermal aging affects the long-term mechanical behavior and strength of polymer-matrix composites," *Compos. Sci. Technol.*, vol. 65, no. 3–4, pp. 395–401, 2005, doi: 10.1016/j.compscitech.2004.09.016.

21. G. B. Mckenna, "On the physics required for prediction of long term performance of polymers and their composites," *J. Res. Natl. Inst. Stand. Technol.*, vol. 99, no. 2, p. 169, 1994, doi: 10.6028/jres.099.014.

22. S. Murray, C. Hillman, and M. Pecht, "Environmental aging and deadhesion of siloxane-polyimide-epoxy adhesive," *IEEE Trans. Compon. Packag. Technol.*, vol. 26, no. 3, pp. 524–531, 2003, doi: 10.1109/TCAPT.2003.817642.

23. G. A. Schoeppner, G. P. Tandon, and E. R. Ripberger, "Anisotropic oxidation and weight loss in PMR-15 composites," *Compos. Part A Appl. Sci. Manuf.*, vol. 38, no. 3, pp. 890–904, 2007, doi: 10.1016/j.compositesa.2006.07.006.

24. M. Mohammed et al., "Effect of kenaf fibre layers on mechanical and thermal properties of kenaf/unsaturated polyester composites," *IOP Conf. Ser. Mater. Sci. Eng.*, vol. 454, no. 1, 2018, doi: 10.1088/1757-899X/454/1/012091.

25. M. Asim et al., "Thermal stability of natural fibers and their polymer composites," *Iran. Polym. J.*, vol. 29, no. 7, pp. 625–648, 2020, doi: 10.1007/s13726-020-00824-6.

26. B. F. Abu-Sharkh and H. Hamid, "Degradation study of date palm fibre/polypropylene composites in natural and artificial weathering: Mechanical and thermal analysis," *Polym. Degrad. Stab.*, vol. 85, no. 3, pp. 967–973, 2004, doi: 10.1016/j.polymdegradstab.2003.10.022.

27. K. Mayandi et al., "An overview of endurance and ageing performance under various environmental conditions of hybrid polymer composites," *J. Mater. Res. Technol.*, vol. 9, no. 6, pp. 15962–15988, 2020, doi: 10.1016/j.jmrt.2020.11.031.

28. V. A. Alvarez, R. A. Ruscekaite, and A. Vazquez, "Mechanical properties and water absorption behavior of composites made from a biodegradable matrix and alkaline-treated sisal fibers," *J. Compos. Mater.*, vol. 37, no. 17, pp. 1575–1588, 2003, doi: 10.1177/0021998303035180.

29. M. Ramesh, L. Rajeshkumar, D. Balaji, and V. Bhuvaneswari, *Influence of Moisture Absorption on Mechanical Properties of Biocomposites Reinforced Surface Modified Natural Fibers*. Springer Singapore.

30. L. Mohammed, M. N. M. Ansari, G. Pua, M. Jawaid, and M. S. Islam, "A review on natural fiber reinforced polymer composite and its applications ," *Int. J. Polym. Sci.*, pp. 1–28, 2015. https://doi.org/10.1155/2015/243947.

31. S. Migneault, A. Koubaa, P. Perré, and B. Riedl, "Effects of wood fiber surface chemistry on strength of wood–plastic composites," *Appl. Surf. Sci.*, vol. 343, pp. 11–18, 2015, doi: 10.1016/j.apsusc.2015.03.010.

32. M. Assarar, D. Scida, A. El Mahi, C. Poilâne, and R. Ayad, "Influence of water ageing on mechanical properties and damage events of two reinforced composite materials: Flax – Fibres and glass – Fibres," *Mater. Des.*, vol. 32, no. 2, pp. 788–795, 2011, doi: 10.1016/j.matdes.2010.07.024.

33. M. Z. Abdullah, Y. Dan-mallam, P. Sri, and M. Megat, "Effect of environmental degradation on mechanical properties of kenaf / polyethylene terephthalate fiber reinforced polyoxymethylene hybrid composite," *Adv. Mater. Sci.*, vol. 2013, pp. 1–8, 2013.

34. M. D. H. Beg and K. L. Pickering, "Mechanical performance of Kraft fibre reinforced polypropylene composites: Influence of fibre length, fibre beating and hygrothermal ageing," *Compos. Part A Appl. Sci. Manuf.*, vol. 39, no. 11, pp. 1748–1755, 2008, doi: 10.1016/j.compositesa.2008.08.003.

35. A. Hodzic, J. K. Kim, A. E. Lowe, and Z. H. Stachurski, "Science and the effects of water aging on the interphase region and interlaminar fracture toughness in polymer – Glass composites," vol. 64, no. 13–14, pp. 2185–2195, 2004, doi: 10.1016/j.compscitech.2004.03.011.

36. K. Imieli and L. Guillaumat, "The effect of water immersion ageing on low-velocity impact behaviour of woven aramid – Glass fibre / epoxy composites," vol. 64, no. 13–14, pp. 2271–2278, 2004, doi: 10.1016/j.compscitech.2004.03.002.

37. M. R. Sanjay and J. Parameswaranpillai, "A comprehensive review of techniques for natural fibers as reinforcement in composites: Preparation, processing and characterization,", *Carbohydr. Polym.*, vol. 207, pp. 108–121, 2019, doi: 10.1016/j.carbpol.2018.11.083.

38. S. Deshpande and T. Rangaswamy, "A comparative study on dry sliding wear characteristics of Al_2O_3 and bone powder filled hybrid composites," *J. Miner. Mater. Charact. Eng.*, vol. 04, no. 2, pp. 164–180, 2016, doi: 10.4236/jmmce.2016.42016.

39. B. C. Ray, "Temperature effect during humid ageing on interfaces of glass and carbon fibres reinforced epoxy composites," *J. Colloid Interface Sci.*, vol. 298, no. 1, pp. 111–117, 2006.

40. A. C. De Albuquerque, K. Joseph, L. Hecker De Carvalho, and J. R. M. D'Almeida, "Effect of wettability and ageing conditions on the physical and mechanical properties of uniaxially oriented jute-roving-reinforced polyester composites," *Compos. Sci. Technol.*, vol. 60, no. 6, pp. 833–844, 2000, doi: 10.1016/S0266-3538(99)00188-8.

41. A. Afzal, M. K. Bangash, A. Hafeez, and K. Shaker, "Aging effects on the mechanical performance of carbon fiber-reinforced composites,", vol. 2023, 2023 https://doi.org/10.1155/2023/4379307.

42. N. Vellguth, M. Shamsuyeva, H. Endres, and F. Renz, "Accelerated ageing of surface modified flax fiber reinforced composites," *Compos. Part C*, vol. 6, p. 100198, 2021, doi: 10.1016/j.jcomc.2021.100198.

43. T. K. Tsotsis, "Thermo-oxidative aging of composite materials," *J. Compos. Mater.*, vol. 29, no. 3, pp. 410–422, 1995, doi: 10.1177/002199839502900307.

44. T. K. Tsotsis, S. Keller, J. Bardis, and J. Bish, "Preliminary evaluation of the use of elevated pressure to accelerate thermo-oxidative aging in composites," *Polym. Degrad. Stab.*, vol. 64, no. 2, pp. 207–212, 1999, doi: 10.1016/S0141-3910(98)00190-6.

45. H. T. Lee and D. W. Levi, "Effect of curing temperature on the thermal degradation of an epoxide resin," *J. Appl. Polym. Sci.*, vol. 13, no. 8, pp. 1703–1705, 1969, doi: 10.1002/app.1969.070130811.

46. T. K. Tsotsis and S. M. Lee, "Long-term thermo-oxidative aging in composite materials: Failure mechanisms," *Compos. Sci. Technol.*, vol. 58, no. 3–4, pp. 355–368, 1998, doi: 10.1016/S0266-3538(97)00123-1.

47. K. J. Bowles, "Thermal and mechanical durability of graphite-fiber-reinforced PMR-15 composites," in *32nd ISTC*, 2000, p. 16.

48. H. Parvatareddy, J. Z. Wang, D. A. Dillard, T. C. Ward, and M. E. Rogalski, "Environmental aging of high-performance polymeric composites: Effects on durability," *Compos. Sci. Technol.*, vol. 53, no. 4, pp. 399–409, Jan. 1995, doi: 10.1016/0266-3538(95)00029-1.

49. J. L. Sullivan, "Creep and physical aging of composites," *Compos. Sci. Technol.*, vol. 39, no. 3, pp. 207–232, 1990.

50. J. L. Sullivan, E. J. Blais, and D. Houston, "Physical aging in the creep behavior of thermosetting and thermoplastic composites," *Compos. Sci. Technol.*, vol. 16, pp. 389–403, 1993.

51. L. R. Bao and A. F. Yee, "Moisture diffusion and hygrothermal aging in bismaleimide matrix carbon fiber composites: Part II - Woven and hybrid composites," *Compos. Sci. Technol.*, vol. 62, no. 16, pp. 2111–2119, 2002, doi: 10.1016/S0266-3538(02)00162-8.

52. Y. Zheng, R. D. Priestley, and G. B. McKenna, "Physical aging of an epoxy subsequent to relative humidity jumps through the glass concentration," *J. Polym. Sci. Part B Polym. Phys.*, vol. 42, no. 11 SPEC. ISS., pp. 2107–2121, 2004, doi: 10.1002/polb.20084.

53. "Materials Vo110 (January 1976) pp 69–78 Ion exchange strengthening of glass used in the manufacturing of glass ceramics Vo132 (January 1976) pp 335–337 of composite materials, Vo110 (January 1976) pp 2–20 physics Vo147 (April 1976) pp 1351–1355 R," no. January, p. 7050, 1977.

54. C. Browning, G. Husman, and J. Whitney, "Moisture effects in epoxy matrix composites," in *100 Barr Harbor Drive, PO Box C700, West Conshohocken, PA 19428-2959: ASTM International*, pp. 481–481–16, 1977.

55. N. G. Gaylord and R. S. Stein, "Physical properties of polymers, F. BUECHE, interscience, New York, 1962. x + 354 pp. $9.50. Mechanical properties of polymers, L. E. NIELSEN, Rheinhold, New York, 1962. 284 pp. $11.00," *J. Polym. Sci. Part A Gen. Pap.*, vol. 1, no. 11, pp. 3519–3520, Nov. 1963, doi: 10.1002/pol.1963.100011122.

56. A. Vedrtnam, D. Gunwant, H. Verma, and K. Kalauni, "Effect of aging and UV exposure on mechanical properties of natural fiber composites." In: *Aging Effects on Natural Fiber-Reinforced Polymer Composites*. Springer Singapore, 2022, pp. 189–217.

57. P. K. Bajpai, D. Meena, S. Vatsa, and I. Singh, "Tensile behavior of nettle fiber composites exposed to various environments," *J. Nat. Fibers*, vol. 10, no. 3, pp. 244–256, 2013, doi: 10.1080/15440478.2013.791912.

58. K. C. C. C. Beninia, H. J. C. Voorwald, and M. O. H. Cioffi, "Mechanical properties of HIPS/sugarcane bagasse fiber composites after accelerated weathering," *Procedia Eng.*, vol. 10, pp. 3246–3251, 2011, doi: 10.1016/j.proeng.2011.04.536.

59. H. F. Brinson, W. I. Griffith, and D. H. Morris, "Creep rupture of polymer-matrix composites," *Exp. Mech.*, vol. 21, no. 9, pp. 329–335, 1981.

60. M. A. Musthaq, H. N. Dhakal, Z. Zhang, A. Barouni, and R. Zahari, "The effect of various environmental conditions on the impact damage behaviour of natural-fibre-reinforced composites (NFRCs)—A critical review," *Polymers (Basel)*, vol. 15, no. 5, 2023, doi: 10.3390/polym15051229.

61. K. J. Bowles, L. McCorkle, and L. Ingrahm, "Comparison of graphite fabric reinforced PMR-15 and avimid N composites after long-term isothermal aging at various temperatures," *J. Adv. Mater.*, vol. 30, no. 1, pp. 27–35, 1998.

62. S. F. Marin, Gerardo (University of San Francisco), Sabogal, Fabio (University of California, San Francisco), Marin, Barbara VanOss (University of California, San Francisco), Otero-Sabogal, Regina (University of California, San Francisco), Perez-Stable, Eliseo, "From the SAGE social science collections: All rights," *Hisp. J. Behav. Sci.*, vol. 9, no. 2, pp. 183–205, 1987, doi: 10.1177/07399863870092005.

63. H. L. Friedman, "Kinetics of thermal degradation of char-forming plastics from thermogravimetry. Application to a phenolic plastic," *J. Polym. Sci. Part C Polym. symp.*, vol. 6, no. 1, pp. 183–195, Mar. 2007, doi: 10.1002/polc.5070060121.

64. J. B. Henderson and M. R. Tant, "A study of the kinetics of high temperature carbon silica reactions in an ablative polymer composite," *Polym. Compos.*, vol. 4, no. 4, pp. 233–237, 1983, doi: 10.1002/pc.750040408.

65. H. A. Papazian and M. M. Corporation, "Prediction of polymer degradation kinetics at moderate temperatures from TGA measurements," *J. Therm. Anal. Calorim.*, vol. 16, pp. 2503–2510, 1972.

3

Evaluation of Hybrid Corn Husk Fiber/Hibiscus tiliaceus Fiber Powders–Reinforced Coconut Shell/Polyester Composites: Effect of Volume Fraction on Mechanical and Morphological Properties

N. H. Sari , S. Suteja, Y. A. Sutaryono, S. Sujita, and E. Syafri

3.1 Introduction

Currently, there is interest in developing the modification of natural fibres with other materials to produce better properties. Natural fibres, such as corn husk fibre (CHF), include agricultural waste that is not utilized, which has very large economic value, as well as fibre from the branch bark of the *Hibiscus tiliaceus* (HT) plant. This HT fibre is very easy to find in Indonesia and has fairly good tensile performance. Considering that these two types of fibres are easy to obtain, inexpensive, harmless to health, renewable, and environmentally friendly, they have great potential to be used as reinforcement for polymer composites.

Characterization of CHF and its composites has been reported by several researchers [1–4]. Sari et al. [1] have reported that after immersion in water for 6 hours, the lowest water absorption value (2.39%) was obtained from the composite with 20% CHF and 80% polyester (PE). They also reported that the mechanical strength of composites decreased when the CHF content was 10%–30% (vol. fraction), and above, and that after immersion in water for 27 hours, the water absorption and swelling properties of the composites reached 0.24%–1.38% and 0.08%–1.04%, respectively [5]. Furthermore, after exposure to ultraviolet, the mechanical ability of polyester composites with a CHF content of 10%–50% showed a small decrease, suitable for applications in outdoor environments [6].

DOI: 10.1201/9781003408215-3

Several researchers have also investigated hibiscus skin fibre–reinforced polymer composites (HT), and their properties have been reported [5, 7–10]. Surata et al. [7] reported that after 5% NaOH treatment, the single tensile strength and elastic modulus of HT reached 44.604 and 365.864 MPa, respectively. This HT fibre is recommended as a substitute for marine canoe synthetic ropes. Purnowidodo et al. [8] reported the behaviour of a single-layer woven bisphenol composite from HT fibre with an angle of 90°/0° made using a vacuum infusion resin and yielding a tensile strength close to 400 MPa. Then, Wirawan et al. [9] investigated HT composites treated with NaOH-silane. They reported that the tensile strength and elongation of the composites were obtained at 65.589 MPa and 2.4%, respectively. Furthermore, Sari et al. [11] reported that the tensile strength of the powder composite of 5% HT fibre increased by 110% after modification with 10% carbon powder.

Coconut shells are a waste product from coconuts, they are cheap and plentiful, and they are found in Indonesia. Reinforcement of polymer composites with coconut shell powder has been known to improve the mechanical properties of polymer composites [12–15]. Sarki et al. [12] reported that coconut shell particles can increase the tensile strength and modulus of elasticity of epoxy composites. Furthermore, due to the reduced costs and ease of processing, Keerthika et al. [13] investigated coconut shell powder (CSP) as a filler for acrylonitrile and butadiene rubber copolymers. They reported that the addition of 50%CSP and Fast Extrusion Furnace (FEF) was satisfactory. The best mechanical properties can be obtained from composites with 25% and 50%CSP. Then, Singh et al. [14] reported that the mechanical properties and density of the composite were strongly influenced by the volume fraction of the CSP. The water absorption properties of the epoxy composites increase when the volume fraction of CSP is 20%–40% [16]. They obtained the highest tensile and bending strength of 19.23 and 86.45 MPa, respectively. Next, Sari et al. [17] reported that composites containing 10%–30% CHF at 10%CSP have higher mechanical properties than 5%CSP; the bending strength reached 24,233–45,844 MPa, which is suitable for structural applications. These previous studies show that research that combines two or three types of these materials has not been carried out. The combination of these many natural materials can produce different properties that will enhance their function as desired.

Therefore, this chapter aims to discuss the CHF-reinforced polymer composite with a hybrid filler modification of carbon powders from coconut shell and *H. tiliaceus* fibre. The effect of the volume fraction of the powder on the mechanical properties of the composite will be addressed in terms of mechanical properties and the fracture morphology of the composite. Furthermore, the properties of corn husk fibre and its composites, *H. tiliaceus* fibre and its properties, carbon powder from coconut shell, and their extraction and the resulting powders will be thoroughly explained.

3.2 Corn Husk Fiber and Its Composites

CHF is made from the husks of corn (see Figure 3.1a). Corn husks are the hard outer layer of corn kernels, which are made up of various layers of protective fibres that protect the kernel's contents. CHFs are frequently extracted by soaking them in 8% NaOH solution for 120 minutes. They are used in a variety of industries, including food, pharmaceuticals, textiles, and others. CHF has advantageous properties such as high strength, biodegradability (natural decomposition), and environmental friendliness. Because this fibre absorbs water well, it is used in cleaning products and as a food additive to increase fibre levels. Aside from that, CHF can be used to make paper, filler material in composite panels, packaging materials, and other products. CHF can help reduce agricultural waste and encourage the use of environmentally friendly materials in a variety of industries [.

CHF is a strong natural fibre with good tensile loading properties. When these fibres are reinforced in a resin matrix or polymer material, they increase the tensile strength of the composite. CHF's stiff and flexible properties can increase fracture resistance and help prevent excessive deformation of the composite. CHF-reinforced composites also exhibit good fatigue resistance in applications requiring repeated loading. Corn husks, based on their specific gravity, are a light natural fibre. When used as a reinforcement in composites, CHF can significantly reduce the specific gravity of the composite when compared with heavier fillers such as metal fibres. One of the most important characteristics of CHF is its ability to naturally degrade in the environment. As a result, the composite is environmentally friendly and can decompose when it reaches the end of its life. Furthermore, CHF is less expensive than synthetic or other fibres. This is why CHF-reinforced composites are a cost-effective and environmentally friendly option in a variety of applications. Although CHF-reinforced composites have advantageous properties, it is important to keep in mind that the mechanical and physical properties of the composite can be influenced by a variety of factors, including fibre treatment

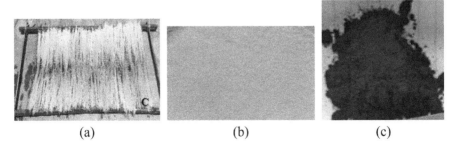

(a)   (b)   (c)

FIGURE 3.1
(a) Corn husk fibre (CHF), (b) *Hibiscus tiliaceus* powder (HTP), (c) coconut shell powder (CSP).

prior to placement, the manufacturing process, and fibre–matrix compatibility. As a result, careful research and testing are required to ensure that the desired properties are achieved in CHF-reinforced composites.

3.3 Hibiscus tiliaceus Powder (HTP)

The bark of the hibiscus tree is used to produce hibiscus leather fibre (HT). The hibiscus tree is a tropical plant that grows in abundance. The bark contains strong and elastic fibres that can be extracted and used in a variety of industries. The physical and mechanical properties of CHF, HTP, CSP, and polyester are shown in Table 3.1. HTP is strong, is resistant to weathering

TABLE 3.1

Physical and Mechanical Properties of CHF, HTP, CSP, and Polyester

Specifications	Polyester	Catalyst (methylethyl ketone peroxide)	CHF	HTP	CSP
Density (g/cm³)	1.2×10^{-6}	N/A	0.61	0.68	N/A
Tensile strength	8.8×10^2 kg/cm²	N/A	368.25 MPa	5144.9 MPa	N/A
Flexural strength (kg/cm²)	2.5×10^2	N/A	N/A	N/A	N/A
Molecular mass (g/mol)	N/A	210.22	N/A	N/A	N/A
Solubility (g/L, 20 °C)	N/A	6.53	N/A	N/A	N/A
Steam pressure (Hg, 20 °C)	N/A	0.01	N/A	N/A	N/A
Water content (%)	4.54	N/A	11.187	7.49	4.54
Elongation (%)	2.3	N/A	N/A	N/A	N/A
Tensile modulus (kg/cm²)	50×10^3	N/A	N/A	N/A	N/A
Viscosity (P)	6–8 (25 °C)	N/A	N/A	N/A	N/A
Cellulose (%)	N/A	N/A	62.87	71.4	N/A
Crystalline index (%)	N/A	N/A	59.49	N/A	N/A
Ash (%)	N/A	N/A	N/A	N/A	3.47
Volatile matter (VM, %)	N/A	N/A	N/A	N/A	39.27
Fixed carbon (%)	N/A	N/A	N/A	N/A	52.72
References	[1,18]	[1]	[3]	[19]	[20]

and insect damage, and has a high water absorption rate. As a result, this fibre is used to make a variety of products, including rope, webbing, paper, fabric, handicrafts, and many more. Aside from that, HT is used in the textile industry as a raw material for the production of yarn, fabric, and other textile products. These fibres are frequently blended with other fibres, such as cotton or silk, to increase the strength and quality of the final product. Because HT trees are easy to grow and can be renewed quickly, using HT can encourage the use of sustainable raw materials. This fibre can also help reduce agricultural waste and add value to the products produced. As well as fibre, HT fibre can also be obtained in powder form (HTP) by grinding it to a fine powder and sieving it through a 200-mesh sieve, as shown in Figure 3.1b.

3.4 Carbon Powder from Coconut Shell

Coconut shells can be used to create carbon powder. Carbonization, or heating the coconut shells under oxygen-reducing conditions to remove non-carbonic substances and leave carbon residue, is used to produce carbon powder from coconut shells. The coconut shell is a hard and rough layer that protects the coconut meat inside. After the coconut meat is extracted or used for another purpose, the coconut shells frequently become waste that can be reused. One of the most common applications for coconut shells is as a raw material for making carbon powder [21–23]. Carbon powder made from coconut shells has a variety of applications. This carbon powder can be used as a filtration medium to remove dissolved substances in water or as an adsorbent to absorb gases or chemicals. Apart from that, carbon powder can be used to make charcoal briquettes or as a component in the creation of craft items. Using carbon powder from coconut shells can help reduce waste from coconut farming while also adding value to this waste. Furthermore, carbon powder has good adsorption properties and low production costs, making it an environmentally friendly alternative to using activated carbon [23–25, 19].

3.5 CSP/HTP/CHF Composites

CSP/HTP/CHF composites can be made by combining three ingredients, namely, CSP and HTP, into polyester resin, stirring until a homogeneous mixture is formed, then pouring into a mould filled with CHF and hot moulding for 20 minutes at 45 MPa and 50 °C. Table 3.2 displays the CSP/HTP/CHF content ratio in the polyester composite. To investigate the mechanical properties of the composite, each mechanical test was repeated at least three times.

TABLE 3.2

Coconut Shell Powder, *Hibiscus tiliaceus* Powder, and Corn Husk Fiber Composition Ratio in Polyester Composite

No.	Specimen Code	Coconut Shell Powder (CSP)	Corn Husk Fibre (CHF)	*Hibiscus tiliaceus* Fibre Powder (HTP)	Polyester Resin (PR)
			Compositions (Vol. fraction, %)		
1	TL	10	25	10	55
2	TR		20	15	55
3	TT		15	20	55
4	TY		10	25	55
5	SL	5	25	10	60
6	SR		20	15	60
7	ST		15	20	60
8	SY		10	25	60

3.5.1 Tensile Strength

The ability of a composite to withstand tensile loads before failure or fracture is referred to as composite tensile strength. These properties are expressed in megapascals (MPa) or pounds per square inch (psi) units of stress or tensile strength. Standard tensile testing was performed on the composites to obtain more accurate tensile strength figures using a Universal Tensile Machine (Instron RTG-1310) tensile testing machine with a crosshead speed of 5 mm/min [26, 27]. Tensile test samples are based on ASTM D-3039 international standards (Figure 3.2a). The results of this test can reveal the maximum amount of stress that the composite can withstand before failing or breaking.

Figure 3.3a shows the tensile strength of the different CSP/CHF/HTP composites. Overall, the tensile strength of the composite decreased with increasing HTP content. The tensile strength of composite 5% CSP shows the minimum and maximum values of 17.24 MPa (SY composite) and 23.56 MPa (SL composite), respectively. For 10% CSP composites, the highest and lowest tensile strength of the composites were obtained from TY (15.32 MPa) and TL composites (21.02 MPa), respectively. An increase in tensile strength is seen when the CHF content increases. On the other hand, the tensile strength decreases because CHF and CSP act as a reinforcement in the composite. This has been confirmed from the tensile strength values of the 5% CSP and 10% CSP composites (22.01 and 23.74 MPa, respectively). In 5% CSP composites, the presence of large amounts of CHF (from 10% to 25%) is known to increase the tensile strength of the composites, because the interfacial bond between the resin and CSP-CHF-HTP is quite strong. On the other hand, the addition of HTP (from 10% to 25%) is thought to decrease the tensile strength

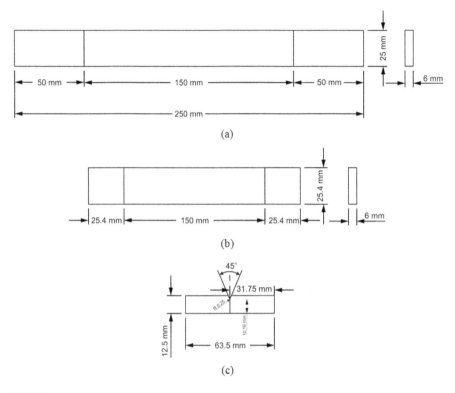

FIGURE 3.2
Composite specimens for (a) tensile strength, (b) bending strength, and (c) impact strength [27, 28].

because HTP does not act as a reinforcement but as a void or defect between the resin and CSP/CHF interface, which causes stress transfer between CSP, CHF, and HTP. This decrease is due to weak interfacial bonds between all constituent materials, and consequently, the tensile strength decreases with increasing HTP. Figure 3.3 also shows that the tensile strength curve of the 10% CSP composite is lower than that of the 5%CSP composite, which is associated with a large amount of CHF, CSP, and HTP not being wetted by polyester, causing the interface bond between CSP, CHF, and HTP to be less strong, resulting in CHF pull-out. The mechanism of the interface strength and volume fraction of the reinforcement lowering the tensile strength of the composites has been reported by Norain et al. [28]; the stress transfer between the reinforcement and the matrix can be reduced if the interface bond is weak, because the volume fraction of the reinforcement exceeds the volume fraction of the matrix. The tensile strength value of the developed CSP/CHF/HTP composite is higher than that of the corn husk fibre–reinforced polypropylene (PP) composite, with a composite tensile strength

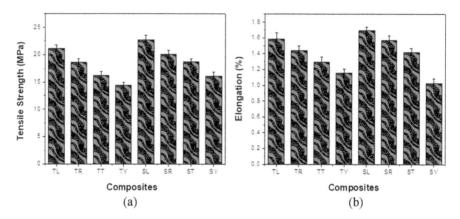

FIGURE 3.3
(a) Tensile strength, (b) elongation of SCP/CHF/HTP composites.

value of 10 MPa [2]. These results suggest that CSP/CHF/HTP composites can be used as an alternative to PP/CHF composites.

3.5.2 Elongation

Figure 3.3b shows the fracture elongation of the CSP/CHF/HTP composite. The elongation curve of the composite shows the same pattern as the tensile strength of the composite. For the 5% CSP composite, the highest elongation was obtained from the SL sample, and the lowest was obtained from the SY sample at the 5% CSP composite. This increase in elongation value was associated with an increase in CHF in the composite. In contrast, a decrease in the elongation value was associated with an increase in the stiffness of the composite as the HTP content increased. This result is in line with the report of Sari et al. [11], who stated that the presence of CSP/HTP tends to increase the stiffness or brittleness of the polyester matrix itself.

Figure 3.3b also shows that the addition of 10%CSP to the CHF/HTP composite decreased the elongation value of the studied composites.

This HTP has been chemically treated with NaOH, bleaching (NaClO), and H_2SO_4, so that it is expected to improve the properties of HTP; the HTP surface becomes rougher and shows perfect wettability of the resin. However, it was observed that the increased content of HTP and CSP caused poor dispersion in the final blending process and eventually, reduced the interaction of the HTP/CSP/CHF interface with polyester.

3.5.3 Modulus of Elasticity

Figure 3.4 shows the modulus of elasticity of the different CSP/CHF/HTP composites. It was observed that for the 10% CSP composite, Young's modulus of the CHF/HT composite tended to be higher than that of the 5% CSP

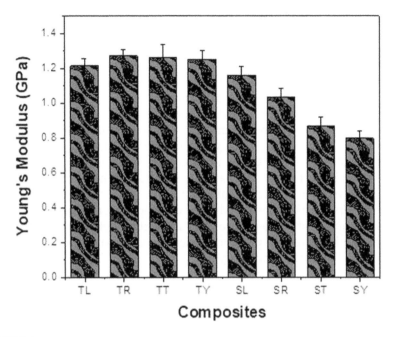

FIGURE 3.4
Young's modulus of SCP/CHF/HTP composites.

composite; this is associated with the increased stiffness of the CHF/HTP composite due to the increase in the content of CHF and 10%CSP. However, when the HTP content increases from 10% to 25%, Young's modulus value of the composite tends to decrease due to voids at the interface between CSP/CHF/HTP and polyester.

There are several factors that can cause the elastic modulus of CSP/HTP/CHF composites to increase, including the following. (1) The high stiffness of CSP/HTP powder significantly contributes to the composite's elastic modulus. Powders with a high modulus of elasticity resist deformation and provide optimal load transfer, increasing the overall stiffness of the composite. (2). Uniform distribution of powder in the matrix can increase the elastic modulus of the composite by allowing load transfer between the powder and the matrix to occur efficiently, increasing the overall modulus of elasticity of the composite. (3). Because the powder density in a powder composite is high, the load transfer between the powder and the matrix is efficient, increasing the elastic modulus of the composite. In contrast, if the powder density is low, load transfer is impaired, and the elastic modulus of the composite is reduced.

To ensure optimal elongation, these factors must be considered during the composite design and manufacturing processes in order to avoid or minimize the causes of decreased elongation in the composite [21, 24, 26].

The elastic modulus of this composite is thought to be decreasing due to several factors, including the following: (1) Weak bonds between the fibres or filler and the matrix. When these bonds are weak, applying force to the composite can cause the fibres and powders to separate from the matrix, resulting in reduced load transfer and a low elastic modulus; (2) Uneven fibre distribution. The elastic modulus of the composite can decrease if the fibre arrangement in the composite is not uniform, such as random orientation or non-uniformity of fibre distribution. Uneven tensile forces in the fibres are caused by non-uniform distribution, reducing the composite's ability to resist deformation and lowering its elastic modulus; (3) The presence of defects in the composite, such as air bubbles, cracks, or inclusions, can weaken the structure and cause a decrease in the elastic modulus. These flaws have an impact on the mechanical properties of the composite and can reduce its stiffness or elastic modulus.

These factors must be considered when developing composites with a high elastic modulus in order to ensure good interaction between the fibres and the matrix and to avoid the presence of defects or weaknesses that can cause a decrease in the elastic modulus [29].

3.5.4 Flexural Strength and Flexural Modulus

Figure 3.5a shows the bending strength of HTP/CSP/CHF composites. It is observed that this bending strength curve shows the same pattern as the tensile strength curve of HTP/CSP/CHF composites. For 5%CSP composites, the bending strengths of JWA, JWB, JWX, and JWY composites were 39.4, 37.8, 33.9, and 22.1 MPa, respectively. For 10%CSP composite, the flexural strength was 33.1, 31.9, 28.4, and 18.2 MPa, respectively. It was observed that there was a decrease in flexural strength from 4.08% to 34.73% successively from JWA. Then, 10%CSP composite showed a decrease of about 3.7%–36.01%. This decrease in flexural strength indicates that the volume fraction of CHF

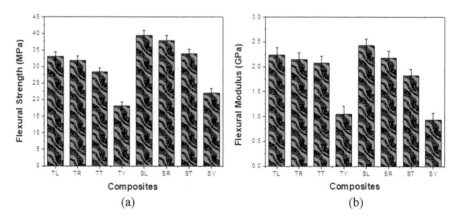

FIGURE 3.5
(a) Flexural strength, (b) Young's modulus of SCP/CHF/HTP composites.

greatly determines the flexural strength of the composite. This is because the presence of CHF in the composite will resist the external forces acting on the composite. Figure 3.5b also shows that the flexural strength of 5%CSP CHF/HTP composites is higher than that for 10%CSP, which is associated with less interaction of CHF/HTP and CSP with polyester, fibre pull-out, and some trapped voids (Figure 3.7). Compiling the results of the tensile and flexural tests (Figure 3.3 and Figure 3.5), a gradual increase was seen from the 5% composite reinforced with unsaturated polyester resin to 25% CHF when compared with the other composites studied.

Figure 3.5b shows the flexural modulus of the composite HTP/CSP/CHF. It was observed that for 10% CSP, the highest flexural modulus of the composite was obtained from the TL sample of 2.24 GPa; it then decreased by 3.994% (TR sample), 7.06% (TT sample), and 53.34% (TY sample), respectively. As for 5%CSP, Young's flexural from SL was 2.45 GPa, and it then tended to decrease by 10.16%, 24.54%, and 61.4% for SR, ST, and SY samples, respectively. The decrease in value occurred because the bond between polyester and CHF became weak, and polyester was more likely to interact with HTP and CSP, with increasing volume fraction of HTP. Islam et al. [30] stated that the weak value of the bending characteristic of the composite is highly dependent on the mismatch between the filler and polyester, the inhomogeneous spread of the filler, and the pores or voids at the filler interface.

3.5.5 Impact Strength

Figure 3.6 displays the impact strength of various composites. For all composites of 10%CSP and 5%CSP, a decrease in the impact strength value was observed of around 6.70%–30.18% and 21.175%–26.245%, respectively, with increasing HTP levels and decreasing CHF levels. The maximum and minimum impact strength values were obtained from SL and SY composites. Figure 3.6 also shows that the 10%CSP composite has a higher impact strength than the 5%CSP (TL sample). These results indicate that CSP and CHF play a role in increasing the ability to absorb energy received by the composite [23, 31]. In other words, when CSP and CHF increase and HTP decreases, the composite impact strength increases; conversely, when CSP and CHF decrease and HTP increases, the composite impact strength decreases. The decrease in impact strength of the composite is thought to be caused by low fibre or powder tensile strength and stiffness. When this happens, the impact energy is distributed through the composite matrix, with the matrix absorbing the majority of the energy rather than the fibres or powder. As a result, the impact strength of powdered or low fibre composites is lower because it is dependent on the mechanical properties of the matrix. Furthermore, the orientation of the fibres or powders in the composite can affect impact strength. When low fibres or powders are randomly arranged, they may not absorb optimal energy during impact [28, 32].

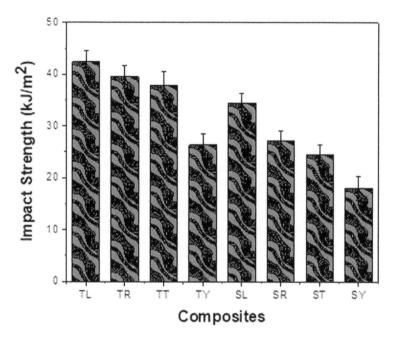

FIGURE 3.6
Impact strength of SCP/CHF/HTP composites.

Even though powder or low fibre composites have lower impact strength, the advantage of this type of composite is that it is easier and less expensive to produce. Powdered or low fibre composites can also be used in applications that do not require high impact strength but instead emphasize stiffness, tensile strength, or corrosion resistance.

3.6 Morphology of CSP/CHF/HTP Composites

The surfaces of the fractured specimens of tensile CSP/CHF/HTP composite were examined directly by scanning electron microscopy (SEM) (model FEI, inspect S-50 type) at different magnifications to conclude the effect of volume fraction of CHF/HTP on CSP/polyester composite through interfacial adhesion of the composites.

Figure 3.7 shows the morphology of the fracture surfaces of the CSP/CHF/HTP composites taken from the tensile test specimens of the 5%CSP composite for the SL and SY specimens (see Figure 3.7A and 3.7B), and the 10% CSP composite represented by the TL and TY composite (see Figure 3.7C and 3.7D). Figure 3.7A shows a strong interface bond between CSP-CHF-HTP and polyester, and a little CHF pull-out, which supports the reason why the tensile strength of the SL composite is higher than other composites. Figure 3.7B

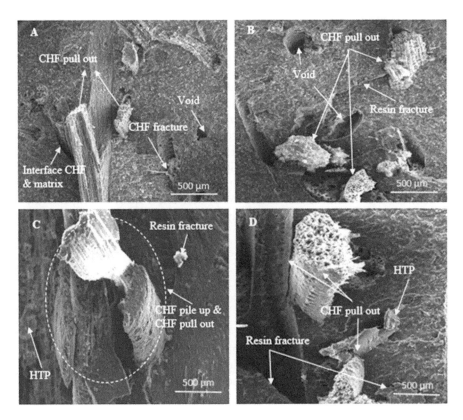

FIGURE 3.7
SEM images of CSP/CHF/HTP composite samples: (a) SL, (b) SY, (c) TL, and (d) TY. Magnification 500×.

shows SEM photos of SY composites; there are a large number of CHF pull-outs and some voids, which are thought to be due to the presence of a large amount of HTP and ultimately, cause the lowest tensile strength of the SY composites. Figure 3.7C presents a fairly tight bond between CSP-CHF-HTP and polyester, CHF pull-out, voids, and fractures along with the resin; the failure rate is also the highest found in Figure 3.7D. This causes the tensile strength of the composite to be very low.

3.7 Conclusions

The evaluation results show that the volume fraction of the hybrid filler CHF and HTP affects the mechanical properties of the polyester/CSP composite. The mechanical properties of the composites were found to decrease with the increase in the volume fraction of HTP; in other words, the mechanical properties increased with the increase in the volume fraction of CHF. The best properties of tensile strength, elongation, flexural strength, and flexural modulus were obtained from 5%CSP composites, with the highest value obtained from SL composites. However, the best impact strength properties were obtained from 10%CSP composite, with the maximum impact value obtained from TL specimens. The increase in tensile and flexural strength was associated with a strong interface bond between CSP-CHF-HTP and polyester with increasing CHF content. The decrease in tensile and flexural strength properties of 5%CSP composite was about 11.5%–13.8% and 4.0%–34.7%, respectively. Because HTP does not act as an amplifier but as a void or defect, the evaluation results from SEM photos show quite strong interface bonds between CHF-CSP-HTP and polyester, CHF pull-out, and voids in the composite.

Acknowledgements

All authors have participated in conception and design, or analysis and interpretation of the data; drafting the article or revising it critically for important intellectual content; and approval of the final version.

References

1. N. H. Sari, C. I. Pruncu, S. M. Sapuan, R. A. Ilyas, A. D. Catur, S. Suteja, Y. A. Sutaryono, and G. Pullen, "The effect of water immersion and fibre content on properties of corn husk fibres reinforced thermoset polyester composite," Polymer Testing, vol. 91, p. 106751, 2020a, https://doi.org/10.1016/j.polymertesting.2020.106751.

2. S. Huda, and Y. Yang, "A novel approach of manufacturing light-weight composites with propylene web and mechanically split cornhusk," *Industrial Crops and Products*, vol. 30, no. 1, pp. 17–23, 2009, https://doi.org/10.1016/j.indcrop.2008.12.007.

3. N. H. Sari, I. N. G. Wardana, Y. S. Irawan, and E. Siswanto, "The effect of sodium hydroxide on chemical and mechanical properties of corn husk fiber," *Oriental Journal of Chemistry*, vol. 33, no. 6, pp. 3037–3042, 2017, http://doi.org/10.13005/ojc/330642.

4. N. H. Sari, and S. Suteja, "Corn husk fibers reinforced polyester composites: Tensile strength properties, water absorption behavior, and morphology," *IOP Conference Series: Materials Science and Engineering*, vol. 722, no. 1, p. 012035, October 2019, https://doi.org/10.1088/1757-899X/722/1/012035.

5. N. H. Sari, J. Fajrin, S. Suteja, and A. Fudholi, "Characterisation of swellability and compressive and impact strength properties of corn husk fibre composites," *Composites Communications*, vol. 18, pp. 49–54, 2020b, https://doi.org/10.1016/j.coco.2020.01.009.

6. N. H. Sari, S. Suteja, R. A. Ilyas, E. Syafri, and S. Indran, "Characterization of the density and mechanical properties of corn husk fiber reinforced polyester composites after exposure to ultraviolet light," *Functional Composites and Structures*, vol. 3, no. 3, p. 034001, 2021a, https://doi.org/10.1088/2631-6331/ac0ed3.

7. I. W. Surata, T. G. T. Nindhia, and D. M. Widagdo, "Promoting natural fiber from bark of Hibiscus tiliaceus as rope to reduce marine pollution from microplastic fiber yield from synthetic rope," *E3S Web of Conferences*, vol. 158, no. 00005, 2019, https://doi.org/10.1051/e3sconf/202015800005.

8. A. Purnowidodo, S. Suteja, D. B. Darmadi, and K. Anam, "Tensile strength and fatigue crack growth behaviour of natural fibre metal laminates," *Journal of Engineering Science and Technology School of Engineering, Taylor's University*, vol. 15, no. 4, pp. 2809–2822, 2020.

9. W. A. Wirawan, M. A. Choiron, E. Siswanto, and T. D. Widodo, "Analysis of the fracture area of tensile test for natural woven fiber composites (hibiscus tiliaceus-polyester)," *Journal of Physics: Conference Series*, vol. 1700, no. 3, p. 012034, October 2020, https://doi.org/10.1088/1742-6596/1700/1/012034.

10. S. Suteja, A. Purnowidodo, D. B. Darmadi, and N. H. Sari, "Perilaku tarik komposit laminat serat kulit waru-aluminium," *Jurnal Rekayasa Mesin*, vol. 10, no. 1, pp. 17–24, 2019.

11. N. H. Sari, S. Suteja, M. N. Samudra, and H. Sutanto, "Composite of hibiscus tiliaceus stem fiber/polyester modified with carbon powder: Synthesis and characterization of tensile strength, flexural strength and morphology properties," *International Journal of Nanoelectronics and Materials*, vol. 14, no. 3, pp. 247–258, July 2021b.

12. J. Sarki, S. B. Hassan, V. S. Aigbodion, and J. E. Oghenevweta, "Potential of using coconut shell particle fillers in eco-composite materials," *Journal of Alloys and Compounds*, vol. 509, pp. 2381–2385, 2011, https://doi.org/10.1016/j.jallcom.2010.11.025.

13. B. Keerthika, M. Umayavalli, T. Jeyalalitha, and N. Krishnaveni, "Coconut shell powder as cost effective filler in copolymer of acrylonitrile and butadiene rubber," *Ecotoxicology and Environmental Safety*, vol. 130, pp. 1–3, 2016, http://doi.org/10.1016/j.ecoenv.2016.03.022.

14. A. Singh, S. Singh, and A. Kumar, "Study of mechanical properties and absorption behaviour of coconut shell powder-epoxy composites," *International Journal of Materials Science and Applications*, vol. 2, no. 5, pp. 157–161, 2013, https://doi.org/10.11648/j.ijmsa.20130205.12.

15. K. Balan, S. M. Parambil, S. Vakyath, J. T. Velayudhan, S. Naduparambath, and P. Etathil, "Coconut shell powder reinforced thermoplastic polyurethane/natural rubber blend-composites: Effect of silane coupling agents on the mechanical and thermal properties of the composites," *Journal of Materials Science*, vol. 52, no. 11, pp. 6712–6725, 2017, https://doi.org/10.1007/s10853-017-0907-y.

16. M. Kumar, K. Rithin, R. Raghuveer, S. Sharun, K. Shanmukh, R. Yathiraj, B. Shreeprakash, and B. Shreeprakash, "Study on effect of stirring parameters on tensile properties of coconut shell powder reinforced epoxy matrix composite," *Applied Mechanics and Materials*, vol. 592–594, pp. 1180–1184, 2014, https://doi.org/10.4028/www.scientific.net/AMM.592-594.1180.

17. N. H. Sari, S. Suteja, A. Fudholi, A. Zamzuriadi, E. D. Sulistyowati, P. Pandiatmi, S. Sinarep, and A. Zainuri, "Morphology and mechanical properties of coconut shell powder-filled untreated cornhusk fibre-unsaturated polyester composites," Polymer, vol. 222, no. 22, p. 123657, April 2021c, https://doi.org/10.1016/j.polymer.2021.123657.

18. N. H. Sari, and S. Suteja, "Polimer thermoset," *Penerbit Deepublish*, ISBN: 978-623-02-2305-1, 2021.

19. N. H. Sari, and Y. A. Padang, "The characterization tensile and thermal properties of hibiscus tiliaceus cellulose fibers," *IOP Conference Series: Materials Science and Engineering*, vol. 539, no. 1, p. 012031, 2019.

20. S. Ojha, S. K. Acharya, and G. Raghavendra, "Mechanical properties of natural carbon black reinforced polymer composites," *Journal of Applied Polymer Science*, vol. 132, no. 1, p. 41211, 2015.

21. L. J. Kumar, V. J. S. Jagadeesh, P. Ganjigatti, G. Irfan, and R. Thara, "Investigation of mechanical properties of Al6061 with reinforcement of coconut shell ash and graphene metal matrix composites," *Materials Today: Proceedings*, 2023, https://doi.org/10.1016/j.matpr.2023.05.529.

22. A. Kumar, A. Nirala, V. P. Singh, B. K. Sahoo, R. C. Singh, R. Chaudhary, A. K. Dewangan, G. K. Gaurav, J. J. Klemeš, and X. Liu, "The utilisation of coconut shell ash in production of hybrid composite: Microstructural characterisation and performance analysis," *Journal of Cleaner Production*, vol. 398, p. 136494, 2023, https://doi.org/10.1016/j.jclepro.2023.136494.

23. L. T. N. Huynh, T. N. Pham, T. H. Nguyen, V. H. Le, T. T. Nguyen, T. D. K. Nguyen, T. N. Tran, P. A. V. Ho, T. T. Co, T. T. T. Nguyen, T. K. A. Vo, T. H. Nguyen, T. T. Vu, V. M. Luong, H. Uyama, G. V. Pham, T. Hoang, and D. L. Tran, "Coconut shell-derived activated carbon and carbon nanotubes composite: A promising candidate for capacitive deionization electrode," *Synthetic Metals*, vol. 265, p. 116415, 2020, https://doi.org/10.1016/j.synthmet.2020.116415.

24. P. Satheeshkumar, and A. I. Selwynraj, "Optimal thermophysical characteristics and uptaking capacity of activated carbon-based composite adsorbent for enhancing the performance of an adsorption cooling system," *Materials Letters*, vol. 351, p. 135016, ISSN 0167-577X, 2023, https://doi.org/10.1016/j.matlet.2023.135016.

25. K. Kang, C. Zheng, Y. Xie, H. Song, Y. Liang, J. Hu, J. Kang, and S. Bai, "The effect of relative humidity on multicomponent organic vapor adsorption on composite beds with micro-fibrous entrapped activated carbon," *Journal of Environmental Chemical Engineering*, p. 111005, 2023, https://doi.org/10.1016/j.jece.2023.111005.

26. N. H. Sari, Sujita Suteja, R. A. Ilyas, E. Sari, M. R. Sanjay, and S. Siengchin, "Fabrication of bio-fiber based Eichhornia crassipes /Al2O3 particles hybrid biocomposites and investigation of important properties," *Proceedings of the Institution of Mechanical Engineers, Part E: Journal of Process Mechanical Engineering*, 2023, https://doi.org/10.1177/09544089231167750 (095440892311677).

27. M. K. Gupta, P. Manimaran, B. Suresha, A. S. S. S. J. sekaran, and M. K. N. Marichelvam, "Investigation of mechanical and dynamic mechanical properties of novel Acacia arabica fiber polyester hybrid composites," *Polymer Composites*, vol. 43, no. 5, pp. 2724–2735, 2022, https://doi.org/10.1002/pc.26569.

28. H. F. Norain, H. Salmah, and M. M. Zakaria, "Properties of all-cellulose composite films from coconut shell powder and microcrystalline cellulose," *Applied Mechanics and Materials*, vol. 754–755, pp. 39–43, 2015, https://doi.org/10.4028/www.scientific.net/AMM.754-755.39.

29. N. H. Sari, S. Suteja, A. Fudholi, Y. A. Sutaryono, M. Maskur, R. Srisuk, S. M. Rangappa, and S. Siengchin, "Evaluation of impact, thermo-physical properties, and morphology of cornhusk fiber-reinforced polyester composites," *Polyester Composites*, vol. 43, no. 5, pp. 2771–2778, May 2022.

30. Md. T. Islam, S. C. Das, J. Saha, D. Paul, M. T. Islam, M. Rahman, and M. A. Khan, "Effect of coconut shell powder as filler on the mechanical properties of coir-polyester composites," *Chemical and Materials Engineering*, vol. 5, no. 4, pp. 75–82, 2017, https://doi.org/10.13189/cme.2017.050401.

31. K. Kang, C. Zheng, Y. Xie, H. Song, Y. Liang, J. Hu, J. Kang, and S. Bai, "The effect of relative humidity on multicomponent organic vapor adsorption on composite beds with micro-fibrous entrapped activated carbon," *Journal of Environmental Chemical Engineering*, p. 111005, 2023, https://doi.org/10.1016/j.jece.2023.111005.

32. K. S. Chun, T. Maimunah, C. M. Yeng, T. K. Yeow, and O. T. Kiat, "Properties of corn husk fibre reinforced epoxy composites fabricated using vacuum-assisted resin infusion," *Journal of Physical Science*, vol. 31, no. 3, pp. 17–31, 2014, https://doi.org/10.21315/jps2020.31.3.2.

4

Tropical Natural Fibres for Packaging Paper and Its Regulation

Ainun Zuriyati Mohamed, Sharmiza Adnan, Mohd
Ashadie Kusno, Nur Hanani Zainal Abedin, Latifah
Jasmani, Zakiah Sobri, A. Atiqah, and E. S. Zainudin

4.1 About Tropical Natural Fibres

Natural fibres can be derived from a wide variety of sources, including wood, grass, fruits, agricultural crops, seeds, water plants, palms, wild plants, leaves, feathers, and skins (Salit 2014). In this chapter, only plant-derived fibres are discussed. Plants native to tropical climates (Southeast and Central Asia, South America, and Africa) that normally have two seasons, wet and dry (sunlight is intense), are the sources of tropical natural fibres. Native people have relied on them for millennia to provide basic necessities, including clothes, shelter, and household goods. There has been a resurgence of interest in tropical natural fibres in recent years as an eco-friendly replacement for synthetic materials (El Nemr 2012). Due to their distinctive properties and relatively easy availability, these fibres have risen in importance as a raw material to produce pulp and paper.

Papermaking is just one of the many ways that tropical natural fibres have been used for centuries. More than 1,500 years ago, for instance, hemp fibres were used to make the first paper in China, followed by mulberry, rattan, and bamboo (Tsien 1973). People in some tropical locations, including Southeast Asia and South America, developed pulp and paper industries in the nineteenth century. The transition from hand-made to machine-made paper began with an emphasis on natural resources, most notably wood pulp. The development of the usage of tropical natural fibres has been greatly aided by the rubber industry in Southeast Asia and the paper industry in Brazil and Indonesia. The rubber tree was initially processed for natural rubber (latex) products, but after extensive research starting in the late 1970s on usage of its timber after felling for replanting, in particular for furniture applications, it quickly became one of the commodities for countries such as Malaysia and Thailand. Countries like Indonesia, Brazil, Malaysia, Thailand,

DOI: 10.1201/9781003408215-4

the Philippines, and India all play significant roles in the manufacturing of products from tropical natural fibres (Kestur et al. 2009).

Successful instances of tropical fibre–based pulp and paper production can be observed in Indonesia, Malaysia, and Brazil. The Indonesian pulp and paper industry uses mostly acacia and eucalyptus trees as raw materials, making it the largest in Southeast Asia (Nirsatmanto et al. 2022). They adopt sustainable forest management to establish and harvest trees for pulp manufacturing. Looking into the use of palm fibres, Malaysia is the world's second-largest producer of palm oil and consequently produces a tremendous quantity of palm fibres (Parthasarathy et al. 2022). Eco Palm Paper Sdn. Bhd., a pulp plant in Malaysia, has the world's largest annual capacity of producing pulp from oil palm empty fruit bunches, at 50,000 tonnes (Ecopalm 2023). This new development shows how switching to a feedstock other than wood fibres can reduce greenhouse gas emissions and slow global warming. Paper made from eucalyptus fibres is popular in Brazil because the trees grow quickly and can be pulped using either mechanical or chemical methods. The Brazilian pulp factory run by Chilean firm Arauco uses eucalyptus from its own farm.

4.2 Types of Tropical Fibres Used for Pulp and Paper

Tropical fibres commonly used for pulp and paper include abaca, acacia, bamboo, coir, eucalyptus, palm, sisal, and bagasse. These fibres have unique properties, such as a high strength-to-weight ratio, low density, and high dimensional stability (Karimah et al. 2021). Compared with temperate wood and non-wood fibres, tropical fibres can be more sustainable, as they grow faster and require less energy in processing (Abd El-Sayed et al. 2020). These fibres can be processed in various ways to produce quality pulp and paper products. Furthermore, these types of fibres are often used in combination with other wood fibres to create a stronger and more durable paper product (Sharma et al. 2020). Tropical fibres such as bamboo and kenaf are becoming increasingly popular in the pulp and paper industry due to their comparable cellulose content to wood and their fast growth attributes.

4.2.1 Eucalyptus

Eucalyptus is native to Australia but has been widely planted in tropical regions around the world due to its fast-growing nature and high fibre content. In comparison to other tree species, which can take up to 20 years to mature, it can achieve maturity in 7 years (Edberg et al. 2022). This allows more frequent harvesting of eucalyptus, making it a more sustainable supply of fibre for the pulp and paper sector. The adaptability of eucalyptus to varied climates and soil types is another advantage of employing it for pulp and paper manufacturing (Amândio et al. 2022). Eucalyptus trees can thrive in a wide range of environments, from arid to humid, and from acidic to alkaline soils. As a result, eucalyptus is a versatile and widely available raw material source for the industry. Yet, there are several drawbacks to using eucalyptus for pulp and paper production. The influence of eucalyptus plantations on water supplies is one of the primary concerns. Eucalyptus trees consume a lot of water, which can cause water scarcity and have an impact on local ecosystems. Eucalyptus can provide a reliable source of raw material for the paper and pulp industry while also supporting economic growth and environmental stewardship with careful management and sustainable sourcing practices (Arnold et al. 2022).

4.2.2 Acacia

Acacia trees are also extensively employed in the pulp and paper industries as a raw material. These trees are native to Africa, Australia, and South Asia, and they are recognised for their rapid growth, adaptation to various soil types, and capacity to grow in arid circumstances. Acacia wood has a cellulose content comparable to other tree species commonly used in the pulp and paper industry, such as pine and spruce. The high growth rate of acacia trees is one of the primary advantages of employing them for pulp and paper production. Acacia trees, like eucalyptus, can mature within five years. The adaptability of acacia trees to varied soil types and climates is another advantage of employing them for pulp and paper manufacturing. Acacia trees may thrive in a wide range of environments, from dry and sandy soils to humid and tropical climates. This results in acacia being a versatile and widely available source of raw material for the pulp and paper industry (Arnold et al. 2022). Despite the benefits, there are some issues about using acacia trees for pulp and paper manufacturing. The influence of acacia plants on local ecosystems and water supplies is one of the key issues. Acacia tree monoculture plantations on a large scale can cause soil degradation, biodiversity loss, and water scarcity in some areas. Recent publications also indicate that monocrop acacia plantations in some regions exhibit signs of fungal pathogen attack that lead to wilt disease (Harsh 2022; Barnes et al. 2023; Lapammu et al. 2023; Wingfield et al. 2023).

4.2.3 Bamboo

Bamboo, belonging to subfamily Bambusoideae of the grass family, is a fast-growing plant, which can have diverse applications throughout the world. The multi-application of bamboo is not only due to its versatility but also due to its environmentally friendly attributes. There are around 1,500 species of bamboo around the world (Ahmad et al. 2021). Bamboo exists in many parts of the world, where it can survive hot and cold climates. It can usually be harvested after three to four years, as it has already reached maturity. Bamboo, for example, can contain up to 70% cellulose, making it an alternative source of raw material for paper production.

4.2.4 Kenaf

Kenaf, a plant related to cotton, is another tropical fibre that is gaining popularity in the industry due to its high-quality fibre and short rotation time (Al-Mamun et al. 2023). Kenaf (*Hibiscus cannabinus*) is a fast-growing crop that can be harvested after between three and four months. It may grow up to 14 feet tall and has a high fibre yield, making it an efficient and cost-effective raw material source for pulp manufacture. The high cellulose content of kenaf is one of its advantages in pulp manufacture. Another advantage of kenaf is that it is a low-cost crop to cultivate. It uses less water than other crops, such as cotton, and can be produced on marginal land that would otherwise be inappropriate for other crops. As a result, kenaf is a sustainable raw material source for pulp manufacture (Xu et al. 2020). Because the plant is inherently resistant to pests and illnesses, pesticides and other chemicals are not required in the growing process. Kenaf is also a renewable resource; therefore, it does not contribute to deforestation.

4.2.5 Bagasse

Bagasse, a surplus product of sugarcane processing, is also high in cellulose and has been used for years in the production of paper and pulp in countries like Brazil and India (Kumar et al. 2021). Bagasse is produced after squeezing the sweet water from the sugar cane. It is a well-known raw material among manufacturers due not only to its bright pulp colour but importantly, to its sustainability (Pydimalla et al. 2023). In India, bagasse is a popular pulp trade; 75 to 90 million tonnes of wet residual lignocellulosic sugarcane bagasse is produced annually from 600 operational sugar mills in India (Quereshi et al. 2020). Recently, research and development has shown encouraging signs in the sugarcane industries that bagasse can be applied in biodegradable material, packaging products, energy and more (Singh et al. 2022).

4.2.6 Abaca

Abaca is one of the most popular tropical fibres used in the packaging industry due to its mechanical, physical, and thermal properties (Kurien et al. 2023). It is a native plant of the Philippines (Salmorin and Gepty 2023) and is characterized by its long and strong fibres. It is used mainly in the production of high-quality paper (Celestino et al. 2016) like paper bags, coffee filters, and shipping sacks.

4.2.7 Jute

Jute is another type of tropical fibre that is widely used in the packaging industry (Samanta et al. 2020). Jute comes from the Mediterranean; however, other countries in Asia, namely, India, China, and Bangladesh, and countries of South America are exporting premium jute (Saleem et al. 2020). Jute fibres are versatile and are used in a wide range of applications, including bags, carpets, and curtains.

4.2.8 Coir

Coir is extracted from the outer husk of coconuts and is used in the production of ropes, mats, and carpets, among other products. It is an attractive option for packaging materials due to its strong and durable nature (Hamouda 2021).

4.2.9 Sisal

Sisal is a plant that is native to Mexico and is characterised by its long and fine white fibres (Trejo-Torres et al. 2018). It is commonly used in the production of twines, ropes, and carpets, among others, due to its strength and durability.

4.2.10 Pineapple Leaf

Pineapple leaf fibre is extracted from the leaves of the pineapple plant. It is used in the production of bags, clothing, and paper, among others. Pineapple leaf fibre is a sustainable and eco-friendly option because it is a by-product of the pineapple industry (Sarkar et al. 2018). Several studies have reported the potential of pineapple leaf fibre for various applications (Waham et al. 2015; Jaafar et al. 2018; Jutarut et al. 2020).

4.2.11 Oil Palm Empty Fruit Bunch

Oil palm empty fruit bunch (EFB) is the waste generated from oil palm production after the oil is squeezed out. It is proven to be suitable for pulp and paper manufacture (Indriati et al. 2020), and a few companies have been producing pulps and moulded paper from EFB, such as Eco Palm Paper Sdn. Bhd. and Ecopremium Packaging Sdn. Bhd., respectively, in Malaysia.

4.3 About Paper Packaging

Paper packaging refers to the use of paper-based materials that involves some technologies for enclosing, protecting, maintaining, and presenting various products in their original state until they arrive at consumers (Ibrahim et al.

2019; Choi and Burgess 2007). It involves the creation of containers, boxes, bags, wrappers, and other packaging forms using different grades and types of paper, such as recycled, kraft, corrugated, and cardboard or paperboard, a lightweight but strong paper-based material made from multiple layers of paper. According to Huang (2017), paper and cardboard constitute more than one-third of various types of packaging and containers all over the world, which is higher than the proportion of packaging using plastic, metals and glass. Other materials used to make paper packaging include plastic, foil, wax, and metal. Paper packaging also serves multiple other purposes, including as information and communication media, for marketing and branding, to demonstrate eco-friendliness, and for convenience.

The introduction of paper packaging can be traced back to ancient times when people discovered the versatility and protective properties of paper. Ancient Mesopotamians used clay tablets to store and transport goods, while ancient Egyptians used papyrus to wrap their mummies. The invention of paper in ancient China around the second century BCE opened up new possibilities for its use. People quickly realised that paper could be folded, wrapped, or formed into containers to hold and protect various items.

The industrial revolution in the nineteenth century brought significant advancements to paper production, making it more accessible and affordable. This led to the rise of paper packaging as a viable solution for industries seeking reliable and cost-effective packaging materials. With the development of paper making machines, such as the Fourdrinier machine in the early nineteenth century, paper production became faster and more efficient. This allowed mass production of paper, including various types of paperboards and cardboard, which are commonly used for packaging.

Paper packaging found applications in multiple industries. It was used for packaging food, beverages, household products, cosmetics, pharmaceuticals, and more. Paperboard and corrugated cardboard were particularly popular choices for shipping boxes, cartons, and other types of packaging. Over time, paper packaging evolved in terms of design, functionality, and customization. Manufacturers introduced different folding techniques, printing capabilities, and innovative structural designs to meet the specific requirements of products and branding needs.

Recently, paper packaging has gained favour due to its biodegradability, recyclability, and renewable nature. This has led to the increased adoption of paper-based packaging as a sustainable alternative to non-recyclable and non-biodegradable materials. The paper packaging industry continues to evolve with ongoing innovations. Over 70% of consumers are opting for sustainable packaging while shopping, according to Cone Communications (Kunam et al. 2022). Furthermore, consumers are ready to spend 10% extra for brands that use sustainable packaging materials (Escursell et al. 2021; Srivastava et al. 2022).

The use of recycled fibres, sourcing from responsibly managed forests, and reducing wastes are some of the practices in developing sustainable paper packaging. One of the great options for producing paper packaging is from tropical fibres. "Tropical fibres" refers to natural plant-based fibres that are primarily sourced from plants and trees native to tropical regions. They are often renewable, sustainable, and biodegradable. Common tropical fibres used for paper packaging include abaca, sisal, coir, jute, and bamboo.

Tropical fibres such as sisal, abaca, and jute are ideal for paper packaging materials due to their strong, durable, and lightweight properties. They are naturally biodegradable, making them a great eco-friendly alternative to plastic and other synthetic packaging materials. These fibres also have good absorbency, making them suitable for use in food and beverage packaging. As they are sourced from renewable resources, they are also cost-effective and sustainable. They can also be blended with other materials to create a range of textured and coloured paper packaging.

Paper packaging made from tropical fibres is a great way to reduce environmental impact while still providing a safe and secure way to store and ship products. Tropical fibres have many advantages over traditional materials like plastic and cardboard, including the fact that they are lightweight, durable, and biodegradable. They are also renewable, which helps reduce the carbon footprint associated with paper production. Additionally, tropical fibres can be used to make custom-shaped paper packaging, which can be tailored to fit each item inside. This makes it easier to pack items securely and also reduces the amount of packaging waste. Finally, paper packaging made from tropical fibres is more aesthetically pleasing than cardboard or plastic packaging. This makes it a great choice for businesses looking to make a statement with their packaging.

Today, paper packaging is widely used across industries, striking a balance between functionality, affordability, and environmental sustainability. For instance, in Europe, during the COVID-19 pandemic, packaging paper consumption increased by 2.1%, and sanitary and household paper also increased by 3.1% (Ramakanth et al. 2021). It remains a versatile and adaptable choice, with ongoing efforts to improve its ecological footprint and meet evolving consumer demands. The future of paper packaging is likely to be a balance of sustainability and convenience. Paper packaging will continue to be a popular choice for many products, and it is likely that these products will be made with sustainable, recyclable, and compostable materials.

4.4 Types of Tropical Fibres for Food Applications

Tropical fibres are types of plant fibres naturally grown in tropical regions for their fibre content (for example, cotton, kenaf, and jute) or as a by-product (pineapple, sugar palm, oil palm, banana, etc.). These fibres contain cellulose as the main composition, followed by hemicellulose, lignin, or waxes, depending on the plant type.

Tropical fibres for food packaging applications have been utilised for decades. Some common materials are natural fibres or leaves from banana trees, coconut palm, pineapple, mango, kenaf, oil palm, and sugar palm, which are used directly without further modifications or processes. The use of these materials was originally due to availability, ease to use, and lack of alternatives until other materials, such as plastics, dominated the applications. Nonetheless, concerns about recyclability, the environmental impact, and the abundance of plastic wastes nowadays have launched a paradigm shift towards environmentally friendly, green, natural, and biodegradable materials. Figure 4.1 shows samples of food packaging made from coconut and banana fibres.

The development of fibres for food packaging applications in the form of paper wrappers, trays, cups, bowls, and straws, including corrugated packaging for groceries, has increased dramatically in recent years, predominantly during the COVID-19 pandemic due to the growth in demand for food delivery and hygienic food presentation. According to a published report from *Materials and Packaging* (2022), the paper and paperboard packaging market in Europe is expected to expand at a compound annual growth rate of 4.6% from 2022 to 2029. Furthermore, a previous study revealed that coronavirus survival rates range from 24 to 72 hours on various packaging surfaces, with the longest being on stainless steel and plastic (Feber et al. 2020).

FIGURE 4.1
Food packaging materials from coconut shell and banana trunk.

4.4.1 Advantages and Challenges of Tropical Fibres as Food Packaging Materials

Tropical fibres are sustainable and abundant. These materials are cheaper and nonabrasive compared with standard fibres. Manufacturing tropical fibres from agricultural crops or waste could also reduce their environmental impact. In addition, the lightweight nature of these natural fibres contributes to weight and transportation cost reductions for packaged foods. The materials have various properties depending on the types, treatments, and modifications.

Generally, natural fibres are hydrophilic due to their hygroscopicity. This drawback affects their mechanical properties. Consequently, to overcome this weakness, various studies have been conducted using treatments, composites, etc. In terms of mechanical properties, tropical fibres exhibit lower strength than synthetic fibres. Yet with modifications, these fibres can be manufactured as food containers (clamshells, trays, plates, etc.), paper wrappers, or paper bags for some food products. Applying tropical fibres as food packaging materials for dry food products is an advantage in reducing plastic usage. Table 4.1 shows a comparison of tropical and synthetic fibre characteristics for food packaging.

4.4.2 Current Trends in Tropical Fibres for Food Packaging

Nowadays, tropical fibres can be incorporated into biopolymers or biocomposites as film fillers to enhance their properties. Mohammed et al. (2023) revealed that sugar palm fibres added to biocomposite films of starch with polyvinyl alcohol (PVA) increased the mechanical and thermal properties compared with films without the sugar palm fibres. The surface of oil palm EFB fibres treated with hot water and alkali also enhanced the tensile strength and water resistance of wheat gluten–based bioplastics (Chaiwong et al. 2019). Coconut shell nanofibres were also incorporated into PVA with essential oil as active food packaging (Arun et al. 2022). These fibres with

TABLE 4.1

Characteristics of Tropical Fibres and Synthetic Fibres for Food Packaging

Characteristics	Tropical Fibres	Synthetic Fibres
Mechanical strength	Low	Stronger
Weight	Low	Height
Moisture barrier	Low	Height
Availability	High	Low
Recycling	High	Poor
Biodegradability	Biodegradable	Non-biodegradable
Cost	Low	High

essential oil helped to improve the hydrophobicity and mechanical properties of PVA films.

Besides composite materials, the coating technique is applied to tropical fibre surfaces to improve the barrier properties and mechanical strength. Beeswax chitosan has been used as a coating on fibrous packaging paper from oil palm fruit fibres to enhance mechanical strength and water barrier properties (Chunsiriporn et al. 2022). Numerous studies are being conducted nowadays aiming to protect the environment. For example, lifecycle analysis comparing bamboo and polypropylene tableware indicated that bamboo fibre tableware has better environmental coordination than polypropylene tableware (Chen et al. 2023). This trend is expected to continue in the near future, with more involvement of the research communities in support of the United Nations Sustainable Development Goals (SDG) 2030. A summary of some tropical fibres used for food packaging applications is shown in Table 4.2.

TABLE 4.2

Summary of Tropical Fibres Developed as Food Packaging Materials

Fibre-Based Materials	Properties	Functionality of Packaging Material	References
Arrowroot fibre + starch	Higher degradation, lower WVP	No effect on *Staphylococcus aureus*, *Escherichia coli*, and *Bacillus subtilis*	Tarique et al. (2022)
Oil palm fibre + coated beeswax chitosan	Lower water susceptibility, increased tensile strength	NA	Chunsiriporn et al. (2022)
Ginger essential oil + chitosan + oil palm trunk waste fibre	Chitosan helped to encapsulate essential oil	NA	Maulidna et al. (2020)
Cellulose nano fibre from coconut waste + polyvinyl alcohol (PVA) + essential oil	Increased mechanical, thermal, and optical properties	Contain antioxidant and antimicrobial properties	Arun et al. (2022)
Palm sprout fibre + polylactic acid (PLA) films	Increased mechanical properties	NA	Vanitha and Kavitha (2021)
Cocoa pod husk cellulose + sugarcane bagasse fibre	Reduced susceptibility towards water	NA	Azmin et al. (2020)

4.5 Regulations on the Usage of Tropical Fibres for Packaging

Tropical timbers have long been associated with illegal logging and deforestation. These issues can adversely impact the environment and both local
and indigenous communities. Although there are no regulations related
to tropical fibres per se, many economies have imposed sustainable forest
management as a part of procurement directives. For example, the European
Union (EU) Timber Regulation, which came into force in 2013, is a piece of
legislation that prohibits illegally harvested timbers from being placed on
the EU market (European Union 2010). This regulation sets out mandatory
procedures for those trading in timber within the EU to minimise the risk
of illegal timbers being sold. It applies to both imported and domestically
produced timbers but does not cover all timber products. The EU Timber
Regulation excludes used timber and timber products that have completed
their lifecycle and would otherwise be disposed of as waste (i.e. recycled or
reclaimed woods). Packing cases, boxes, crates, drums, and similar packing
of wood; cable-drums of wood; pallets, box pallets, and other load boards, of
wood; pallet collars of wood, pulp, paper, and paperboard, except bamboo-
based and recovered paper products, are included in this regulation.

In the United States, the Lacey Act (USDA 2008) which was first endorsed
in 1900, prohibits the trafficking of illegally taken, possessed, transported, or
sold fish, wildlife, or plants. In 2008, this act was amended to prohibit trade
of plants and plant products (e.g. timber and paper) that are harvested in
violation of foreign laws. This act requires importers to provide a basic declaration, known as PPQ 505, to accompany every shipment of plants or plant
products. The Lacey Act also excludes declaration for used, recycled, and
reclaimed wooden products. Thus, sustainable forest management practices
such as Programme for the Endorsement of Forest Certification (PEFC) and
Forest Stewardship Council (FSC) Certification play a critical role for tropical
timbers and timber products industry players to ensure that their products
can be accepted in the global market.

PEFC was initiated in 1999 in Europe by a group of family forest own-
ers. The aim was to demonstrate excellence in sustainable forest management. Since its initiation, the PEFC has continued to grow, with 55 national
members and 48 endorsed national forest certification systems in 2022 (PEFC
2022a) Examples of PEFC endorsed national certification are shown in
Table 4.3. A total of 280 million hectares of forest have so far been PEFC
certified. As for Chain of Custody (CoC) product certification by PEFC, more
than 12,000 products have been certified up to December 2022 (PEFC 2022b)
From the statistics, certified tropical forest is associated with Asia, Oceania,
Africa, and Central and South American regions. Currently, Malaysia has the
largest tropical forest area certified with PEFC (5,719,108 hectares), followed
by Brazil (4,706,347 hectares) and Indonesia (3,773,023 hectares). Figure 4.2
shows the areas of PEFC certified forest and the number of certified PEFC

TABLE 4.3

Examples of Nationally Endorsed Forest Certification Systems

Country/ Economy	PEFC Governing Body	Number of Certification Bodies Accredited	
		Forest Management	Chain of Custody
North America	American Tree Farm System (ATFS)	7	17
China	China Forest Certification Council (CFCC)	6	23
Oceania	Responsible Wood	3	11
	New Zealand Forest Certification Association (NZFCA)	3	7
Malaysia	Malaysian Timber Certification Council	2	16
Brazil	Instituto Pró Manejo Florestal (Pro Forest Management Institute)	4	9
Indonesia	Indonesian Forest Certification Cooperation (IFCC)	2	11
Thailand	Federation of Thai Industries (FTI)	1	5
Vietnam	Vietnamese Academy of Forest Sciences (VAFS)	3	4

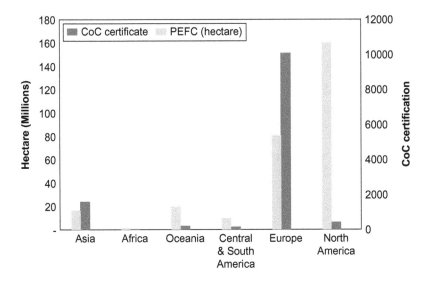

FIGURE 4.2
Status of PEFC certified forest and CoC products.

CoC products for various regions as of December 2022. North America has the largest PEFC certified forest, while Europe has certified most CoC products. In 2021, PEFC started a sustainable forest management programme for rubberwood, particularly in Southeast Asia (PEFC 2021). This will guarantee sustainable products from rubber plantations, such as latex and wood. PEFC works with national certification councils of Thailand, Vietnam, and Malaysia to certify rubber plantations, smallholders, and group planters. As of 2023, one certified rubber plantation in Thailand and 15 rubber plantations and groups in Vietnam are under this programme.

The FSC was founded in 1993 by a group of environmentalists, businesses, and community leaders in order to create a revolutionary market-based approach to improve forestry practices worldwide after the failure of the 1992 Earth Summit in Rio. The FSC is a voluntary certification for sustainable forestry, promoting environmentally sound, socially beneficial, and economically viable management of the world's forests. As of 2023, 193 million hectares of forest in 81 countries are certified by FSC (FSC 2023). FSC also operates CoC certification and has issued 53,343 certificates in 137 countries.

Besides sustainable forest management certification, a restriction on endangered and threatened timber species according to Convention on International Trade in Endangered Species of Wild Fauna and Flora (CITES) (Appendices I, II, and III) is applicable for wood-based products (CITES 1973). Among tropical timbers listed in these appendices are rosewood, Madagascar ebony, ramin, and some mahogany species.

Previously, for other biomass, such as oil palm fibres, no specialised certification or testing requirements have been available. However, some companies utilising oil palm fibres use certified sustainable palm oil (CSPO) as a means of demonstrating the sustainable source of their fibres. CSPO is a commonly used term for palm oil certified by the Roundtable on Sustainable Palm Oil (RSPO). RSPO was formed in 2004 by a few organizations, namely, the World Wildlife Fund (WWF), the Malaysian Palm Oil Association (MPOA), Unilever, AAK, and Migros. RSPO is a global, not-for-profit organisation with voluntary members that operates based on consensus. The RSPO stakeholders consist of oil palm producers, palm oil processors or traders, consumer goods manufacturers, retailers, banks and investors, environmental or nature conservation non-governmental organizations (NGOs), and social or developmental NGOs. Currently, RSPO represents 5,000 global member organisations (RSPO 2022). There are several other palm oil certification schemes, with the main ones being the Indonesia Sustainable Palm Oil (ISPO) and the Malaysian Sustainable Palm Oil (MSPO).

The ISPO was introduced in 2011 and currently certifies 5.45 million hectares of oil palm plantation, contributing to 40% of global oil palm production (Indonesia Palm Oil 2023). ISPO certification is managed by ISPO Committee (*Komite ISPO*) under the jurisdiction of Ministry of Agriculture, Indonesia. ISPO is currently mandatory for all plantations in Indonesia, including plantations supplying for bioenergy production. Smallholders have a grace period until 2025 to be certified.

The MSPO Certification Scheme is the national scheme in Malaysia for oil palm plantations, independent and organised smallholdings, and palm oil processing facilities. The scheme is developed and operated by the Malaysian Palm Oil Certification Council (MPOCC), and assessment is made in conformance with MS 2530 standards (Standards Malaysia 2013a; Standards Malaysia 2013b; Standards Malaysia 2013c; Standards Malaysia 2013d). The MSPO certification covers oil palm plantations, independent smallholders, and palm oil mills and has been made mandatory for all Malaysian oil palm since 2019. Currently, there are 25 Department of Standards Malaysia accredited certification bodies (CBs) for MSPO. A total of 6,406,242.88 hectares of palm oil plantation areas and 442 oil palm mills in Malaysia have valid certification (MPOCC 2023). A summary of RSPO, ISPO, and MSPO certification data is presented in Table 4.4. To cater for the growing demand for oil palm biomass for energy and fibres, MS 2751: 2022 was published in 2022 for the MSPO Chain of Custody for Oil Palm Biomass (Standards Malaysia 2022; MPOCC 2022). Based on the 20% EFB yield from fresh fruit bunch (FFB) (Syed Mahdhar 2011) and the FFB yield ranging between 12 and 38.8 tonnes/hectare/year (Lim et al. 2011), an estimated amount of 346 million to 1.12 billion metric tonnes of EFB can be collected annually around the world. This can be a significant source of sustainable fibre for various applications.

A comparison of principles for the three sustainable palm oil certifications is presented in Table 4.5. Each certification scheme has its own strengths and drawbacks. For example, both ISPO and MSPO rely heavily on local laws and legislation, and are thus less flexible for change compared with RSPO (Abdul Majid et al. 2021). Prior to 2022, the MSPO certification was granted based on seven principles outlined in MS 2530: 2013 Parts 1 to 4

TABLE 4.4

RSPO, ISPO, and MSPO Certification Data

	RSPO (RSPO 2022)	ISPO	MSPO (MPOCC 2023)
Global plantation area certified (%)	16.9	20.0	22.2
Palm oil plantation area certified (hectares)	4,854,915	5,770,000	6,406,242.88
Smallholder areas certified (hectares)	65,603	12,809	1,427,284.45
Number of certificates granted	3,785	682	607
Number of smallholders certified	22,017	17	181,392
Number of oil palm facilities certified	6,484	n/a	5,814
Certification cost per hectare	$87–215[a]	n/a	$30–35[a]

[a]*Kong Lingyu (2021).*

TABLE 4.5

Comparison of RSPO, ISPO, and MSPO

Principles	RSPO	ISPO	MSPO
1. Transparency	✓	x	✓
2. Compliance with law and regulations	✓	✓	✓
3. Long-term economic and financial viability	✓	x	x
4. Best practices by growers and millers	✓	✓	Excluded in MS2530:2022
5. Environmental responsibility and conservation	✓	✓	✓
6. Employees' rights and communities	✓	✓	✓
7. Responsible cultivation of new plant materials	✓	x	Excluded in MS2530:2022
8. Continuous improvement	✓	✓	✓
9. Certification by third party Certification Body	✓	✓	✓

Source: Abdul Majid et al. 2021, Ahmad and Hospes 2018, and Rosearnida (2019).

(Standards Malaysia 2013a; Standards Malaysia 2013b; Standards Malaysia 2013c; Standards Malaysia 2013d). In 2022, these standards were revised to include only five principles.

Other fibres, such as cotton, jute, sisal, abaca, and coconut coir, are planted commercially but at a smaller scale compared with oil palm. These fibres are an important source of income for many communities in tropical regions, and they play a key role in the global textile industry. Applications for other fibre crops include speciality paper, mats and rope, twine, and handicrafts. There are several certification schemes for these types of fibres. A good reference for operators involved in the biomass production and supply chain to choose a suitable certification scheme is published by NL Agency (2012). In their work, a total of 18 certification schemes were assessed, including International Sustainability and Carbon Certification (ISCC), Global GAP, Sustainable Agriculture Network (SAN), and Roundtable on Sustainable Biomaterials (RSB), besides RSPO. Due to the diversified biomass supply chains, most of the available certification schemes are able to certify biomass coming from all over the world. A simplified summary of suitable certification schemes for tropical fibres is depicted in Table 4.6.

In Malaysia, commodity crop planters can apply for Malaysian Good Agriculture Practice certification (MyGAP) managed by the Department of Agriculture Malaysia at the respective district office where the plantation area is located. MyGAP certification is based on Malaysian Standard MS 1784:2005 Crop Commodities – Good Agricultural Practice.

Some issues related to fibre-based packaging are the content of dangerous substances and biodegradability. Fibre-based packaging meant for food contact has to be tested for dangerous substances such as heavy metals. The two most common regulations used are the Food and Drug Agency (FDA) Title 21 of the U.S. Code of Federal Regulations (21 CFR) (FDA 2023) and the

European Parliament and Council Directive 94/62/EC (EU 1994). The regulations pertaining to paper packaging in the 21 CFR are listed in Table 4.7.

The European Parliament and Council Directive 94/62/EC on packaging and packaging waste includes some regulations pertaining to packaging, as indicated in Table 4.8. The primary concerns for the EU are the heavy metal content, packaging recyclability, and biodegradability, although packaging decomposition can also be in the form of physical, chemical, or thermal and not limited to biological means.

TABLE 4.6

Selected Certification Schemes for Tropical Fibres

Name	Chain Coverage	Type of Biomass	Region	Scheme Owner
2BSvs	All stages	All	Global	Control Union Certifications
Biograce	All stages	All	Global	Agency NL (Netherlands)
BSI	All stages	Sugarcane	Global	Bonsucro BSI
Global GAP	Biomass production	Agricultural biomass	Global	Euro-Retailer Produce Working Group (EUREP)
ISCC	All stages	All	Global	ISCC Association (ISCC e.V.)
NTA8080	All stages	All	Global	Netherlands Standardization Institute
RSB	All stages	All	Global	Energy Center (CEN) of the Ecole Polytechnique Fédérale de Lausanne, Switzerland
SAN	Biomass production	Agricultural biomass	Global	Sustainable Agriculture Network (SAN)

TABLE 4.7

FDA Regulation for Food Packaging

Title 21 of the U.S. Code of Federal Regulations (21 CFR)		Substance of Concern
Part 170	(Section 170.39)(a)(1) The substance has not been shown to be a carcinogen	Carcinogenic substance with TD_{50} values greater than 6.25 milligrams per kilogram bodyweight per day
Part 176	Indirect Food Additives: Paper and Paperboard Components	Acrylamide-acrylic acid resins, alkyl ketene dimers, chelating agents
Part 177	Indirect Food Additives: Polymers	Some modified industrial starch
Part 178	Indirect Food Additives: Adjuvants, Production Aids, and Sanitizers	Polymers
Part 179	(Section 179.45) Irradiation in the Production, Processing and Handling of Food	Packaging materials for use during the irradiation of prepackaged foods, including nitrocellulose-coated paper, glassine paper, kraft paper, and vegetable parchment

TABLE 4.8

EU Council Directive 94/62/EC Regulation on Packaging

Section	Requirement	Main Concern
Article 11	The sum of heavy metals in packaging	Lead, cadmium, mercury, and hexavalent chromium
Annex II Section 2.	Composition and the reusable and recoverable, including recyclable, nature of packaging	Packaging shall be designed, produced, and commercialised in such a way as to permit its reuse or recovery, including recycling, and to minimise its impact on the environment when packaging waste or residues from packaging waste management operations are disposed of. Packaging shall be so manufactured that the presence of noxious and other hazardous substances and materials as constituents of the packaging material or of any of the packaging components is minimised with regard to their presence in emissions, ash, or leachate when packaging or residues from management operations or packaging waste are incinerated or landfilled.
Annex IISection 3 (d)	Biodegradability	Packaging is capable of undergoing physical, chemical, thermal, or biological decomposition such that most of the finished compost ultimately decomposes into carbon dioxide, biomass, and water

For bio-based packaging, researchers have reported compostability/ biodegradability test methods meant for plastics and bioplastics, such as ASTM D5511, ASTM D5526, ASTM D5338, ASTM D5209, ASTM D-6400, and EN 14995 (ASTM 2018a, ASTM 2018b, ASTM 2021, ASTM 1992, ASTM 2023). A more suitable standard method for pulp or paper-based packaging is BS EN 13432: 2000 (European Standard 2016). This standard was harmonised with European Parliament and Council Directive 94/62/EC.

4.6 Conclusions

The use of sustainable and eco-friendly raw materials is crucial for the long-term viability of the pulp and paper industry. Tropical fibres offer several environmental and economic advantages compared with conventional wood fibres. For instance, tropical forests are a source of carbon sequestration, and their preservation can contribute to mitigating climate change. The processing of tropical fibres can also create job opportunities for communities in rural areas. However, challenges still exist in the implementation of sustainable production practices for tropical fibres, including certification and traceability, and social issues related to land use policies and indigenous rights.

In packaging terms, the use of tropical fibres has high, increasing acceptance in the market. Whether the fibres come from timber sources, commodity crops, or agricultural wastes, tropical fibres can be a major contributor to sustainable packaging. The current regulations and standard methods for packaging do not specifically address the use of tropical fibres; hence, it is timely that standard methods and certification schemes for them are developed. Tropical fibres from agricultural wastes are especially a guaranteed renewable resource for the future; thus, their traceability based on good agricultural practice is critically important in addressing market concerns.

References

Abd El-Sayed, E.S., El-Sakhawy, M. and El-Sakhawy, M.A.M., 2020. Non-wood fibres as raw material for pulp and paper industry. *Nordic Pulp and Paper Research Journal*, 35(2), 215–230.

Ahmad, Z., Upadhyay, A., Ding, Y., Emamverdian, A. and Shahzad, A., 2021. Bamboo: Origin, habitat, distributions and global prospective. In *Biotechnological Advances in Bamboo: The "Green Gold" on the Earth*, Zishan Ahmad, Yulong Ding and Anwar Shahzad, Eds. (pp. 1–31).

Al-Mamun, M., Rafii, M.Y., Misran, A.B., Berahim, Z., Ahmad, Z., Khan, M.M.H., Oladosu, Y. and Arolu, F., 2023. Kenaf (Hibiscus Cannabinus L.): A promising fibre crop with potential for genetic improvement utilizing both conventional and molecular approaches. *Journal of Natural Fibres*, 20(1), 2145410.

Amândio, M.S., Pereira, J.M., Rocha, J.M., Serafim, L.S. and Xavier, A.M., 2022. Getting value from pulp and paper industry wastes: On the way to sustainability and circular economy. *Energies*, 15(11), 4105.

Arnold, R., Midgley, S., Stevens, P., Phimmavong, S., Kien, N.D. and Chen, S., 2022. Profitable partnerships: Smallholders, industry, eucalypts and acacias in Asia. *Australian Forestry*, 85(1), 38–53.

Arun, R., Shruthy, R., Preetha, R. and Sreejit, V., 2022. Biodegradable nano composite reinforced with cellulose nano fibre from coconut industry waste for replacing synthetic plastic food packaging. *Chemosphere*, 291(1), 132786.

ASTM., 1992. ASTM D5209 Standard test method for determining the aerobic biodegradation of plastic materials in the presence of municipal sewage sludge (Withdrawn 2004).

ASTM., 2018. ASTM D5511 Standard test method for determining anaerobic biodegradation of plastic materials under high-solids anaerobic-digestion conditions.

ASTM., 2018. ASTM D5526 Standard test method for determining anaerobic biodegradation of plastic materials under accelerated landfill conditions.

ASTM., 2021. ASTM D5338 Standard test method for determining aerobic biodegradation of plastic materials under controlled composting conditions, incorporating thermophilic temperatures.

ASTM., 2023. ASTM D6400 Standard specification for labeling of plastics designed to be aerobically composted in municipal or industrial facilities.

Azmin, S.N.H.M., Hayat, N.A.M. and Nor, M.S.M., 2020. Development and characterization of food packaging bioplastic film from cocoa pod husk cellulose incorporated with sugarcane bagasse fibre. *Journal of Bioresources and Bioproducts*, 5(4), 248–255.

Barnes, I., Abdul Rauf, M.R., Fourie, A., Japarudin, Y. and Wingfield, M.J., 2023. Ceratocystis Manginecans and not C. Fimbriata a threat to propagated Acacia spp. in Sabah, Malaysia. *Journal of Tropical Forest Science (JTFS)*, 35, 16–26.

Celestino, E.R., Sarmiento, G.O. and Bencio, J.T., 2016. Value chain analysis of abaca (musa textiles) fiber in Northern Samar, Philippines. *International Journal of Innovative Science, Engineering & Technology*, 3(8), 166–168.

Chaiwong, W., Samoh, N., Eksomtramage, T. and Kaewtatip, K., 2019. Surface-treated oil palm empty fruit bunch fibre improved tensile strength and water resistance of wheat gluten-based bioplastic. *Composites Part B Engineering*, 176, 107331.

Chen, X., Chen, F., Yang, Q., Gong, W., Wang, J., Li, Y. and Wang, G., 2023. An environmental food packaging material part I: A case study of life-cycle assessment (LCA) for bamboo fibre environmental tableware. *Industrial Crops and Products*, 194, 116279.

Choi, S.J. and Burgess, G., 2007. Practical mathematical model to predict the performance of insulating packages. *Packaging Technology and Science*, 20(6), 369–380.

Chungsiriporn, J., Khunthngkaew, P., Wongnoipla, Y., Sopajarn, A., Karrila, S. and Lewkitayakorn, J., 2022. Fibrous packaging paper made of oil palm fibre with beeswax-chitosan solution to improve water resistance. *Industrial Crops and Products*, 177, 114541.

CITES., 1973. Convention on international trade in endangered species of wild fauna and flora. part of the Endangered Species Act (PL 93–205, 93rd Congress) and in 50 appendices. *Code Fed. Reg.*, p.23.

Dermawan, A. and Hospes, O., 2018. When the state brings itself back into GVC: The case of the Indonesian palm oil pledge. *Global Policy*, 9(S2).

Ecopalm. 2023. http://ecopalmpaper.com/about-us/ (Accessed 24 March 2023).

Edberg, S., Tigabu, M. and Odén, P.C., 2022. Commercial Eucalyptus plantations with taungya system: Analysis of tree root biomass. *Forests*, 13(9), 1395.

El Nemr, A., 2012. From natural to synthetic fibres. *Textiles: Types, Uses and Production Methods; El Nemr, A., Ed*, pp. 1–152.

Escursell, S., Llorach-Massana, P. and Roncero, M.B., 2021. Sustainability in e-commerce packaging: A review. *Journal of Cleaner Production*, 280, 124314. https://doi.org/10.1016/J.JCLEPRO.2020.124314.

European Parliament and Council., 1994. European Parliament and Council directive 94/62/EC on packaging and packaging waste.

European Union., 2010. Regulation (EU) No 995/2010 of The European Parliament and of the council laying down the obligations of operators who place timber and timber products on the market.

U.S. Food and Drug Administration. 2023 Title 21 of the U.S. Code of Federal Regulations, Chapter 1, Parts 170 - 179. Office of the Federal Register National Archives and Records Administration..

Feber, D., Lingqvist, O. and Nordigården, D., 2020. Paper, forest products & packaging practice: How the packaging industry can navigate the coronavirus pandemic. *McKinsey & Company*, April 2020, pp. 1–7.

Forest Stewardship Council (FSC)., 2023. Facts & figures. https://connect.fsc.org/impact/facts-figures. Accessed 27 March 2023.

Hamouda, T., 2021. Sustainable packaging from coir fibres. In *Biopolymers and Biocomposites from Agro-Waste for Packaging Applications*, Naheed Saba, Mohammad Jawaid and Mohamed Thariq, Eds. (pp. 113–126). Woodhead Publishing.

Harsh, N.S.K., 2022. A new wilt disease of *Acacia nilotica* caused by *Fusarjum oxysporum*. *Journal of Tropical Forest Science (JTFS)*, 16(4), 453–462.

Huang, J., 2017. Sustainable development of green paper packaging. *Environment and Pollution*, 6(2), 1. https://doi.org/10.5539/ep.v6n2p1.

Ibrahim, S., El Saied, H., & Hasanin, M., 2019. Active paper packaging material based on antimicrobial conjugated nano-polymer/amino acid as edible coating. *Journal of King Saud University - Science*, 31. https://doi.org/10.1016/j.jksus.2018.10.007.

Indonesia Palm Oil., 2023. ISPO by the numbers. https://www.indonesiapalmoilfacts.com/ispo/. Accessed 23 March 2023.

Indriati, L., Elyani, N. and Dina, S.F., 2020, December. Empty fruit bunches, potential fibre source for Indonesian pulp and paper industry. In *IOP Conference Series. Materials Science and Engineering* (Vol. 980, No. 1, p. 012045). IOP Publishing.

Jaafar, J., Siregar, J.P., Mat Piah, M.B.M., Cionita, T., Sharmiza, A. and Teuku, R., 2018. Influence of selected treatment on tensile properties of short pineapple leaf fiber reinforced tapioca resin biopolymer composites. *Journal of Polymers and the Environment*, 26(11), 4271–4281.

Jutarut, I., Piyaporn, K., Yutthawee, W., Kaewta, K., Panumas, S. and Arrisa, S., 2020. Biodegradable plates made of pineapple leaf pulp with biocoatings to improve water resistance. *Journal of Materials Research and Technology*, 9(3), 5056–5066.

Karimah, A., Ridho, M.R., Munawar, S.S., Adi, D.S., Damayanti, R., Subiyanto, B., Fatriasari, W. and Fudholi, A., 2021. A review on natural fibres for development of eco-friendly bio-composite: Characteristics, and utilizations. *Journal of Materials Research and Technology*, 13, 2442–2458.

Kestur, S.G., Ramos, L.P. and Wypych, F., 2009. Comparative study of Brazilian natural fibres and their composites with other. *Natural Fibre Reinforced Polymer Composites: from Macro to Nanoscale*. Old City Publishing, Inc. Philadelphia.

Kumar, A., Kumar, V. and Singh, B., 2021. Cellulosic and hemicellulosic fractions of sugarcane bagasse: Potential, challenges and future perspective. *International Journal of Biological Macromolecules*, 169, 564–582.

Kunam, P.K., Ramakanth, D., Akhila, K. et al., 2022. Bio-based materials for barrier coatings on paper packaging. *Biomass Conv. Bioref.* https://doi.org/10.1007/s13399-022-03241-2.

Kurien, R.A., Selvaraj, D.P., Sekar, M., Koshy, C.P., Paul, C., Palanisamy, S., Santulli, C. and Kumar, P., 2023. A comprehensive review on the mechanical, physical, and thermal properties of abaca fibre for their introduction into structural polymer composites. *Cellulose*, 30(14), 8643–8664.

Lapammu, M., Warburton, P.M., Japarudin, Y., Boden, D., Wingfield, M.J. and Brawner, J.T., 2023. Verification of tolerance to infection by *Ceratocystis manginecans* in clones of *Acacia mangium*. *Journal of Tropical Forest Science (JTFS)*, 35, 42–50.

Lim, K.H., Goh, K.J., Kee, K.K. and Henson, I.E. 2011. Climatic requirements of oil palm. In Goh, K.J., Chiu, S.B. and S Paramananthan (eds.), *Agronomic Principles and Practices of Oil Palm Cultivation*, pp. 3–48. Agricultural Crop Trust (ACT).

Lingyu, K., 2021. Who gets to define sustainable palm oil? China dialogue. https://chinadialogue.net/en/food/who-gets-to-define-sustainable-palm-oil/. Accessed 24 March 2023.

Majid, N.A., Ramli, Z., Sum, S. Md. and Awang, A.H., 2021. Sustainable palm oil certification scheme frameworks and impacts: A systematic literature review. *Sustainability*, 13(6), 3263.

Materials & Packaging., 2022. Europe paper and paperboard packaging market-industry trends and forecast to 2029. Published Report, https://www.databridgemarketresearch.com/reports/europe-paper-paperboard-packaging-market accessed 18 March 2023.

Maulidna., Wirjosentono, B., Tamrin. and Marpaung, L., (2020). Microencapsulation of ginger-based essential oil (Zingiber cassumunar roxb) with chitosan and oil palm trunk waste fiber prepared by spray-drying method. *Case Studies in Thermal Engineering*, 18, 100606. https://doi.org/10.1016/j.csite.2020.100606.

Mohammed, A.A.B.A., Hasan, Z., Omran, A.A.B., Elfaghi, A.M., Ali, Y.H., Ilyas, R.A. and Sapuan, S.M., 2023. Effect of sugar palm fibres on the properties of blended wheat starch/polyvinyl alcohol (PVA)- based biocomposite films. *Journal of Materials Research and Technology*, In press. https://doi.org/10.1016/j.jmrt.2023.02.027.

MPOCC., 2023. MSPO trace. https://mspotrace.org.my/ Accessed 27 March 2023.

MPOCC., 2022. Development process of MS2751: 2022 Malaysian Sustainable Palm Oil (MSPO) Chain of Custody (CoC) of Oil Palm Biomass. Paper presented at CSPO Forum and Launching Ceremony of the MSPO 2022. 22 March 2022, Kuala Lumpur.

Nirsatmanto, A., Sunarti, S., Kartikaningtyas, D., Handayani, B.R., Setyaji, T., Pudjiono, S., Kartikawati, N.K., Kardiansyah, T.E.D.D.Y., Indrawan, D.A., Pari, R.O.H.M.A.H. and Razoki, M., 2022. Evaluation of the characteristics of Eucalyptus pellita and Acacia hybrid superior clones selected from breeding program in Indonesia as materials for pulp and papermaking. *Wood Research*, 67(5), 847–865.

NL Agency., 2012. Selecting a biomass certification system – A benchmark on level of assurance, costs and benefits. Report NL Energy and Climate Change, Netherlands.

Parthasarathy, P., Alherbawi, M., Shahbaz, M., Mackey, H.R., McKay, G. and Al-Ansari, T., 2022. Conversion of oil palm waste into value-added products through pyrolysis: A sensitivity and techno-economic investigation. *Biomass Conversion and Biorefinery*, 1–21. Edited in Chief by Martin Kaltschmitt & Hermann Hofbauer

PEFC., 2021. PEFC is supporting sustainable rubber: ASEAN countries demonstrate best practice in natural rubber production & supply. https://rubber.pefc.org/news/pefc-is-supporting-sustainable-rubber-asean-countries-demonstrate-best-practice-in-natural-rubber-production-supply. Accessed 27 March 2023.

PEFC., 2022a. Find certification bodies. https://www.pefc.org/find-certification-bodies. Accessed 23 March 2023.

PEFC., 2022b. Facts and figures. https://pefc.org/discover-pefc/facts-and-figures. Accessed 20 March 2023.

Pydimalla, M., Chirravuri, H.V. and Uttaravalli, A.N., 2023. An overview on non-wood fibre characteristics for paper production: Sustainable management approach. *Materials Today: Proceedings*. https://doi.org/10.1016/j.matpr.2023.08.278.

Quereshi, S., Naiya, T.K., Mandal, A. and Dutta, S., 2020. Residual sugarcane bagasse conversion in India: Current status, technologies, and policies. *Biomass Conversion and Biorefinery*, 1–23.

Ramakanth, D., Singh, S., Maji, P.K., Lee, Y.S. and Gaikwad, K.K., 2021. Advanced packaging for distribution and storage of COVID-19 vaccines: A review. *Environmental Chemistry Letters*, 19(5), 3597–3608. https://doi.org/10.1007/S10311 -021-01256-1/FIGURES/8.

Rosearnida, S., 2019. Transformation of oil palm independent smallholders through Malaysian sustainable palm oil. *Journal of Oil Palm Research*, 31(3), 496–507.

Roundtable on Sustainable Palm Oil (RSPO)., 2022. Impact report 2022.

Saleem, M.H., Ali, S., Rehman, M., Hasanuzzaman, M., Rizwan, M., Irshad, S., Shafiq, F., Iqbal, M., Alharbi, B.M., Alnusaire, T.S. and Qari, S.H., 2020. Jute: A potential candidate for phytoremediation of metals—A review. *Plants*, 9(2), 258.

Salit, M.S., 2014. Tropical natural fibres and their properties. In *Tropical Natural Fibre Composites: Properties, Manufacture and Applications*, Mohd Sapuan Salit, Eds. (pp. 15–38). Springer.

Salmorin, D.E. and Gepty, V., 2023. Cultural practices & beliefs in abaca farming of the indigenous people. *Journal of Humanities and Social Sciences Studies*, 5(2), 22–32.

Samanta, A.K., Mukhopadhyay, A. and Ghosh, S.K., 2020. Processing of jute fibres and its applications. In *Handbook of Natural Fibres*, Ryszard M. Kozłowski and Maria Mackiewicz-Talarczyk Eds. (pp. 49–120). Woodhead Publishing.

Sarkar, T., Nayak, P. and Chakraborty, R., 2018. Pineapple [Ananas comosus (L.)] product processing techniques and packaging: A review. *Iioabj*, 9(4), 6–12.

Sharma, N., Godiyal, R.D. and Thapliyal, B.P., 2020. A review on pulping, bleaching and papermaking processes. *Journal of Graphic Era University*, 8(2), 95–112.

Singh, R.V., Sharma, P. and Sambyal, K., 2022. Application of sugarcane bagasse in chemicals and food packaging industry: Potential and challenges. *Circular Economy and Sustainability*, 2(4), 1479–1500.

Srivastava, P., Ramakanth, D., Akhila, K. et al., 2022. Package design as a branding tool in the cosmetic industry: Consumers' perception vs. reality. *SN Bus Econ*, 2(6), 58. https://doi.org/10.1007/ s43546-022-00222-5.

Standards Malaysia, 2013a. MS 2530-1:2013. Malaysian Sustainable Palm Oil (MSPO): Part 1: General principles.

Standards Malaysia, 2013b. MS 2530-2:2013 Malaysian Sustainable Palm Oil (MSPO): Part 2: General principles for independent smallholders.

Standards Malaysia, 2013c. MS 2530-3:2013 Malaysian Sustainable Palm Oil (MSPO): Part 3: General principles for oil palm plantations and organised smallholders.

Standards Malaysia, 2013d. MS 2530-4:2013 Malaysian Sustainable Palm Oil (MSPO): Part 4: General principles for palm oil mills.

Standards Malaysia, 2022. MS 2751:2022 Malaysian Sustainable Palm Oil (MSPO) chain of custody of oil palm biomass.

Syed-Mahdhar, S.H. 2011 *Compost and Organic Fertilizer from by Product in Oil Palm Plantation, National Seminar on Oil Palm* (pp. 87–93). Kuala Lumpur.

Tarique, J., Sapuan, S.M., Khalina, A., Ilyas, R.A. and Zainudin, E.S., (2022). Thermal, flammability, and antimicrobial properties of arrowroot (Maranta arundinacea) fiber reinforced arrowroot starch biopolymer composites for food packaging applications. *International Journal of Biological Macromolecules*, 213, 1–10. https://doi.org/10.1016/j.ijbiomac.2022.05.104.

Trejo-Torres, J.C., Gann, G.D. and Christenhusz, M.J., 2018. The Yucatan Peninsula is the place of origin of sisal (Agave sisalana, Asparagaceae): Historical accounts, phytogeography and current populations. *Botanical Sciences*, 96(2), 366–379.

Tsien, T.H., 1973. Raw materials for old papermaking in China. *Journal of the American Oriental Society*, 510–519.

U.S. Department of Agriculture (USDA). Lacey Act. 2008 (Amendment). 16 USC 3371-3378.

Vanitha, R. and Kavitha, C., 2021. Development of natural cellulose fibre and its food packaging application. *Materialstoday: Proceedings*, 36(4), 903–906.

Waham, A.L. and Wan Aizan, W.A.R., 2015. Chemical pulping of waste pineapple leaves fiber for kraft paper production. *Journal of Materials Research and Technology*, 4(3), 254–261.

Wingfield, M.J., Wingfield, B.D., Warburton, P., Japarudin,Y., Lapammu, M., Abdul Rauf, M.R., Boden, D. and Barnes, I., 2023. Ceratocystis wilt of *Acacia mangium* in Sabah: Understanding the disease and reducing its impact. *Journal of Tropical Forest Science (JTFS)*, 35, 51–66.

Wirjosentono, B., & Marpaung, L., 2020. Microencapsulation of ginger-based essential oil (Zingiber cassumunar roxb) with chitosan and oil palm trunk waste fibre prepared by spray-drying method. *Case Studies in Thermal Engineering*, 18, 100606.

Xu, J., Tao, A., Qi, J. and Wang, Y., 2020. Bast fibres: Kenaf. In *Handbook of Natural Fibres*, Ryszard M. Kozłowski and Maria Mackiewicz-Talarczyk, Eds. (pp. 71–92). Woodhead Publishing.

5

Sustainable Substrate Based Biocomposites for Electronic Applications: A Review

Nik Athirah Azzra, A. Atiqah, A. Jalar, R. A. Ilyas,
and Ainun Zuriyati Mohamed

5.1 Introduction

Electronics potting or encapsulation is a method of polymer filling for electronic compartments that protect electronic components (Kokkila 2019). Some polymers in plastic encapsulation for electronic packaging consist of synthetic polymers. The common polymer matrices used for plastic encapsulation are classified into thermosetting resin and thermoplastic resin, such as polyesters, vinyl esters, epoxy, phenolics, polyamides (PA), and bismaleimides (BMI) (Nguyen, Zatar, and Mutsuyoshi 2017).

Thermoset encapsulations are in high demand due to their high linking density, high durability, great chemical adhesion, and thermal properties (Raquez et al. 2010; Yang et al. 2019). Epoxy and polyurethane are some thermosets that attract a lot of industrial attention due to their excellent chemical and heat resistance (Sun and Yi 2021; Nguyen, Zatar, and Mutsuyoshi 2017), and good mechanical (Utekar et al. 2021) and physical properties (Sun and Yi 2021). These resins are widely applied in electrical appliances for harsh environments, infrastructure, and household products (Yang et al. 2019). However, the downside of these polymers is their severe environmental impact. They are non-biodegradable and not recyclable. After polymerization, they can only be disposed of by burning them with precautions under toxic waste substances (Raquez et al. 2010; Naheed Saba et al. 2016; Nguyen, Zatar, and Mutsuyoshi 2017

Conversely, thermoplastics are recyclable (Muhammad Ameerul Atrash Mohsin, Lorenzo Iannuccii 2021). However, it is advisable not to reuse the re-melted thermoplastics, as there are changes in the physical properties due to the breakage of the polymeric chain (Asim et al. 2017). One study reported that Malaysia produced 1,600,000 tons of toxic waste yearly, resulting in numerous critical complaints correlated to environmental issues (Ibrahim et al. 2021). A few adjustments could be made to cope with this situation; some researchers

TABLE 5.1

Physical and Mechanical Properties of Synthetic Polymer of Composites

Synthetic Polymer	Density (g/cm³)	Tensile Strength (MPa)	Young's Modulus (GPa)	Elongation (%)	References
Epoxy	1.171.156	39.3920.575	2.50.51232.38	0.915.3	(Cavalcanti et al. 2021; N. Saba et al. 2019; T. Sun et al. 2021; Verma et al. 2019)
Teflon	2.16	N/A	N/A	N/A	(Struchkova et al. 2022)
Polyester	1.5	34.78	4.97	N/A	(Cavalcanti et al. 2021)
Nylon	1.14	52.3	N/A	N/A	(X. Li et al. 2020)

improve the recycling method, and others may focus on the disposal technique. Another practical method to investigate is replacing synthetic fibre with natural fibre as reinforcement. Therefore, a few studies have taken the initiative to produce biodegradable composites compared with synthetic polymers. Table 5.1 summarizes different types of polymers with their excellent physical properties. Hence, this chapter begins by introducing the critical need for environmentally friendly alternatives to the polymers used in electronic applications, highlighting the growing concerns about electronic waste and the detrimental impact of traditional materials. Then, we delve into the fundamentals of sustainable substrates and biocomposites, elucidating the unique properties of nanocellulose. Throughout the chapter, we will highlight the properties of existing sustainable substrates from biocomposites and future prospects and directions of this exciting field, making it a valuable resource for researchers, engineers, and enthusiasts interested in the intersection of sustainability and electronics.

5.1.1 Encapsulation Biodegradable Sensor Devices

The application of nanocellulose in wearable sensors with multi-functional electronic uses has gained a lot of attention, as it can monitor human conditions by detecting human movement (Wan et al. 2021). The characteristic of the sensor model is the important part of creating flexible and human-friendly devices (Zhu et al. 2021). The latest research into synthesizing a hybrid nanocomposite by fabricating renewable materials and inorganic functional materials attracted attention due to their unique properties and the uses of renewable materials (Brakat and Zhu 2021; Rajinipriya et al. 2018). Encapsulation plays a crucial role in the development of biodegradable sensor devices. The encapsulation layer provides stability and protection to the sensors, allowing them to function effectively in various environments. Several studies have focused on different encapsulation materials and techniques to enhance the performance of these devices. For example, Zou et al. (2023) developed an adhesive hydrogel encapsulation layer that improved the stability and long-term performance of hydrogel-based sensors. Zhang et al. explored the use of polymer adhesives and metal tubes to create a more robust sensor for temperature and salinity sensing (Zhang et al. 2021). Choi et al. investigated the use of polyanhydride-based polymers as hydrophobic encapsulation layers for bioresorbable electronic systems (Choi et al. 2020). Valavan et al. studied the selectivity of thermoplastic encapsulants for gas sensors, ensuring that the encapsulation did not reduce sensor sensitivity (Valavan et al. 2019). Schmitz-Hertzberg et al. developed a capsule system with permeable and sealed compartments for protecting miniaturized analytical chip devices (Schmitz-Hertzberg et al. 2014).

5.1.2 Cellulose

Cellulose has received massive interest from academicians and industrials due to its excellent mechanical properties and has a huge opportunity to be commercialized in a large-scale industry (Ye et al. 2018; Mohammad Padzil et al. 2020). Every year, 170–200 billion tons of lignocellulosic materials are harvested, indicating that cellulose is the most abundant biopolymer (Asim et al. 2017; Banwell et al. 2021). It has been used extensively as an energy source due to its biodegradability, biocompatibility, non-toxicity, non-pollution, and facile modification and renewability properties (Xue, Mou, and Xiao 2017; Otoni et al. 2018; Li et al. 2021). Cellulose is usually made up of a long-chain polymer with repeating units of D-glucose, which represent a simple sugar chain (Huang et al. 2019). Figure 5.1 depicts the cellulose structure, which consists of several glucose molecules linked by β-1,4 glycosidic bonds and a repetitive unit known as cellobiose dimer (Mahardika et al. 2018; Islam et al. 2018; Ouarhim et al. 2018). By referring to the cellulose chemical formula $(C_6 H_{10} O_6)_n$, the number of glucose molecules are represented as n-number, which depends on the type of plant for cellulose of different origin. The polymeric chain of cellulose leads to two ends, which are reducing and non-reducing. The reducing part refers to the free hydroxyl group, while the non-reducing end is part of the hydroxyl group (Islam et al. 2018). Nevertheless,

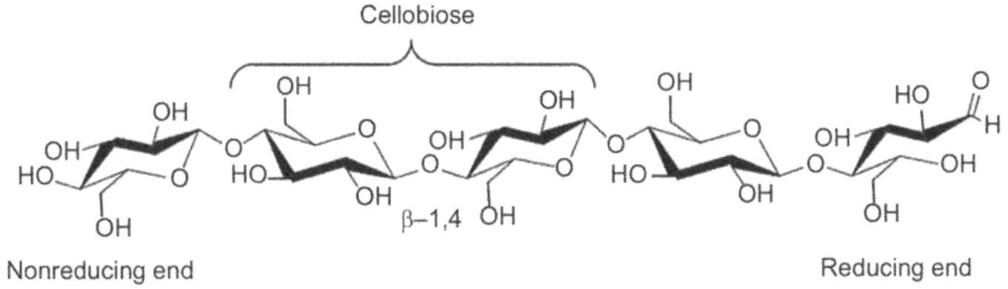

FIGURE 5.1
Cellulose chain structure with repetitive cellobiose. (From Ouarhim, W. et al., *Mechanical and Physical Testing of Biocomposites, Fibre-Reinforced Composites and Hybrid Composites,* 43–60, 2018.)

the content is different in wood, plant leaves, and stalks, which are made up of lignin, hemicelluloses, and cellulose (Borrero-López et al. 2018). Even though it is common to find cellulose in plants, cellulose can also be produced by bacteria (Kalia et al. 2011).

5.2 Substrate Based Biocomposites

There are various plants and biodegradable materials that are high-quality candidates in the production of thin film composites that do not pose a threat to humans and the environment. Over the past five years, industry and researchers have been seen to be more inclined towards the production of biodegradable and more environmentally friendly products. For example, we can see more advertising and daily "green" product usage campaigns for products such as yam-based straws, starch, and recycled paper. In June 2022, an investigation from India chaired by Surendra Kumar made a biodegradable film composite for the use of "Sound Absorbers" using seaweed and nanocrystalline cellulose. Based on studies that have been carried out, porous films with the highest content of agar act as a brilliant sound absorber. However, the addition of nanocellulose material does not change any acoustic properties (Kumar et al. 2022). There have been various efforts to fabricate biodegradable materials, thus making the goal of creating a green environment no longer impossible to achieve. Therefore, further details of biodegradable materials used for different applications will be discussed.

5.2.1 Nanocellulose

Nanocellulose refers to cellulose on a nanometre scale (Kharbanda et al. 2019). There are three different types of nanocellulose: cellulose nanofibres (CNF), cellulose nanocrystals (CNC), and bacterial cellulose (BC) (Trache et al. 2020; Blanco et al. 2018). They have an indistinguishable chemical composition but are different in morphology, particle size, crystallinity, and properties due to differences in extraction methods (Phanthong et al. 2018). Cellulose nanofibres are fibrils that are made up of crystalline and amorphous parts of the cell wall and isolated by mechanical treatment with 13–22 μm diameter and crystallinity range up to 44–65% (Heise et al. 2021; de Amorim et al. 2020). Meanwhile, cellulose nanocrystals are produced by chemical treatment such as bleaching, acid hydrolysis, and alkali treatment to remove the non-cellulose and amorphous part of cellulose. They have a range of 4–25 nm in diameter and 100-1000 nm in length (Pereira et al. 2020; Leão et al. 2017; Ilyas et al. 2018). Also, Gluconacetobacter is able to produce a cellulose form of bacterial cellulose that is chemically similar to plant cellulose with diameter 10–100 nm, 90% crystallinity (de Amorim et al. 2020), and size of 70–80 nm in width (Azeredo, Rosa, and Mattoso 2017). Generally, nanocellulose fibre is light in weight and has a low density of around 1.6 g/ cm^3 (Barbash, Yashchenko, and Vasylieva 2019), with high stiffness, around 220 GPa of elastic modulus, which is higher than Kevlar fibre (Rai and Singh 2020). In addition, these nanomaterials are widely applied in many fields, like electronic devices (Xing et al. 2019), cosmetics (Bongao et al. 2020), texturing agents in food (Oun, Shankar, and Rhim 2020), biodegradable packaging, and as a filler of special textiles (Phanthong et al. 2018). Nanocellulose is also frequently applied in the medical field as a medical packaging membrane for haemodialysis, vascular grafts, wound dressing, tissue engineering, and drug delivery (Xue, Mou, and Xiao 2017). However, there are some disadvantages of the fibre, for instance, weak interfacial adhesion (Lee, Khalina, and Lee 2021), low melting point, moisture absorption (Kahavita et al. 2020), and poor compatibility with hydrophobic substances (Kalia et al. 2011).

Various types of plants that are used to extract nanocellulose. Table 5.2 depicts different plants resulting in different mechanical properties. In order to replace current synthetic polymers, these nanocelluloses were suggested to act as a substrate that enhanced the mechanical properties of natural polymers. Nonetheless, research found that nanocellulose fibrils need to undergo chemical treatment and modification, since they contain components (lignin, hemicellulose, and pectin) that cause dimensional instability. According to

TABLE 5.2

Properties of Nanocellulose

Plants	Type of Cellulose	Diameter (nm)	Hardness (MPa)	Elastic Modulus (GPa)	References
Oil palm empty fruit bunches (OPEFB)	Cellulose Nanocrystal	43	N/A	4,758–7,951.4	(Soetaredjo et al. 2022)
Amorpha fruticosa Linn.	Cellulose Nanofibers	10	36.74	N/A	(Kong et al. 2019)
Oil palm mesocarp fibre	N/A	7–21	N/A	N/A	(Abu Bakar et al. 2021)
Switchgrass	N/A	N/A	18,500	0.43	(Wan et al. 2021)
Hardwoods	N/A	N/A	400–560	14.2–35.4	(Changjie Chen et al. 2017)
Softwoods	N/A	N/A	340–530	14.2–18	(Changjie Chen et al. 2017)
Crops	N/A	N/A	480–850	16.3–20.8	(Changjie Chen et al. 2017)
Bamboo	N/A	12–55	290–507	4.5–8.6	(Wang et al. 2019)
Lyocell fibre	N/A	N/A	250	25	(Adekunle et al. 2012)

Fahma et al. (2021), the significance of undergoing alkali treatment is believed to increase the tensile strength of plant fibres. Table 5.2 summarizes that the nanocelluloses that come from different plants are different in diameter and mechanical properties.

5.2.2 Thermoplastic Starch

The compatibility between thermoplastic starch (TPS) and cellulose fibres as biocomposites may be limited due to their poor adhesion as matrix and reinforcer. The mechanical properties of starch-based polymers are themselves a disadvantage; therefore, adding natural plant fibres from various species with different types of chemical composition, morphological characteristics, and size does improve their mechanical properties (Fazeli, Florez, and Simão 2019). Additionally, by adding glycerol as a plasticizer, the produced composites are more applicable to high heat and shear stress (Ahmadi et al. 2018).

The impact of the reinforcement on physical–chemical properties between natural fibres, hemp, and sisal with TPS has been analysed by Gironès et al. (2012). The addition of reinforcing fibres did not disturb the equilibrium of water content in TPS. However, it did alter the water absorption kinetics. Scanning electron microscopy (SEM) microphotographs show that both matrices and reinforcements have good adhesion. In terms of fibre dispersion, hemp reinforcement dispersed better than sisal fibres. The amount of fibre dispersion does affect the mechanical properties of the composite. The mechanical properties of TPS and natural fibre-reinforced TPS were tested, proving that there is an improvement in Young's modulus (YM) and tensile strength (TS) with fibres as reinforcer.

Hajar et al. (2021) conducted a test on the effects of NCF with different amounts of thymol to study their mechanical, thermal, and barrier properties with corn starch film via the solvent casting method. The results showed that the addition of NCF improved the TS and YM. However, the higher content of NCF results in a decrease in the elongation at break (EAB), oxygen permeability, and water vapour permeability of the Chitosan (CS) films.

5.3 Substrate Conductivity

Biodegradable substrates usually consist of starch and nanocellulose, both of which have a very weak conductivity property. Thus, the addition of conductor materials such as graphene, metal ion, metal oxide, and carbon are highly recommended to increase the material's conductivity. The electrical properties of a substance need to be understood through the state of the electron arrangement and the ability of the valence electrons to go beyond the gap band. In this case, starch and nanocellulose are classified as conductors, which means

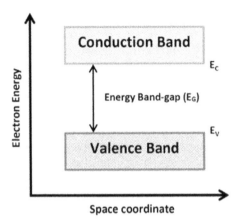

FIGURE 5.2
Electron energy for different types of material. (From Fayyaz, A., PhD thesis, University of Nottingham, 2019.)

that in the form of an energy-level graph, they tend to have a big gap between valence and conduction bands, as depicted in Figure 5.2 (Fayyaz 2019).

5.4 Mechanical Properties

The mechanical properties of substrate biocomposites refer to the physical characteristics and behaviour of these materials under applied forces or loads (N. Saba, Jawaid, and Sultan 2018). Substrate biocomposites typically comprise a substrate material combined with a biocompatible matrix or reinforcement (Balaji, Karthikeyan, and Sundar Raj 2015), often used in biomedical applications. The mechanical properties of substrate biocomposites are crucial to ensure their structural integrity, durability, and suitability for specific applications (Arunachalam and Henderson 2023). Here are some important mechanical properties of substrate biocomposites. The strength of a substrate biocomposite refers to its ability to withstand applied forces or loads without failure. The ultimate tensile, compressive, or flexural strength typically characterizes this. Higher strength values indicate greater resistance to deformation or fracture (Frómeta et al. 2020). Stiffness, also known as modulus of elasticity, describes the material's resistance to deformation under an applied load. It determines how much a material will stretch or deform under stress. Higher stiffness values indicate a more rigid material (Norman et al. 2021). Toughness measures the ability of a material to absorb energy before fracturing. It represents the material's resistance to fracture under impact or sudden loading conditions (Shirvanimoghaddam et al. 2021). Substrate biocomposites with higher toughness are less prone to brittle failure. Hardness refers to the material's resistance to indentation or scratching (Naveen et al. 2021). It is an important property when considering substrate biocomposites' wear resistance and surface durability. Fatigue resistance refers to the ability of a material to withstand repeated cyclic loading without failure (Valandro et al. 2023). It is particularly relevant in biomedical applications, where materials may experience cyclic stresses due to motion or physiological forces. Creep is the gradual deformation of a material over time under a constant load or stress. Substrate biocomposites with good creep resistance maintain their shape and mechanical properties over extended periods, even under constant stress (Peixoto et al. 2020). Fracture toughness measures the material's ability to resist crack propagation or fracture when subjected to a sharp flaw or defect. It is important for assessing the material's resistance to crack initiation and propagation (Bucci and Christensen 2019). In some cases, substrate biocomposites may be designed to degrade over time within the biological environment. Biodegradability is an essential property to ensure that the material breaks down safely without causing adverse effects (Pires et al. 2022)

The mechanical properties of NC can be proved through testing such as TS, EAB, and nanoindentation testing. Srivastava et al. (2021) conducted a mechanical test on PVA–banana pseudostem fibre (BPF) lignocellulosic composite films with increasing amounts of NC. They tested TS and EAB values for different sets of films. Films with 3% NC achieved the highest TS with 43.8 MP; elongation break was recorded as 139.5%. This indicates that the addition of NC improves the properties of PVA-BPF films because NC increases the bond between polymer matrix and the lignocellulosic BPF. Meanwhile films with highest NC scored lowest for both. The sudden drop for 5% NC was due to agglomeration in the presence of non-homogenized films; therefore, the threshold limit of NC content for this composite film is lower than 5%. On the other hand, Kong et al. (Kong et al. 2019) fabricated a low release of volatile organic compounds (VOCs) with the idea of nanocellulose as reinforcement in polyurethane for waterborne wood coating.

Nanoindentation is another mechanical test that is conducted to observe the elastic modulus, testing hardness, and fracture toughness (Lutpi et al. 2012). The conventional indenter test is for bulk material and is not suitable for a small material like thin film. Therefore, a specific type of indenter was used to conduct a mechanical test on nano-size material, with exceptional repeatability on various tests (Lutpi et al. 2012).

5.5 Electrical Properties

Substrate biocomposites refer to materials that combine a substrate or matrix material with biocompatible components, such as biomolecules or cells, to enhance their functionality for biomedical applications (Liu et al. 2023).

While the electrical properties of substrate biocomposites can vary depending on the specific components and fabrication techniques used, here are some common electrical properties to consider.

The electrical conductivity of a substrate biocomposite determines its ability to conduct electric current (Sharma et al. 2021). Conductivity is typically influenced by the composition, structure, and concentration of conductive materials within the biocomposite (Syrový et al. 2019). For example, incorporating conductive nanoparticles like carbon nanotubes or graphene can enhance the electrical conductivity of the substrate.

The dielectric constant, also known as relative permittivity, measures the ability of a material to store electrical energy in an electric field. It characterizes how a material responds to an applied electric field. The dielectric constant of a substrate biocomposite can be influenced by the dielectric properties of its components and the concentration of electrically insulating materials within the matrix (Bonardd et al. 2019). Meanwhile, capacitance is a measure of the ability of a material to store an electric charge. Substrate biocomposites with high capacitance can be used in applications where energy storage or charge separation is required (Ratajczak et al. 2019). Capacitance depends on the dielectric constant and the geometry of the biocomposite.

Impedance is a measure of the opposition to the flow of alternating current (AC). It includes both resistance and reactance components (Chen et al. 2019). The impedance of a substrate biocomposite is influenced by the conductivity and dielectric properties of its constituents (Huang et al. 2019). Impedance measurements can be used to analyse the electrical behaviour of biocomposite materials.

The surface charge of a substrate biocomposite refers to the net charge on its surface, which can be influenced by the presence of charged biomolecules or functional groups (Alves, Ferraz, and Gamelas 2019). The zeta potential, a measure of the electrical potential at the slipping plane between the biocomposite and a surrounding medium, provides information about the surface charge and stability of the biocomposites (Bystrov et al. 2019).

Another study was conducted by Chen et al. (2021) on producing a flexible sensor based on carbonized bacterial nanocellulose and wood-derived nanofibril composite aerogels. Carbonized bacteria were used due to their excellent conductivity properties, which can act as a sensor; meanwhile, cellulose nanofiber based on wood-derived dissolving pulp was used to control the deformation resistance of the aerogel composite. The effects of adding different amounts of cellulose nanofibre were observed in terms of their sensitivity corresponding to the pressure applied. The higher cellulose nanofibre contents lead to higher sensitivity, which allows them to achieve significant deformation in a large pressure loading but slight deformation in a small pressure loading. Figure 5.3 visualizes the structure of the flexible pressure sensor with carbonized bacterial cellulose acting as a conductor (Chen et al. 2021).

Martinez-crespiera et al. (2022) conducted research on nanocellulose to produce water-based conductive inks with silver nanoparticles (Ag NPs)

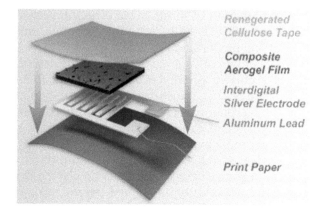

Renegerated
Cellulose Tape

Composite
Aerogel Film

Interdigital
Silver Electrode

Aluminum Lead

Print Paper

FIGURE 5.3
Visualization of flexible sensor with composite aerogel film. (From Chen, S. et al., *ACS Appl. Mater. Interfaces*, 2021, 13, 7, 8754–8763, 2021.)

for printed electronics. Two types of nanocellulose were used to fabricate printed electronics: cellulose nanofibres (CNF) and cellulose nanocrystals (CNC) with Ag NP inks. The conductivity and resistivity measured by the Cylindrical Four-Point probe show that Ag NP/CNC obtained the highest electrical conductivity of 2×10^6 S/m with only 1.1 wt.% CNC and 53.9 wt.% Ag NPs. CNC are more conductive than CNF due to the better interconnection and differences in diameter size.

Other studies conducted by Zheng et al. (2019) provided evidence of the capability of CNF to conduct electricity in the presence of graphene (GN) and PVA hydrogels. GN-CNP@PVA nanocomposite hydrogel specimens with different masses of GN were examined, and they found that by adding 0.7 wt.% of GN, the composite depicts the highest ideal behaviour with conductivity of 3.55 ± 0.1 S/m. The electrical properties of thermoplastic starch biocomposites as presented in Table 5.3.

5.6 Conclusions

It is important to note that the electrical properties of substrate biocomposites can be tailored by varying the composition, processing parameters, and fabrication techniques. Hence, the outcomes are varied, as different composites achieve better strength and conductivity. These properties play a crucial role in determining the performance and functionality of the biocomposites for specific applications such as biosensors, tissue engineering scaffolds, or implantable devices. Therefore, through this chapter, we were able to analyse that the addition of conductive materials has potential in the production of sustainable substrates.

TABLE 5.3

Electrical Properties of Thermoplastic Starch Biocomposites

Matrix	Reinforcement	Conducting Materials	Resistivity	Conductivity	Refs.
TPS	N/A	Calcium chloride, $CaCl_2$	$3.7–9.2\ \Omega m$	N/A	(Peng Liu et al. 2020)
TPS	N/A	Waste iron filings	N/A	7.42×10^{-3} S/m	(Battistelli et al. 2020)
TPS	N/A	Reduced graphene oxide	N/A	1.7×10^{-5} S/cm	(Ferreira et al. 2018)
TPS/ Poly-lactic acid (PLA)	N/A	Reduced graphene oxide	N/A	1.1×10^{-5} S/cm	(Ferreira et al. 2018)
TPS	CNC	Ag NP	N/A	2×10^6 S/m	(Martinez-crespiera et al. 2022)
PVA hydrogels	CNC	Graphene	N/A	3.55 ± 0.1 S/m	(Zheng et al. 2019)

Acknowledgements

The authors would like to acknowledge the financial support provided by Universiti Kebangsaan Malaysia under GUP-2022-072 and collaborators for providing valuable knowledge throughout the completion of this book chapter.

References

Abu Bakar, N. F., N. Abd Rahman, M. B. Mahadi, S. A. Mohd. Zuki, K. N. Mohd. Amin, M. Z. Wahab, en I. Wuled Lenggoro. 2021. "Nanocellulose from oil palm mesocarp fiber using hydrothermal treatment with low concentration of oxalic acid". *Materials Today: Proceedings* 48: 1899–904. https://doi.org/10.1016/j.matpr .2021.09.357.

Adekunle, K. F., S. W. Cho, C. Patzelt, T. Blomfeldt, en M. Skrifvars. 2012. "Impact and flexural properties of flax fabrics and lyocell fiber reinforced bio-based thermoset for automotive and structural applications". *ECCM 2012 - Composites at Venice, Proceedings of the 15th European Conference on Composite Materials*, June: 24–8.

Ahmadi, Mobina, Tayebeh Behzad, Rouhollah Bagheri, en Pejman Heidarian. 2018. "Effect of cellulose nanofibers and acetylated cellulose nanofibers on the properties of low-density polyethylene/thermoplastic starch blends". *Polymer International* 67(8): 993–1002. https://doi.org/10.1002/pi.5592.

Alves, L., E. Ferraz, en J. A. F. Gamelas. 2019. "Composites of nanofibrillated cellulose with clay minerals: A review". *Advances in Colloid and Interface Science* 272: 101994. https://doi.org/10.1016/j.cis.2019.101994.

Amorim, Julia Didier Pedrosa de, Karina Carvalho de Souza, Cybelle Rodrigues Duarte, Izarelle da Silva Duarte, Francisco de Assis Sales Ribeiro, Girlaine Santos Silva, Patrícia Maria Albuquerque de Farias, et al. 2020. "Plant and bacterial nanocellulose: Production, properties and applications in medicine, food, cosmetics, electronics and engineering: A review". *Environmental Chemistry Letters* 18(3): 851–69. https://doi.org/10.1007/s10311-020-00989-9.

Arunachalam, Krishna Prakash, en Jane Helena Henderson. 2023. "Experimental study on mechanical strength of vibro-compacted interlocking concrete blocks using image processing and microstructural analysis". *Iranian Journal of Science and Technology, Transactions of Civil Engineering.* https://doi.org/10.1007/s40996-023-01194-8.

Asim, Mohammad, Mohammad Jawaid, Ramengmawii Naheed Saba, Mohammad Nasir, en Mohamed Thariq Hameed Sultan. 2017. *Processing of Hybrid Polymer Composites-A Review: Hybrid Polymer Composite Materials: Processing.* Elsevier Ltd. https://doi.org/10.1016/B978-0-08-100789-1.00001-0.

Azeredo, Henriette M. C., Morsyleide F. Rosa, en Luiz Henrique C. Mattoso. 2017. "Nanocellulose in bio-based food packaging applications". *Industrial Crops and Products* 97: 664–71. https://doi.org/10.1016/j.indcrop.2016.03.013.

Balaji, A., B. Karthikeyan, en C. Sundar Raj. 2015. "Bagasse fiber – The future bio-composite material: A review". *International Journal of ChemTech Research* 7(1): 223–33.

Banwell, Martin G., Brett Pollard, Xin Liu, en Luke A. Connal. 2021. "Exploiting nature's most abundant polymers: Developing new pathways for the conversion of cellulose, hemicellulose, lignin and chitin into platform molecules (and beyond)". *Chemistry - An Asian Journal* 16(6): 604–20. https://doi.org/10.1002/asia.202001451.

Barbash, Valerii A., Olha V. Yashchenko, en Olesia A. Vasylieva. 2019. "Preparation and properties of nanocellulose from Miscanthus x giganteus". *Journal of Nanomaterials* 2019. https://doi.org/10.1155/2019/3241968.

Battistelli, Danilo, Diana P. Ferreira, Sofia Costa, Carlo Santulli, en Raul Fangueiro. 2020. "Conductive thermoplastic starch (TPS) composite filled with waste iron filings". 4(3): 136–47.

Blanco, Angeles, M. Concepcion Monte, Cristina Campano, Ana Balea, Noemi Merayo, en Carlos Negro. 2018. "Nanocellulose for industrial use: Cellulose nanofibers (CNF), cellulose nanocrystals (CNC), and bacterial cellulose (BC)". In *Handbook of Nanomaterials for Industrial Applications.* Elsevier Inc. https://doi.org/10.1016/B978-0-12-813351-4.00005-5.

Bonardd, Sebastián, Viviana Moreno-Serna, Galder Kortaberria, David Díaz Díaz, Angel Leiva, en César Saldías. 2019. "Dipolar glass polymers containing polarizable groups as dielectric materials for energy storage applications. A minireview". *Polymers* 11(2): 1–10. https://doi.org/10.3390/polym11020317.

Bongao, Harveen C., Ryan Russel A. Gabatino, Christlyn Faith H. Arias, en Eduardo R. Magdaluyo. 2020. "Micro/nanocellulose from waste Pili (Canarium ovatum) pulp as a potential anti-ageing ingredient for cosmetic formulations". *Materials Today: Proceedings* 22: 275–80. https://doi.org/10.1016/j.matpr.2019.08.117.

Borrero-López, A. M., E. Masson, A. Celzard, en V. Fierro. 2018. "Modelling the reactions of cellulose, hemicellulose and lignin submitted to hydrothermal treatment". *Industrial Crops and Products* 124(July): 919–30. https://doi.org/10.1016/j.indcrop.2018.08.045.

Brakat, Abdelrahman, en Hongwei Zhu. 2021. "Nanocellulose-graphene hybrids: Advanced functional materials as multifunctional sensing platform". *Nano-Micro Letters* 13. Springer Singapore. https://doi.org/10.1007/s40820-021-00627-1.

Bucci, Giovanna, en Jake Christensen. 2019. "Modeling of lithium electrodeposition at the lithium/ceramic electrolyte interface: The role of interfacial resistance and surface defects". *Journal of Power Sources* 441(September): 227186. https://doi.org/10.1016/j.jpowsour.2019.227186.

Bystrov, Vladimir, Anna Bystrova, Yuri Dekhtyar, Igor A. Khlusov, Vladimir Pichugin, Konstantin Prosolov, en Yurii Sharkeev. 2019. "Electrical functionalization and fabrication of nanostructured hydroxyapatite coatings". *Bioceramics and Biocomposites*: 149–90. https://doi.org/10.1002/9781119372097.ch7.

Cavalcanti, D. K. K., M. D. Banea, J. S. S. Neto, en R. A. A. Lima. 2021. "Comparative analysis of the mechanical and thermal properties of polyester and epoxy natural fibre-reinforced hybrid composites". *Journal of Composite Materials* 55(12): 1683–92. https://doi.org/10.1177/0021998320976811.

Chen, Changjie, Guicui Chen, Xin Li, Hongyun Guo, en Guohe Wang. 2017. "The influence of chemical treatment on the mechanical properties of windmill palm fiber". *Cellulose* 24(4): 1611–20. https://doi.org/10.1007/s10570-017-1205-1.

Chen, Chen, Zhen Wen, Aimin Wei, Xinkai Xie, Ningning Zhai, Xuelian Wei, Mingfa Peng, Yina Liu, Xuhui Sun, en John T. W. Yeow. 2019. "Self-powered on-line ion concentration monitor in water transportation driven by triboelectric nanogenerator". *Nano Energy* 62(April): 442–48. https://doi.org/10.1016/j.nanoen.2019.05.029.

Chen, Sheng, Yanglei Chen, Deqiang Li, Yanglei Xu, en Feng Xu. 2021. "Flexible and sensitivity-adjustable pressure sensors based on carbonized bacterial nanocellulose/wood-derived cellulose nano fi bril composite aerogels". https://doi.org/10.1021/acsami.0c21392.

Choi, Yeon Sik, Jahyun Koo, Young Joong Lee, Geumbee Lee, Raudel Avila, Hanze Ying, Jonathan Reeder, et al. 2020. "Biodegradable polyanhydrides as encapsulation layers for transient electronics". *Advanced Functional Materials* 30(31): 1–10. https://doi.org/10.1002/adfm.202000941.

Fahma, Farah, Faiza Ayu Lestari, Ika Amalia Kartika, Nurmalisa Lisdayana, en Evi Savitri Iriani. 2021. "Nanocellulose sheets from oil palm empty fruit bunches treated with NaOH solution". *Karbala International Journal of Modern Science* 7(1): 10–7. https://doi.org/10.33640/2405-609X.1892.

Fayyaz, Asad. 2019. "Performance and robustness characterisation of SiC power MOSFETs". PhD thesis, University of Nottingham, October.

Fazeli, Mahyar, Jennifer Paola Florez, en Renata Antoun Simão. 2019. "Improvement in adhesion of cellulose fibers to the thermoplastic starch matrix by plasma treatment modification". *Composites Part B: Engineering* 163: 207–16. https://doi.org/10.1016/j.compositesb.2018.11.048.

Ferreira, Willian H., Karim Dahmouche, Cristina T. Andrade, Eloisa Mano, Universidade Federal, en Centro De. 2018. "Tuning the mechanical and electrical conductivity properties of graphene-based thermoplastic starch / poly (lactic acid) hybrids". https://doi.org/10.1002/pc.24902.

Frómeta, D., S. Parareda, A. Lara, S. Molas, D. Casellas, P. Jonsén, en J. Calvo. 2020. "Identification of fracture toughness parameters to understand the fracture resistance of advanced high strength sheet steels". *Engineering Fracture Mechanics* 229(February): 106949. https://doi.org/10.1016/j.engfracmech.2020 .106949.

Gironès, J., J. P. López, P. Mutjé, A. J. F. Carvalho, A. A. S. Curvelo, en F. Vilaseca. 2012. "Natural fiber-reinforced thermoplastic starch composites obtained by melt processing". *Composites Science and Technology* 72(7): 858–63. https://doi.org/10 .1016/j.compscitech.2012.02.019.

Hajar, Siti, Norhazirah Nordin, Nur Ayuni, Aziera Azman, Intan Sya, Mohamed Amin, en Roseliza Kadir. 2021. "International journal of biological macromolecules effects of nanocellulose fiber and thymol on mechanical, thermal, and barrier properties of corn starch films". 183: 1352–61. https://doi.org/10.1016/j .ijbiomac.2021.05.082.

Heise, Katja, Gwendoline Delepierre, Alistair W. T. King, Mauri A. Kostiainen, Justin Zoppe, Christoph Weder, en Eero Kontturi. 2021. "Chemical modification of reducing end-groups in cellulose nanocrystals". *Angewandte Chemie - International Edition* 60(1): 66–87. https://doi.org/10.1002/anie.202002433.

Huang, Chih Feng, Cheng Wei Tu, Rong Ho Lee, Cheng Han Yang, Wei Chen Hung, en Kun Yi Andrew Lin. 2019. "Study of various diameter and functionality of TEMPO-oxidized cellulose nanofibers on paraquat adsorptions". *Polymer Degradation and Stability* 161: 206–12. https://doi.org/10.1016/j.polymdegrad-stab.2019.01.023.

Huang, Yao, Semen Kormakov, Xiaoxiang He, Xiaolong Gao, Xiuting Zheng, Ying Liu, Jingyao Sun, en Daming Wu. 2019. "Conductive polymer composites from renewable resources: An overview of preparation, properties, and applications". *Polymers* 11(2): 1–32. https://doi.org/10.3390/polym11020187.

Ibrahim, Mohd. Faiz, Rozita Hod, Mazrura Sahani, Azmawati Mohammed Nawi, Idayu Badilla Idris, Hanizah Mohd Yusoff, en Haidar Rizal Toha. 2021. "The impacts of illegal toxic waste dumping on children's health: A review and case study from pasir gudang, Malaysia". *International Journal of Environmental Research and Public Health* 18(5): 1–17. https://doi.org/10.3390/ijerph18052221.

Ilyas, R. A., S. M. Sapuan, M. R. Ishak, E. S. Zainudin, en M. S. N. Atikah. 2018. "Characterization of sugar palm nanocellulose and its potential for reinforcement with a starch-based composite". *Sugar Palm Biofibers, Biopolymers, and Biocomposites*. https://doi.org/10.1201/9780429443923-10.

Islam, M. S., L. Chen, J. Sisler, en K. C. Tam. 2018. "Cellulose nanocrystal (CNC)-inorganic hybrid systems: Synthesis, properties and applications". *Journal of Materials Chemistry B* 6(6): 864–83. https://doi.org/10.1039/c7tb03016a.

Kahavita, K. D. H. N., A. M. P. B. Samarasekara, D. A. S. Amarasinghe, en L. Karunanayake. 2020. "Nanofibrillated cellulose reinforced polypropylene composites: Influence of silane (SI-69) surface modification". *Cellulose Chemistry and Technology* 54(7–8): 789–97. https://doi.org/10.35812/CelluloseChemTechnol .2020.54.78.

Kalia, Susheel, Alain Dufresne, Bibin Mathew Cherian, B. S. Kaith, Luc Avérous, James Njuguna, en Elias Nassiopoulos. 2011. "Cellulose-based bio- and nanocomposites: A review". *International Journal of Polymer Science* 2011. https://doi .org/10.1155/2011/837875.

Kharbanda, Yashu, Mateusz Urbańczyk, Ossi Laitinen, Kirsten Kling, Sakari Pallaspuro, Sanna Komulainen, Henrikki Liimatainen, en Ville Veikko Telkki. 2019. "Comprehensive NMR analysis of pore structures in superabsorbing cellulose nanofiber aerogels". *Journal of Physical Chemistry C* 123(51): 30986–95. https://doi.org/10.1021/acs.jpcc.9b08339.

Kokkila, Toni. 2019. "Effects of PDMS potting compound outgassing and other properties on measurement device design".

Kong, Linglong, Dandan Xu, Zaixin He, Fengqiang Wang, Shihan Gui, Jilong Fan, Xiya Pan, et al. 2019. "Nanocellulose-reinforced polyurethane for waterborne wood coating". *Molecules* 24(17): 1–13. https://doi.org/10.3390/molecules24173151.

Kumar, Surendra, Kousar Jahan, Abhishek Verma, Manan Agarwal, en C. Chandraprakash. 2022. "Agar-based composite films as effective biodegradable sound absorbers". *ACS Sustainable Chemistry and Engineering* 10(26): 8242–53. https://doi.org/10.1021/acssuschemeng.2c00168.

Leão, Rosineide Miranda, Patrícia Câmara Miléo, João M. L. L. Maia, en Sandra Maria Luz. 2017. "Environmental and technical feasibility of cellulose nanocrystal manufacturing from sugarcane bagasse". *Carbohydrate Polymers* 175: 518–29. https://doi.org/10.1016/j.carbpol.2017.07.087.

Lee, Ching Hao, Abdan Khalina, en Seng Hua Lee. 2021. "Importance of interfacial adhesion condition on characterization of plant-fiber-reinforced polymer composites: A review". *Polymers* 13(3): 1–22. https://doi.org/10.3390/polym13030438.

Li, Xiaoheng, Yucheng He, Xia Dong, Xiaoning Ren, Hongxu Gao, en Wenbing Hu. 2020. "Effects of hydrogen-bonding density on polyamide crystallization kinetics". *Polymer* 189(January): 122165. https://doi.org/10.1016/j.polymer.2020.122165.

Li, Zhangdi, Fengxian Qiu, Xuejie Yue, Qiong Tian, Dongya Yang, en Tao Zhang. 2021. "Eco-friendly self-crosslinking cellulose membrane with high mechanical properties from renewable resources for oil/water emulsion separation". *Journal of Environmental Chemical Engineering* 9(5): 105857. https://doi.org/10.1016/j.jece.2021.105857.

Liu, Peng, Cong Ma, Ying Li, Liming Wang, Linjie Wei, Yinlei Yan, en Fengwei Xie. 2020. "Facile preparation of eco-friendly, flexible starch-based materials with ionic conductivity and strain-responsiveness". *ACS Sustainable Chemistry and Engineering* 8(51): 19117–28. https://doi.org/10.1021/acssuschemeng.0c07473.

Liu, Yingyu, Haiyan Liu, Susu Guo, Yifan Zhao, Jin Qi, Ran Zhang, Jianing Ren et al. 2023. "A review of carbon nanomaterials/bacterial cellulose composites for nanomedicine applications". *Carbohydrate Polymers*, 121445.

Lutpi, H. A., H. Anuar, N. Samat, S. N. Surip, en N. N. Bonnia. 2012. "Evaluation of elastic modulus and hardness of polylactic acid-based biocomposite by nanoindentation". *Advanced in Materials Research* 576: 446–49. https://doi.org/10.4028/www.scientific.net/AMR.576.446.

Mahardika, Melbi, Hairul Abral, Anwar Kasim, Syukri Arief, en Mochamad Asrofi. 2018. "Production of nanocellulose from pineapple leaf fibers via high-shear homogenization and ultrasonication". *Fibers* 6(2): 1–12. https://doi.org/10.3390/fib6020028.

Martinez-Crespiera, Sandra, Belén Pepió-Tàrrega, Rosa M. González-Gil, Francisco Cecilia-Morillo, Javier Palmer, Ana M. Escobar, Sirio Beneitez-Álvarez et al. 2022. "Use of nanocellulose to produce water-based conductive inks with Ag NPs for printed electronics". *International Journal of Molecular Sciences* 23(6): 2946.

Padzil, Mohammad, Farah Nadia, Seng Hua Lee, Zuriyati Mohamed Asa ari Ainun, Ching Hao Lee, en Luqman Chuah Abdullah. 2020. "Potential of oil palm empty fruit bunch resources in nanocellulose hydrogel production for versatile applications: A review". *Materials* 13(5). https://doi.org/10.3390/ma13051245.

Mohsin, Muhammad Ameerul Atrash, Lorenzo Iannuccii, en Emile S. Greenhalgh. 2021. "Experimental and numerical analysis of low-velocity impact thermoplastic composites". https://doi.org/10.3390/polym13213642.

Naveen, R., M. Kumar, A. Mathan, en D. Dhushyanath. 2021. "Investigation on the effect of stacking sequence on mechanical properties of a basalt and carbon fiber hybrid composite". *Journal of Engineering Research (Kuwait)* 9: 1–14. https://doi.org/10.36909/jer.ICMMM.15803.

Nguyen, Hai, Wael Zatar, en Hiroshi Mutsuyoshi. 2017. "Mechanical properties of hybrid polymer composite". In *Hybrid Polymer Composite Materials: Properties and Characterisation*. Elsevier Ltd. https://doi.org/10.1016/B978-0-08-100787-7.00004-4.

Norman, Michael D. A., Silvia A. Ferreira, Geraldine M. Jowett, Laurent Bozec, en Eileen Gentleman. 2021. "Measuring the elastic modulus of soft culture surfaces and three-dimensional hydrogels using atomic force microscopy". *Nature Protocols* 16(5): 2418–49. https://doi.org/10.1038/s41596-021-00495-4.

Otoni, Caio G., Beatriz D. Lodi, Marcos V. Lorevice, Renato C. Leitão, Marcos D. Ferreira, Márcia R.de Moura, en Luiz H. C. Mattoso. 2018. "Optimized and scaled-up production of cellulose-reinforced biodegradable composite films made up of carrot processing waste". *Industrial Crops and Products* 121(March): 66–72. https://doi.org/10.1016/j.indcrop.2018.05.003.

Ouarhim, Wafa, Nadia Zari, Rachid Bouhfid, en Abou El Kacem Qaiss. 2018. "Mechanical performance of natural fibers-based thermosetting composites". *Mechanical and Physical Testing of Biocomposites, Fibre-Reinforced Composites and Hybrid Composites*: 43–60. https://doi.org/10.1016/B978-0-08-102292-4.00003-5.

Oun, Ahmed A., Shiv Shankar, en Jong Whan Rhim. 2020. "Multifunctional nanocellulose/metal and metal oxide nanoparticle hybrid nanomaterials". *Critical Reviews in Food Science and Nutrition* 60(3): 435–60. https://doi.org/10.1080/10408398.2018.1536966.

Peixoto, Tânia, Maria Conceição Paiva, António T. Marques, en Maria A. Lopes. 2020. "Potential of graphene–polymer composites for ligament and tendon repair: A review". *Advanced Engineering Materials* 22(12): 1–20. https://doi.org/10.1002/adem.202000492.

Pereira, Paulo Henrique Fernandes, Heitor Luiz Ornaghi Júnior, Luana Venâncio Coutinho, Benoit Duchemin, en Maria Odila Hilário Cioffi. 2020. "Obtaining cellulose nanocrystals from pineapple crown fibers by free-chlorite hydrolysis with sulfuric acid: Physical, chemical and structural characterization". *Cellulose* 27(10): 5745–56. https://doi.org/10.1007/s10570-020-03179-6.

Phanthong, Patchiya, Prasert Reubroycharoen, Xiaogang Hao, Guangwen Xu, Abuliti Abudula, en Guoqing Guan. 2018. "Nanocellulose: Extraction and application". *Carbon Resources Conversion* 1(1): 32–43. https://doi.org/10.1016/j.crcon.2018.05.004.

Pires, João Ricardo Afonso, Victor Gomes Lauriano Souza, Pablo Fuciños, Lorenzo Pastrana, en Ana Luísa Fernando. 2022. "Methodologies to assess the biodegradability of bio-based polymers—Current knowledge and existing gaps". *Polymers* 14(7): 1–24. https://doi.org/10.3390/polym14071359.

Rai, Gangesh Kumar, en V. P. Singh. 2020. "Study of fabrication and analysis of nano-cellulose reinforced polymer matrix composites". *Materials Today: Proceedings* 38(xxxx): 85–8. https://doi.org/10.1016/j.matpr.2020.06.018.

Rajinipriya, Malladi, Malladi Nagalakshmaiah, Mathieu Robert, en Saïd Elkoun. 2018. "Importance of agricultural and industrial waste in the field of nanocel-lulose and recent industrial developments of wood based nanocellulose: A review". *ACS Sustainable Chemistry and Engineering* 6(3): 2807–28. https://doi.org /10.1021/acssuschemeng.7b03437.

Raquez, J. M., M. Deléglise, M. F. Lacrampe, en P. Krawczak. 2010. "Thermosetting (bio) materials derived from renewable resources: A critical review". *Progress in Polymer Science (Oxford)* 35(4): 487–509. https://doi.org/10.1016/j.progpolymsci.2010.01.001.

Ratajczak, Paula, Matthew E. Suss, Friedrich Kaasik, en François Béguin. 2019. "Carbon electrodes for capacitive technologies". *Energy Storage Materials* 16: 126–45. https://doi.org/10.1016/j.ensm.2018.04.031.

Saba, N., Othman Y. Alothman, Zeyaddddb Almutairi, M. Jawaid, en Waheedullah Ghori. 2019. "Date palm reinforced epoxy composites: Tensile, impact and mor-phological properties". *Journal of Materials Research and Technology* 8(5): 3959–69. https://doi.org/10.1016/j.jmrt.2019.07.004.

Saba, N., M. Jawaid, en M. T. H. Sultan. 2018. "An overview of mechanical and physical testing of composite materials". In *Mechanical and Physical Testing of Biocomposites, Fibre-Reinforced Composites and Hybrid Composites*. Elsevier Ltd. https://doi.org/10.1016/B978-0-08-102292-4.00001-1.

Saba, Naheed, Mohammad Jawaid, Othman Y. Alothman, M. T. Paridah, en Azman Hassan. 2016. "Recent advances in epoxy resin, natural fiber-reinforced epoxy composites and their applications". *Journal of Reinforced Plastics and Composites* 35(6): 447–70. https://doi.org/10.1177/0731684415618459.

Schmitz-Hertzberg, Sebastian Tim, Rick Liese, Carsten Terjung, en Frank F. Bier. 2014. "Towards a smart encapsulation system for small-sized electronic devices: A new approach". *International Journal of Polymer Science*. https://doi. org/10.1155/2014/713603.

Sharma, S., P. Sudhakara, A. Abdoulhdi, B. Omran, S. Jujhar, en R. A. Ilyas. 2021. "Recent trends and developments in conducting polymer nanocomposites for multifunctional applications". *Polymers* 13(17). https://doi.org/10.3390/ polym13172898.

Shirvanimoghaddam, Kamyar, K. V. Balaji, Ram Yadav, Omid Zabihi, Mojtaba Ahmadi, Philip Adetunji, en Minoo Naebe. 2021. "Balancing the toughness and strength in polypropylene composites". *Composites Part B: Engineering* 223(March): 109121. https://doi.org/10.1016/j.compositesb.2021.109121.

Soetaredjo, Felycia Edi, Shella Permatasari Santoso, Gladdy L. Waworuntu, en Farida Laniwati Darsono. 2022. "Cellulose nanocrystal (Cnc) capsules from oil palm empty fruit bunches (opefb)". *Biointerface Research in Applied Chemistry* 12(2): 2013–21. https://doi.org/10.33263/BRIAC122.20132021.

Srivastava, K. R., S. Dixit, D. B. Pal, P. K. Mishra, P. Srivastava, N. Srivastava, A. Hashem, A. A. Alqarawi, en E. F. Abd_Allah. 2021. "Effect of nanocellulose on mechanical and barrier properties of PVA–banana pseudostem fiber compos-ite films". *Environmental Technology and Innovation* 21: 101312. https://doi.org/10 .1016/j.eti.2020.101312.

Struchkova, Tatyana S., Andrey P. Vasilev, Aleksey G. Alekseev, Aitalina A. Okhlopkova, en Sakhayana N. Danilova. 2022. "Mechanical and tribological properties of polytetrafluoroethylene composites modified by carbon fibers and zeolite". *Lubricants* 10(1). https://doi.org/10.3390/lubricants10010004.

Sun, Lei, en Wenjun Yi. 2021. "Analysis of the influence of shrinkage tensile stress in potting material on the anti-overload performance of the circuit board". *Sensors* 21(7). https://doi.org/10.3390/s21072316.

Sun, Tao, Hongyu Fan, Xin Liu, en Zhanjun Wu. 2021. "Sheets for improved mechanical and thermal properties of epoxy resin". *Composites Science and Technology* 207(October 2020): 108671. https://doi.org/10.1016/j.compscitech.2021.108671.

Syrový, Tomáš, Stanislava Maronová, Petr Kuberský, Nanci V. Ehman, María E. Vallejos, Silvan Pretl, Fernando E. Felissia, María C. Area, en Gary Chinga-Carrasco. 2019. "Wide range humidity sensors printed on biocomposite films of cellulose nanofibril and poly(ethylene glycol)". *Journal of Applied Polymer Science* 136(36): 1–10. https://doi.org/10.1002/app.47920.

Trache, Djalal, Ahmed Fouzi Tarchoun, Mehdi Derradji, Tuan Sherwyn Hamidon, Nanang Masruchin, Nicolas Brosse, en M. Hazwan Hussin. 2020. "Nanocellulose: From fundamentals to advanced applications". *Frontiers in Chemistry* 8. https://doi.org/10.3389/fchem.2020.00392.

Utekar, Shubham, V. K. Suriya, Neha More, en Adarsh Rao. 2021. "Comprehensive study of recycling of thermosetting polymer composites – Driving force, challenges and methods". *Composites Part B: Engineering* 207(December 2020): 108596. https://doi.org/10.1016/j.compositesb.2020.108596.

Valandro, Luiz Felipe, Ana Carolina Cadore-Rodrigues, Kiara Serafini Dapieve, Renan Vaz Machry, en Gabriel Kalil Rocha Pereira. 2023. "A brief review on fatigue test of ceramic and some related matters in Dentistry". *Journal of the Mechanical Behavior of Biomedical Materials* 138: 105607. https://doi.org/https. https://doi.org/10.1016/j.jmbbm.2022.105607.

Valavan, Ashwini, Abiodun Komolafe, Nick Harris, en Steve Beeby. 2019. "Encapsulation process and materials evaluation for e-textile gas sensor". *Multidisciplinary Digital Publishing Institute Proceedings* 32(1): 8. https://doi.org/10.3390/proceedings2019032008.

Verma, Akarsh, Laxmi Budiyal, M. R. Sanjay, en Suchart Siengchin. 2019. "Processing and characterization analysis of pyrolyzed oil rubber (from waste tires) -Epoxy polymer blend composite for lightweight structures and coatings applications": 1–11. https://doi.org/10.1002/pen.25204.

Wan, Caichao, Luyu Zhang, Ken-Tye Yong, Jian Li, en Yiqiang Wu. 2021. "Recent progress in flexible nanocellulosic structures for wearable piezoresistive strain sensors". *Journal of Materials Chemistry C* 9(34): 11001–29.

Wang, Jie, Xin Liu, Tao Jin, Haifeng He, en Lei Liu. 2019. "Preparation of nanocellulose and its potential in reinforced composites: A review". *Journal of Biomaterials Science, Polymer Edition* 30(11): 919–46. https://doi.org/10.1080/09205063.2019.1612726.

Xing, Jinghao, Peng Tao, Zhengmei Wu, Chuyue Xing, Xiaoping Liao, en Shuangxi Nie. 2019. "Nanocellulose-graphene composites: A promising nanomaterial for flexible supercapacitors". *Carbohydrate Polymers* 207: 447–59. https://doi.org/10.1016/j.carbpol.2018.12.010.

Xu, Kaimeng, Zhengjun Shi, Jianhua Lyu, Qijun Zhang, Tuhua Zhong, Guanben Du, en Siqun Wang. 2020. "Effects of hydrothermal pretreatment on nano-mechanical property of switchgrass cell wall and on energy consumption of isolated lignin-coated cellulose nanofibrils by mechanical grinding". *Industrial Crops and Products* 149(December 2019): 112317. https://doi.org/10.1016/j.indcrop.2020.112317.

Xue, Yan, Zihao Mou, en Huining Xiao. 2017. "Nanocellulose as a sustainable biomass material: Structure, properties, present status and future prospects in biomedical applications". *Nanoscale* 9(39): 14758–81. https://doi.org/10.1039/c7nr04994c.

Yang, Guozhen, Wanting Xie, Mengfei Huang, Victor K. Champagne, Jae Hwang Lee, John Klier, en Jessica D. Schiffman. 2019. "Polymer particles with a low glass transition temperature containing thermoset resin enable powder coatings at room temperature". *Industrial and Engineering Chemistry Research* 58(2): 908–16. https://doi.org/10.1021/acs.iecr.8b04698.

Ye, Hanzhou, Yang Zhang, Yu Zhiming, en Jun Mu. 2018. "Effects of cellulose, hemicellulose, and lignin on the morphology and mechanical properties of metakaolin-based geopolymer". *Construction and Building Materials* 173: 10–6. https://doi.org/10.1016/j.conbuildmat.2018.04.028.

Zhang, Li Hui, Jing Wang, Ji Chao Liu, Jun Cheng Zhang, Yun Fei Hou, en Shan Shan Wang. 2021. "Encapsulation research of microfiber mach-zehnder interferometer temperature and salinity sensor in seawater". *IEEE Sensors Journal* 21(20): 22803–13. https://doi.org/10.1109/JSEN.2021.3110789.

Zheng, Chunxiao, Yiying Yue, Lu Gan, Xinwu Xu, Changtong Mei, en Jingquan Han. 2019. "Highly stretchable and self-healing strain sensors based on nanocellulose-supported graphene dispersed in electro-conductive hydrogels". *Nanomaterials* 9(7): 937. https://doi.org/10.3390/nano9070937.

Zhu, Enwen, Haiyu Xu, Yuanyuan Xie, Yiheng Song, Dongning Liu, Yujiao Gao, Zhuqun Shi, Quanling Yang, en Chuanxi Xiong. 2021. "Antifreezing ionotronic skin based on flexible, transparent, and tunable ionic conductive nanocellulose hydrogels". *Cellulose* 28(9): 5657–68. https://doi.org/10.1007/s10570-021-03878-8.

Zou, Jian, Zhuo Chen, Sheng-Ji Wang, Zi-Hao Liu, Yue-Jun Liu, Pei-Yong Feng, en Xin Jing. 2023. "A flexible sensor with excellent environmental stability using well-designed encapsulation structure". *Polymers* 15(10): 2308.

6

A Review on Functionalization and Mechanical Properties of Multiwalled Carbon Nanotubes/ Natural Fibre–Reinforced Epoxy Composites

M. M. Huzaifa, M. K. Nur Fitrah, N. M. Nurazzi,
M. H. M. Kassim, and M. R. Nurul Fazita

6.1 Introduction

Composite materials are made up of two or more different types of materials that are mixed to create a brand-new material with enhanced features. Due to their great strength, longevity, and versatility, composite materials have been extensively used in a range of industries, including aircraft, construction, and transportation. Polymer matrix composites and metal matrix composites are the two categories into which composite materials can be divided. A polymer matrix is filled with reinforcing fibres, such as glass or carbon fibres, to form polymer matrix composites. Metal matrix composites combine a metal matrix with another material, like fibres or ceramic particles. The type and loading of reinforcing fibres, as well as the matrix material, all affect the performance of composite structures. The matrix material offers toughness and shields the fibres from damage, while the reinforcing fibres give the composite material its strength and stiffness [1]. These features are further improved by the addition of nanofillers to composite materials, creating composite hybrids that have better mechanical, thermal, and electrical properties. For instance, research has demonstrated that adding nanofillers like CNTs or graphene to composite materials can improve their mechanical and thermal characteristics [2, 3]. This improvement in mechanical properties is due to the unique characteristics of nanofillers, such as their high surface area, small size, and strong interfacial interaction with the matrix material. These properties allow the nanofillers to effectively transfer loads

DOI: 10.1201/9781003408215-6

between the matrix and the reinforcement, leading to an overall improvement in the mechanical performance of the composite material.

Despite its high strength and rigidity, which make it a popular choice for reinforcing composite materials, carbon fibre comes with a set of challenges. These include a higher cost compared with other reinforcing materials such as glass fibre or natural fibres. The manufacturing process of carbon fibre composites requires specific machinery and expertise, adding to the complexity and cost. Furthermore, the production of carbon fibre is energy-intensive, produces waste, and results in emissions. Carbon fibre composites are also prone to absorbing moisture, affecting their performance. Thus, despite its advantages, the use of carbon fibre as a reinforcing material in composites presents several issues that need careful consideration [4–6]. The production of carbon fibre can pose environmental challenges due to its high energy consumption, waste generation, and emission of gases. The process, often powered by fossil fuels, may lead to the release of carbon dioxide (CO_2), a potent greenhouse gas. The creation of carbon fibre precursors, such as polyacrylonitrile, can also contribute to air pollution through the emission of volatile organic compounds (VOCs). Furthermore, the propensity of carbon fibre composites to absorb moisture can compromise their mechanical properties and long-term durability. Therefore, while carbon fibre offers significant benefits, its production and use also present considerable environmental and performance-related concerns [4–6].

Natural fibres, originating from plant and animal sources, find extensive use in various products such as clothing, textiles, and composite materials. Their popularity is due to their easy availability, cost-effectiveness, and long-lasting durability. A significant benefit of natural fibres is their sustainable and renewable characteristics. Their biodegradability lessens the environmental impact of composite materials, positioning them as an eco-friendlier substitute for synthetic fibres [7, 8]. Ideal for applications where end-of-life disposal is crucial, natural fibres, unlike synthetic ones that depend on finite fossil fuels, are derived from plentiful and renewable resources. This helps reduce our reliance on limited resources and fosters a more sustainable future. Such fibres include cotton, hemp, flax, jute, and sisal. Recent studies underscore the remarkable mechanical performance of these fibres, including their low-velocity and thermal performance [9, 10]. These fibres are highly desirable for various applications due to their strength, flexibility, and biodegradability. Therefore, natural fibres present an environmentally friendly alternative and excel in performance parameters. Due to their strength, rigidity, and environmental friendliness, natural fibres are significantly used in composite materials, serving as an ideal choice for reinforcement. These composites are employed in various products, including consumer goods, construction materials, and automobile parts. A prime example is the use of jute fibre composites as reinforcement material. Their high strength and cost-effectiveness make them suitable for diverse applications such as packaging and insulation materials. Therefore, natural fibres not only enhance the performance of composite materials but also foster sustainability [11–13].

CNTs are a unique form of carbon that possesses a cylindrical nanostructure. They are constructed from graphene, which is a single layer of carbon atoms arranged in a hexagonal pattern, and this sheet of graphene is rolled into a tube-like shape to form CNTs. The properties of CNTs are exceptional, boasting impressive mechanical strength, excellent electrical conductivity, and superior thermal properties. These attributes make CNTs highly desirable for a wide range of applications across various fields. For instance, in the field of electronics, the electrical properties of CNTs can enhance the performance and efficiency of devices. In energy storage, CNTs contribute to the development of high-capacity batteries and supercapacitors. Furthermore, in the fabrication of composite materials, the robust mechanical strength of CNTs can improve their durability and resilience [14]. Therefore, with their unique structure and superior properties, CNTs are paving the way for advancements in numerous fields, highlighting their significance in the landscape of materials science.

There are two main types of CNTs: single-walled carbon nanotubes (SWCNTs) and MWCNTs. CNT filler material is added to a polymer or matrix material to improve its mechanical, thermal, and electrical properties. CNTs act as reinforcement, strengthening the overall material structure and enhancing its performance. This composite material is often referred to as a CNT-filled polymer. MWCNTs are a type of CNTs consisting of multiple concentric layers of graphene. They have unique physical and mechanical properties, such as high thermal conductivity, high mechanical strength, and high electrical conductivity, which make them attractive for various applications in fields such as electronics, energy storage, and composite materials. The multiwall structure of MWCNTs gives them improved stability and resistance to thermal and chemical degradation compared with SWCNTs. For instance, MWCNTs have Young's modulus of over 1 TPa, which is significantly higher than that of traditional reinforcing fibres like glass or carbon fibres, according to a 2006 study by Zhu et al. MWCNTs are the perfect reinforcement material for composites, the study also discovered, having high tensile strengths of over 100 GPa. The excellent thermal stability of MWCNTs has also been discovered; they can withstand high temperatures without significantly degrading their mechanical properties [15, 16]. These findings imply that MWCNTs are the best reinforcement materials because they have the potential to significantly enhance the mechanical properties of composites. They are usually synthesized through chemical vapour deposition methods using a metal catalyst and a carbon-rich gas as the precursors.

CNTs have ushered in a new era for natural fibre composites. When these tiny tubes are woven into natural fibres such as flax, hemp, or jute, the result is a dramatic enhancement in their mechanical properties. This includes an increase in tensile strength, modulus, and impact resistance. But, the benefits continue beyond there. CNTs also boost the thermal stability of these composites, making them more resilient under varying temperature conditions. The unique mechanical and thermal properties of CNTs have led to the development of high-performance natural fibre composites [17]. Research indicates that integrating CNTs with these composites can significantly improve their mechanical attributes, such as tensile strength and stiffness. This is achieved by strengthening the bond between the fibres and the matrix material, resulting in a strong and durable composite.

The nanoscale diameter and high aspect ratio of CNTs enable them to deeply penetrate natural fibres, creating a strong bond with the matrix material. CNTs vary in size from nanometres to millimetres in length. This unique property enhances the mechanical strength and durability of natural fibre composites [14, 18]. The diameter of CNTs can vary depending on synthesis techniques and other factors. SWCNTs typically have an average diameter of 1–2 nm, while MWCNTs typically range from 5 to 30 nm in diameter. This variation in diameter is essential for tailoring CNT properties to specific applications, making them a versatile material in nanotechnology [19, 20]. Additionally, the addition of CNTs can lessen the need for synthetic fibres and improve the sustainability of composite materials by blending them with natural fibres. According to a study by Kumar et al. [21], the addition of CNTs to natural fibre composites led to an increase in the tensile strength and modulus by up to 25% and 55%, respectively.

6.2 Thermosetting Polymer Matrices

Polymers are widely used in various applications due to their exceptional qualities, such as flexibility, strength, and light weight. Polymers can be used as a matrix to improve the mechanical, thermal, and electrical characteristics of composites. The polymer matrix's ability to evenly distribute the reinforcing material throughout the composite construction is one of its essential functions [22, 23]. The reinforcing material is held in position by the polymer matrix, which prevents agglomeration and ensures that it is distributed uniformly throughout the composite structure. As a result, mechanical properties like stiffness and strength are enhanced, along with durability and deformation resistance. Protecting the reinforcing substance from environmental deterioration, such as moisture and UV radiation, is another important function of the polymer matrix. The polymer matrix is a barrier, keeping out moisture and other damaging elements that could degrade the reinforcing material. As a result, the composite material's condition is preserved over time, increasing its longevity and resilience. The type of polymer matrix can also affect the electrical and thermal characteristics of the composite material [14, 24, 25].

Thermosetting and thermoplastic polymer matrices are the two primary polymer types frequently used in composites. Crosslinking of thermosetting polymers during curing produces a rigid and long-lasting material. Due to their superior mechanical and thermal characteristics, they are frequently employed in high-performance uses like aerospace, automotive, and construction. On the other hand, thermoplastic polymers are readily processed and recyclable due to their ability to melt and solidify numerous times without degrading. Due to their cheap cost and simplicity of processing, they are extensively used in many applications, including packaging, medical devices, and consumer goods. Due to their distinctive qualities, including high strength, toughness, and dimensional stability, thermosetting polymers are extensively used in numerous industrial applications. Thermosetting polymers are networks of crosslinked polymer chains that are chemically cured to form a three-dimensional network structure. Epoxy, polyester, and phenolic resins are the thermosetting polymers that are most frequently employed in composite fabrications. These composites often use glass fibres, carbon fibres, or ceramic particles as reinforcing components. The kind of polymer matrix, the kind and quantity of reinforcement material, and the manufacturing method all affect these composites' performance [26–28].

In aerospace and marine applications, epoxy resin matrix composites are the most often utilized thermosetting polymer matrix composites because of their superior mechanical characteristics, chemical resistance, and minimal shrinkage while curing. Research has been done to incorporate various reinforcement elements, including carbon fibres, glass fibres, and nanoparticles, to improve the properties of these composites [29, 30]. Vacuum infusion was used to create carbon fibre–reinforced epoxy composites. When the composites' mechanical characteristics were examined, it was discovered that adding carbon fibres increased the tensile and flexural strength of the composites. Also, the researchers found that the nanoparticle addition enhanced the interfacial adhesion between the epoxy matrix and the carbon fibres, leading to higher mechanical characteristics and thermal stability [29–32]. Table 6.1 shows an example of the thermosetting polymer matrix, its properties, and its applications, and Table 6.2 shows some examples of the structure of the thermosetting polymer matrix, while Table 6.3 presents the mechanical properties of the thermosetting polymer matrix.

TABLE 6.1

Example of Thermosetting Polymer Matrix, Properties, and Applications

Polymer Matrix	Properties	Applications	Relevant Citation
Epoxy Resin	High strength, toughness, chemical resistance, low shrinkage during curing	Aerospace, automotive, electronics, construction	[33, 34]
Polyester Resin	Good water resistance, low cost	Marine, construction	[35, 36]
Phenolic Resin	Excellent thermal stability, fire resistance	Aerospace, automotive, construction	[28, 37, 38]
Vinyl Ester	Good chemical resistance, excellent toughness and fatigue resistance, low shrinkage, excellent adhesion to various substrates	Chemical processing, marine, aerospace, and automotive industries	[39–41]

TABLE 6.2

Structure of Thermosetting Polymer Matrix

Polymer Matrix	Structure	Relevant Citation
Epoxy Resin		[42]
Polyester Resin	Unsaturated polyester resin (UPR)	[43]
Phenolic Resin	Resole	[44]
Vinyl Ester		[45]

TABLE 6.3

Mechanical Properties of Thermosetting Polymer Matrix

Polymer Matrix	Tensile Strength (MPa)	Flexural Strength (MPa)	Impact Strength (kJ/m^2)	Relevant Citation
Epoxy Resin	60 to 120	80 to 150	15 to 25	[46–48]
Polyester Resin	30 to 80	60 to 120	8 to 16	[49–51]
Phenolic Resin	50 to 150	80 to 200	10 to 20	[52–54]
Vinyl Ester	50 to 100	60 to 120	20 to 50	[55–57]

6.3 Epoxy Resin

Epoxy resin is a thermosetting polymer frequently utilized as a matrix material in composite applications due to its good mechanical, thermal, and chemical properties. Epoxy resins are manufactured by reacting epoxide monomers with curing agents such as amines, anhydrides, or acid catalysts. The curing process creates a strongly crosslinked and rigid three-dimensional network structure. Epoxy resins are well-suited for use in the aerospace, marine, and automotive industries because of their high strength, stiffness, and fatigue resistance [42, 58]. In addition, they are ideal for use as coatings and adhesives because they have excellent adhesion properties. An oxygen-bonded chain of carbon atoms makes up the fundamental basic structure of epoxy resin. The epoxide groups at the chain's end act as the polymer's crosslinking reactive sites [59]. Figure 6.1 shows the structure of a monomer for epoxy resin.

One of epoxy resins' outstanding characteristics is their ease of modification with different additives, including fillers, fibres, and nanoparticles, to tailor their properties for specific uses [60]. For instance, the mechanical properties of an epoxy matrix can be considerably improved by adding carbon fibres, and the addition of nanoparticles can enhance its thermal and electrical properties. Table 6.4 presents the properties of epoxy resin. Due to its superior mechanical performance, chemical resistance, and minimal shrinkage while curing, epoxy resin is a frequently utilized thermosetting polymer matrix in composite materials. Epoxy resin matrix composites are

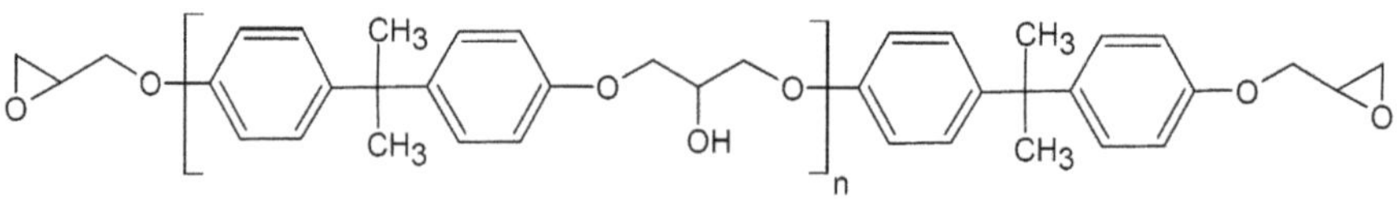

FIGURE 6.1
Structure of common monomer for epoxy resin [59].

TABLE 6.4

Properties of Epoxy Resin

Properties	Description	References
High mechanical strength and stiffness	Epoxy resin is ideal for load-bearing applications because of its high tensile and compressive strength. Its high rigidity also makes it the best choice for applications requiring dimensional stability.	[61]
Excellent chemical resistance	Epoxy resin is extremely chemically resistant, including acids, bases, and solvents.	[62]
Good adhesion properties	Epoxy resin adheres well to many substrates, such as metals, ceramics, and composites.	[63]
Low shrinkage and good dimensional stability	Little shrinkage of epoxy resin during curing ensures superior dimensional stability of the finished product.	[64]

employed in many industries, including construction, automotive, electronics, and aerospace.

In the year 2018, a detailed study was conducted by Martully et al. [65], which meticulously examined the complex effects of the type and concentration of carbon fibres on the mechanical properties of carbon fibre–reinforced epoxy composites. The research findings highlighted a distinct correlation between the type of carbon fibre chosen and the resulting mechanical characteristics. Additionally, it was observed that an increase in the concentration of carbon fibres corresponded to an improvement in the mechanical properties of these composites. A study conducted by Kumar et al. [66] explored the effects of CNTs on the mechanical attributes of epoxy composites that are reinforced with carbon fibre. The researchers found that the incorporation of CNTs resulted in an enhancement of the interfacial bonding between the epoxy matrix and the carbon fibres. This improvement led to a noticeable increase in the mechanical properties of the composites.

The manufacturing process of epoxy resin matrix composites, which includes techniques such as hand lay-up, vacuum infusion, and resin transfer moulding, plays a crucial role in determining the final properties of the composite. A study by Demir et al. [67] investigated the influence of these manufacturing techniques on the mechanical properties of carbon fibre–reinforced epoxy composites. The findings revealed that the vacuum infusion method resulted in composites with superior mechanical properties compared with those produced by manual lay-up and resin transfer moulding. In a separate study, Wu et al. [68] explored the effect of curing temperature and pressure on the mechanical properties of carbon fibre–reinforced epoxy composites. The study concluded that curing at elevated temperatures and pressures yielded composites with enhanced mechanical performance.

MWCNTs, known for their superior mechanical, electrical, and thermal properties, have been the subject of extensive research as potential reinforcing fillers for epoxy resin matrix composites. The mechanical characteristics of the composite material, such as tensile strength, flexural strength, and impact strength, can be significantly improved by the integration of MWCNTs into the epoxy resin matrix. Zhou et al. [69] conducted a study that delved into the effects of MWCNTs on the mechanical properties of an epoxy resin matrix composite. They developed composites with different weight fractions of MWCNTs (0.1%, 0.3%, and 0.5%) and assessed the mechanical properties of the materials. The study found that the inclusion of MWCNTs led to an enhancement in the mechanical properties of the composites, with the composite containing a 0.3% weight fraction of MWCNTs showing the most substantial improvement. When compared with the neat epoxy matrix, there was a 32.6% and 46.2% increase in tensile and flexural strength, respectively. Moreover, the composite with 0.3% MWCNTs exhibited a 56.1% increase in impact strength.

In a similar vein, Nguyen et al. [70] explored the influence of MWCNTs on the mechanical properties of an epoxy resin matrix composite. They fabricated composites with varying weight fractions of MWCNTs (0.1%, 0.5%, and 1.0%) and evaluated the mechanical properties of the materials. The research concluded that the addition of MWCNTs led to an improvement in the mechanical properties of the composites, with the composite containing a 0.5% weight fraction of MWCNTs demonstrating the most significant enhancement. In comparison to the pure epoxy resin matrix, there was a 26.2% and 29.7% increase in tensile and flexural strength, respectively. Furthermore, the composite with 0.5% MWCNTs showed a 41.7% increase in impact strength. Table 6.5 presents the different types of epoxy resin commonly used in structural applications.

TABLE 6.5

Different Types of Epoxy Resin

Epoxy Resin Type	Description	References
Bisphenol A Epoxy Resin	Most commonly used type of epoxy resin, made from the reaction of Bisphenol A and epichlorohydrin	[71, 72]
Novolac Epoxy Resin	Made from the reaction of phenolic novolac and epichlorohydrin, known for its high temperature resistance	[73, 74]
Cycloaliphatic Epoxy Resin	Made from the reaction of cycloaliphatic compounds and epichlorohydrin, known for its excellent electrical properties	[75, 76]
Glycidyl amine Epoxy Resin	Made from the reaction of polyamine and epichlorohydrin, known for its excellent adhesion properties	[77, 78]
Diglycidyl Ether of Tetrabromobisphenol A Epoxy Resin	Made from the reaction of tetrabromobisphenol A and epichlorohydrin, known for its flame retardant properties	[57, 79]

Bisphenol A diglycidyl ether (DGEBA) is a popular epoxy resin in various applications due to its outstanding mechanical properties and chemical resistance. It is produced by combining Bisphenol A with epichlorohydrin, then dehydrochlorination, resulting in a diglycidyl ether. Figure 6.2 shows the synthesis of DGEBA from Bisphenol A and epichlorohydrin.

DGEBA epoxy resin, due to its high tensile strength, flexural strength, and elastic modulus, is frequently employed in various industries such as aerospace, automotive, and construction. It also exhibits outstanding chemical resistance to a variety of solvents and acids. The curing agents most commonly used for DGEBA epoxy resin are amine compounds, such as aliphatic amines and aromatic amines. These compounds are chosen due to their strong reactivity and ability to form three-dimensional networks during curing. Diethylenetriamine (DETA), known for its low viscosity and strong reactivity, is one of the most widely used catalysts for DGEBA epoxy resin [80–83].

Another frequently used catalyst for DGEBA epoxy resin is triethylenetetramine (TETA). TETA, an aliphatic amine with four primary amino groups, has a high curing rate, making it suitable for applications that require rapid curing. The curing of DGEBA epoxy resin with amines is a complex process involving several chemical reactions. These include epoxy-amine addition, amine homopolymerization, and amine–amine crosslinking. It has been demonstrated that a higher curing temperature and a longer curing time can lead to greater crosslinking, thereby improving the mechanical properties of the DGEBA epoxy resin [84–87]. Figure 6.3 shows the chemical structure of TETA.

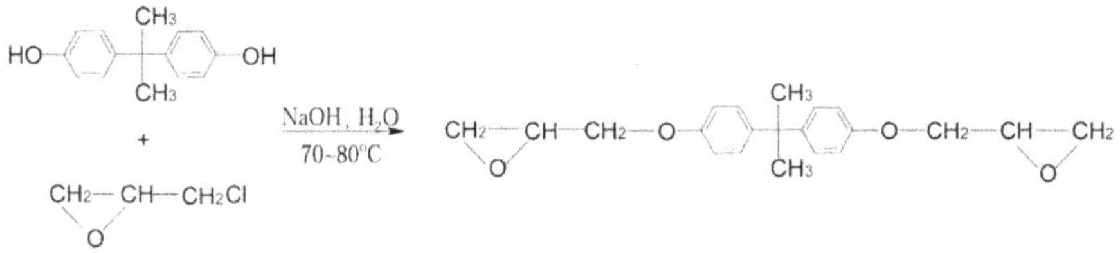

FIGURE 6.2
Synthesis of DGEBA from bisphenol A and epichlorohydrin [88].

$$NH_2-C_2H_4-NH-C_2H_4-NH-C_2H_4-NH_2$$

FIGURE 6.3
Chemical structure of TETA.

6.4 Functionalization and Dispersion of CNTs in Epoxy Resin

CNTs have attracted a lot of interest because of their unusual mechanical, electrical, and thermal properties, making them a potential material for various applications, including reinforcement in polymer composites. However, due to their high tendency to form agglomerates, it might be difficult for CNTs to disperse evenly throughout the polymer matrix, which is necessary for their efficient use in composites. Thus, it is essential to functionalize and disperse CNTs in epoxy resin to achieve uniform dispersion and enhance the mechanical properties of the final composite material [89–91].

CNTs are functionalized by changing their surface chemistry to improve their compatibility with the polymer matrix. Covalent functionalization, non-covalent functionalization, and hybrid functionalization are some of the techniques that have been developed [92]. Covalent functionalization involves adding chemical groups to their surface through chemical processes to improve the CNTs' interfacial contact with the epoxy resin matrix. Covalent functionalization creates chemical bonds through various chemical interactions between the functional groups, such as carboxyl, amino, or hydroxyl groups, and the CNT surface [93]. As a result, the interfacial bonding and mechanical properties of the resultant composite can be improved because of these functional groups' ability to create covalent bonds with the epoxy matrix. Strong acid treatment, however, has the potential to damage CNTs and lessen their aspect ratio, which would lower the effectiveness of the reinforcement.

Non-covalent functionalization involves altering the surface chemistry of CNTs and increasing their solubility in the polymer matrix using surfactants or polymers. The adsorption of surfactants, polymers, or other molecules onto the surface of CNTs through van der Waals forces, hydrogen bonds, or electrostatic interactions is known as non-covalent functionalization [92]. In addition to acting as a dispersing agent, the surfactants or polymers can prevent CNTs from adhering together and enhance their dispersion inside the epoxy matrix. Although non-covalent functionalization is straightforward and reasonably priced, the final composite may have poor mechanical properties due to the weak interaction between the CNTs and the surfactant or polymer [92].

To produce better dispersion and compatibility, hybrid functionalization blends covalent and non-covalent methods. In this technique, the CNTs are first covalently functionalized with functional groups, and then, surfactants or polymers are adsorbed onto the surface of the functionalized CNTs [92, 94]. As a result, the CNTs and the polymer matrix can interact strongly thanks to the combination of covalent and non-covalent functionalization, which enhances the mechanical properties. The finished composite's specific application and required qualities determine the benefits and drawbacks of the various functionalization techniques. For example, strong interfacial bonding and improved mechanical properties can be achieved through covalent functionalization; however, the CNTs may suffer damage and have their aspect ratio reduced. Although non-covalent functionalization is quick and inexpensive, it may result in subpar mechanical properties due to the weak interaction between the CNTs and the surfactant or polymer. The strongest contact between the CNTs and the polymer matrix can be achieved through hybrid functionalization, but this can also be more time-consuming and expensive than other approaches [92–94].

The effective incorporation of CNTs into an epoxy resin matrix is a critical factor in optimizing the properties of the resulting composite material. Various techniques have been developed to ensure uniform dispersion of CNTs within the epoxy matrix. These include sonication, shear mixing, and high-pressure homogenisation. Sonication, which employs ultrasonic vibrations, is commonly used for dispersing and de-agglomerating CNTs within the polymer matrix. Despite its advantages in terms of dispersion efficiency, cost-effectiveness, and simplicity, sonication can also lead to the degradation of CNTs, resulting in defects and a consequent loss of their inherent properties. On the other hand, shear mixing leverages mechanical shear forces to break up CNT agglomerates of CNTs. Another technique, high-pressure homogenisation, utilizes high-pressure fluid to disperse CNTs throughout the polymer matrix. Notably, the functionalization of CNTs can significantly influence their dispersion within the epoxy resin. Functional groups can alter the colloidal behaviour of CNTs in the precursor epoxy resin, thereby affecting the synthesis process and the subsequent interfacial interactions between the polymer matrix and the CNTs [95–97].

The remarkable increases in mechanical characteristics seen in composites, including well-dispersed and functionalized CNTs, demonstrate the significance of functionalization and dispersion of CNTs in epoxy resin. Together with greater stiffness, strength, and toughness, these enhancements also feature improved thermal and electrical conductivity. However, the difficulties in generating uniform CNT dispersion and functionalization in epoxy resin underscore the need for more study in this field. The functionalization and dispersion of CNTs in epoxy resins have been studied in much research [95, 98, 99]. For instance, CNTs were functionalized with carboxyl groups using a three-step process, and the functionalized CNTs were successfully mixed with epoxy resin using high-shear sonication. Furthermore, the mechanical properties of the composites were significantly improved using a combination of non-covalent and covalent functionalization to enhance CNT dispersion in the epoxy resin [95, 97, 98].

6.5 Mechanism of CNTs in Epoxy Resin

Functionalization and dispersion are the two critical steps involved in the mechanism of CNTs in epoxy resin. While dispersion entails obtaining a homogeneous distribution of CNTs inside the polymer matrix, functionalization entails altering the surface of CNTs to improve their compatibility with the epoxy resin matrix [100, 101]. Therefore, the desirable properties of the resultant nanocomposite depend on the effective functionalization and dispersion of CNTs in epoxy resin. Both van der Waals forces and hydrogen bonds play a significant role in the physical interactions between CNTs and epoxy resin. Temporary dipoles cause van der Waals forces to form between two molecules, which are proportional to the interacting molecules' surface areas and inversely proportional to their separation. Due to the enormous surface area of CNTs, many van der Waals interactions can effectively disperse the CNTs throughout the epoxy resin matrix. In addition, a hydrogen atom and a single pair of electrons on an electronegative atom, such as oxygen or nitrogen, can interact electrostatically to form hydrogen bonds. Oxygen-containing functional groups, such as carboxyl and hydroxyl groups, can make it easier for epoxy resin and CNTs to form hydrogen bonds, increasing the adhesion of the two materials [102–104].

CNTs and epoxy resin can interact chemically through covalent bonding and π–π stacking. When two atoms share electrons, a chemical bond is created, which is known as covalent bonding. The functionalization of CNTs with reactive groups, such as epoxide and amine groups, which can react with the epoxy resin matrix, can result in covalent bonds between CNTs and epoxy resin [104–106]. Aromatic rings, like the ones found in CNTs and epoxy resin, can interact non-covalently by forming π–π stacks. Due to the high density of aromatic rings in both the CNTs and the epoxy resin, π–π stacking can result in powerful interactions between the materials. The interactions between CNTs and epoxy resin significantly influence the properties of the resulting CNTs/epoxy nanocomposites. By physical interactions, including van der Waals forces and hydrogen bonding, CNTs can be effectively dispersed inside the epoxy resin matrix, improving interfacial adhesion and the resulting nanocomposite's mechanical properties, such as stiffness, strength, and toughness [104].The mechanical properties of the resultant nanocomposite can also be improved by covalent bonding between CNTs and epoxy resin [107]. The π–π stacking interactions can also increase the mechanical properties of the resultant nanocomposite, particularly its rigidity [108, 109]. It can be challenging to disperse CNTs evenly inside the epoxy resin matrix, and this might cause agglomeration and lower mechanical properties [14].

CNTs can considerably increase epoxy resin's stiffness, strength, and toughness, improving its mechanical properties. CNTs' high aspect ratio and strong tensile strength make them ideal for use in the epoxy resin matrix for load transmission and reinforcing [110, 111]. Furthermore, several studies have shown that CNTs/epoxy nanocomposites have significantly better mechanical performance than neat epoxy resin [102]. The studies revealed that CNTs/epoxy nanocomposites had stiffness and strength that were 50% and 30% higher, respectively, than those of neat epoxy resin.

The integration of CNTs into epoxy resin has been observed to considerably bolster the material's toughness. Serving as nucleating agents, CNTs contribute to the formation of an epoxy matrix that is denser and more structured. The incorporation of CNTs also mitigates the number and size of voids within the composite and retards the development of microcracks, thereby enhancing the interconnection between the matrix and the reinforcement [95, 112]. The inherent high conductivity of CNTs provides a significant boost to the electrical conductivity of the epoxy matrix [113, 114]. This attribute renders CNT-reinforced epoxy composites apt for electrical applications, including electromagnetic interference (EMI) shielding and conductive coatings [95, 114]. The expansive surface area and high aspect ratio of CNTs establish additional conductive paths within the epoxy resin matrix, thereby augmenting electrical conductivity. It has been substantiated that CNT–epoxy nanocomposites possess electrical properties superior to those of pure epoxy resin [95, 113, 114].

CNTs also markedly improve the thermal properties of epoxy resin by enhancing its thermal conductivity and stability. The extensive surface area and high aspect ratio of CNTs facilitate efficient heat transfer within the epoxy resin matrix, thereby elevating thermal conductivity [95, 115]. Additionally, the exceptional thermal stability of CNTs can amplify the thermal stability of the resultant nanocomposite when incorporated into epoxy resin. It has been demonstrated that CNT–epoxy nanocomposites possess thermal properties that are superior to those of pure epoxy resin [95, 115]. Extensive research has been conducted on the reinforcing mechanism of CNTs in epoxy resins [95, 112]. It has been established that the inclusion of CNTs escalates the glass transition temperature and storage modulus of the epoxy matrix, indicating enhanced mechanical and thermal stability [95]. Noteworthy improvements in the thermal conductivity and flame retardancy of CNT-reinforced epoxy composites have been reported, which are attributed to the synergistic effects of the CNTs and other additives [95]. Differences, advantages, and disadvantages of MWCNTs and SWCNTs are presented in Table 6.6.

While CNTs integrated with epoxy resin offer numerous advantages, they also present certain challenges and drawbacks. One of the primary issues is achieving a uniform dispersion of CNTs within the epoxy resin matrix. The strong propensity of CNTs to agglomerate can limit the formation of reinforcing and conducting pathways in the epoxy resin matrix [95]. This poor dispersion can result in suboptimal performance of the composites due to weak interfacial bonding in certain sections of the nanocomposite [95]. Furthermore, CNTs, due to their high aspect ratio, are classified as potentially hazardous substances, posing risks to respiratory health and other health issues [126]. Therefore, it is crucial to implement stringent occupational

TABLE 6.6

Differences, Advantages, and Disadvantages of MWCNTs and SWCNTs

Property	MWCNTs	SWCNTs	References
Diameter	2–100 nm	0.4–2 nm	[116–118]
Length	Several micrometres to millimetres	Several nanometres to micrometres	[116, 117]
Structure	Multiple layers of graphene sheets	Single layer of graphene sheet	[119, 120]
Strength	High	Extremely high	[121, 122]
Stiffness	High	Extremely high	[121, 122]
Electrical conductivity	Good	Excellent	[121, 122]
Thermal conductivity	Good	Excellent	[121, 122]
Dispersion in epoxy resin	Difficult	Relatively easy	[123]
Cost	Relatively low	Relatively high	[124, 125]

health and safety measures to minimize exposure and ensure safe handling procedures during the processing of CNTs [126].

Another limiting factor for the widespread industrial application of CNT–epoxy nanocomposites is the high production cost of CNTs. The complex and energy-intensive manufacturing processes required for CNT production contribute significantly to their high cost [127]. Moreover, the use of CNTs in epoxy resin can complicate processing techniques such as extrusion and injection moulding. Lastly, the potential adverse environmental impacts of CNT–epoxy nanocomposites cannot be overlooked. Given that CNTs are not biodegradable, their persistence in the environment could pose threats to ecosystems and living organisms [128, 129]. Therefore, it is imperative to consider these challenges and drawbacks when developing and utilizing CNT–epoxy nanocomposites.

6.6 Mechanical Performance of Hybrid MWCNT/ Natural Fibre–Reinforced Epoxy Composites

In recent years, interest in hybrid composites made of MWCNTs and natural fibres has grown due to their potential to improve the mechanical properties of composites while preserving their biodegradability and sustainability. The creation of hybrid composites involves combining a polymer matrix, such as epoxy resin, with two or more other types of reinforcements, such as natural fibres and MWCNTs. Hybrid composites are used to acquire higher mechanical properties than those of individual components. The type, amount, and orientation of reinforcements, the degree of interfacial bonding between the reinforcements and the matrix, and the curing conditions of the polymer matrix are only a few of the variables that affect a composite material's mechanical properties. MWCNTs can be hybridized with hybrid fibre systems to provide several advantages and benefits, such as enhanced mechanical and electrical characteristics, higher thermal conductivity, and increased resistance to deterioration from the environment [130, 131]. In addition, MWCNTs can significantly improve the mechanical characteristics of fibre composites, such as improved strength, stiffness, and toughness. For example, according to a study by Ali et al. [132], adding MWCNTs to carbon fibre–reinforced polymer composites increased their flexural and impact strength.

The characteristics of hybrid composites can be affected by numerous factors. The properties of the matrix, the processing methods, and the nature and concentration of the filler materials are the most significant factors. The type and amount of filler components significantly impact the characteristics of hybrid composites. In the case of MWCNTs/fibre composites, the type and surface chemistry of the fibres and the concentration and aspect ratio of MWCNTs can significantly impact the composite's ultimate properties. The glass transition temperature, degree of curing, and molecular weight of the matrix all play a significant part in determining the properties of hybrid composites [133–136]. The properties of hybrid composites can be influenced by the processing methods to produce them, and reinforcement and matrix compatibility. According to the fabrication method, the MWCNTs should be evenly distributed throughout the matrix and between the fibres. MWCNT/fibre composites have been manufactured using various processing methods, including electrospinning, melt blending, and solution mixing. Another critical factor in regulating the characteristics of hybrid composites is the interfacial bonding between the MWCNTs and fibres. Poor stress transfer between the fibre and the matrix due to weak interfacial adhesion can result in subpar mechanical properties. The interfacial interaction between MWCNTs and fibres has been improved using several methods, including surface modification, functionalization, and chemical coupling [133–136].

One study examined the effects on kenaf fibre/epoxy composites of MWCNT addition [137]. With a modest MWCNT loading of 0.5 wt.%, the inclusion of MWCNTs increased the composites' tensile strength and modulus by up to 19% and 57%, respectively. Nevertheless, due to inadequate interfacial interaction between the MWCNTs and the kenaf fibres, further increasing the MWCNT loading to 1.5 wt.% caused a drop in the tensile strength and modulus. With the addition of MWCNTs, the composites' flexural strength and modulus also rose, with the most significant gains shown at a loading of 0.5 wt.% [138–140]. Another study studied the effects of mixing pineapple leaf fibres and MWCNTs in epoxy composites The tensile and flexural characteristics of the composites were significantly improved by the inclusion of MWCNTs, with the most significant gains seen at a loading of 0.5 wt.%. Compared with the pure epoxy and pineapple leaf fibre composites, the hybrid composites showed up to 30% improvement in tensile strength and modulus and up to 70% improvement in flexural strength and modulus [138–140].

The effects of combining oil palm empty fruit bunch fibres with MWCNTs in epoxy composites were also examined in a study. The findings demonstrated that the inclusion of MWCNTs greatly enhanced the tensile and flexural characteristics of the composites [141–143]. Compared with the pure epoxy and oil palm empty fruit bunch fibre composites, the hybrid composites showed up to 66% improvement in tensile strength and modulus and up to 67% improvement in flexural strength and modulus [142, 143]. In general, the types and characteristics of the natural fibres, the loading and dispersion of MWCNTs, and the interfacial bonding between the MWCNTs and the natural fibres all significantly impact the mechanical properties of hybrid MWCNTs/natural fibre–reinforced epoxy composites. High-performance hybrid composites are achieved by adequately functionalizing and dispersing MWCNTs in the epoxy matrix and optimizing MWCNT loading and hybridization with natural fibres. Table 6.7 presents the mechanical properties of hybrid MWCNT/natural fibre-reinforced epoxy composites.

6.7 Conclusions

This review on epoxy composites reinforced with MWCNTs and natural fibres showcases their adaptability for a wide range of applications such as packaging, electronics, structural, automotive, and aerospace applications. Including MWCNTs enhances these composites' mechanical properties, including their robustness, rigidity, and durability. CNTs, known for their superior mechanical attributes, such as high tensile strength and toughness, are increasingly used as reinforcement fillers in biocomposites. The fabrication conditions for these composites are optimal, and the impact of MWCNTs on the microstructure and mechanical behaviour of the composites is significant. The mechanical performance of CNTs in polymer composites can be influenced by several factors, such as the aspect ratio of CNTs, the dispersion of CNTs in the polymer composites' matrix, the concentration of CNTs in the polymer composites, the selection of matrix material in the polymer composites, and the fabrication method employed to produce the polymer composites. Understanding these factors is crucial for designing and fabricating high-performance biocomposites using CNTs as reinforcing fillers. This knowledge could lead to novel advancements in material science and engineering, potentially revolutionizing various industries.

TABLE 6.7

Mechanical Properties of Hybrid MWCNT/Natural Fibre–Reinforced Epoxy Composites

Polymer	Reinforcements	MWCNT Content (wt.%)	Tensile Strength (MPa)	Compressive Strength (MPa)	Flexural Strength (MPa)	Impact Strength (kJ/m^2)	Relevant Studies
Epoxy	Jute fibre	1.0	63.6	–	96.8	19.7	[144]
Epoxy	Kenaf fibre	1.0	28.2	57.2	55.2	3.38	[145]
Epoxy	Coir fibre	1.0	34.2	–	49.3	–	[146]
Epoxy	Banana fibre	0.5	53.7	–	82.1	18.2	[147]
Epoxy	Flax fibre	1.0	48.3	63.6	66.8	11.7	[11]
Polyester	Jute fibre	1.0	35.2	–	49.1	6.0	[146]
Polyester	Kenaf fibre	1.0	23.4	40.6	39.5	1.53	[148]
Polyester	Coir fibre	1.0	22.1	–	29.8	–	[149]
Polyester	Banana fibre	0.5	41.2	–	54.6	8.6	[150]
Polyester	Flax fibre	0.5	36.7	49.5	54.2	5.6	[151]

References

1. Wang, Z., Y. Sun, X.-Q. Chen, C. Franchini, G. Xu, H. Weng, X. Dai, and Z. Fang, Dirac semimetal and topological phase transitions in ${A}_{3}$Bi ($A=\text{Na}$, K, Rb). *Physical Review B*, 2012. **85**(19): p. 195320.
2. Prabhuram, T., S.P. Singh, D.E. Raja, J.I. Durairaj, M.C. Das, and P. Ravichandran, Development and mechanical characterization of jute fibre and multi-walled carbon nanotube-reinforced unsaturated polyester resin composite. *Materials Today: Proceedings*, 2023.
3. Ponnamma, D., K.K. Sadasivuni, M. Strankowski, Q. Guo, and S. Thomas, Synergistic effect of multi walled carbon nanotubes and reduced graphene oxides in natural rubber for sensing application. *Soft Matter*, 2013. **9**(43): p. 10343–10353.
4. Chen, C., Y. Yang, Y. Zhou, C. Xue, X. Chen, H. Wu, L. Sui, and X. Li, Comparative analysis of natural fiber-reinforced polymer and carbon fiber reinforced polymer in strengthening of reinforced concrete beams. *Journal of Cleaner Production*, 2020. **263**: p. 121572.
5. Khandai, S., R. Nayak, A. Kumar, D. Das, and R. Kumar, Assessment of mechanical and tribological properties of flax/kenaf/glass/carbon fiber reinforced polymer composites. *Materials Today: Proceedings*, 2019. **18**: p. 3835–3841.
6. Moudood, A., A. Rahman, H.M. Khanlou, W. Hall, A. Öchsner, and G. Francucci, Environmental effects on the durability and the mechanical performance of flax fiber/bio-epoxy composites. *Composites Part B Engineering*, 2019. **171**: p. 284–293.
7. Gupta, M., M. Ramesh, and S. Thomas, Effect of hybridization on properties of natural and synthetic fiber-reinforced polymer composites (2001–2020): A review. *Polymer Composites*, 2021. **42**(10): p. 4981–5010.
8. Kumar, P.S.S. and K.V. Allamraju, A review of natural fiber composites [Jute, Sisal, Kenaf]. *Materials Today: Proceedings*, 2019. **18**: p. 2556–2562.
9. Mulla, M.H., M.N. Norizan, C.K. Abdullah, N.F.M. Rawi, M.H.M. Kassim, N. Abdullah, M.N.F. Norrrahim, and M.E.M. Soudagar, Low velocity impact performance of natural fibre reinforced polymer composites: A review. *Functional Composites and Structures*, 2023.
10. Nurazzi, N.M., M.R.M. Asyraf, S. Fatimah Athiyah, S.S. Shazleen, S.A. Rafiqah, M.M. Harussani, S.H. Kamarudin, M.R. Razman, M. Rahmah, E.S. Zainudin, R.A. Ilyas, H.A. Aisyah, M.N.F. Norrrahim, N. Abdullah, S.M. Sapuan, and A. Khalina, A review on mechanical performance of hybrid natural fiber polymer composites for structural applications. *Polymers*, 2021. **13**(13): p. 2170.
11. Gholampour, A. and T. Ozbakkaloglu, A review of natural fiber composites: Properties, modification and processing techniques, characterization, applications. *Journal of Materials Science*, 2020. **55**(3): p. 829–892.
12. Ashraf, M.A., M. Zwawi, M. Taqi Mehran, R. Kanthasamy, and A. Bahadar, Jute based bio and hybrid composites and their applications. *Fibers*, 2019. **7**(9): p. 77.
13. Mohammed, L., M.N. Ansari, G. Pua, M. Jawaid, and M.S. Islam, A review on natural fiber reinforced polymer composite and its applications. *International Journal of Polymer Science*, 2015. **2015**.

14. Nurazzi, N., F. Sabaruddin, M. Harussani, S. Kamarudin, M. Rayung, M. Asyraf, H. Aisyah, M. Norrrahim, R. Ilyas, and N. Abdullah, Mechanical performance and applications of cnts reinforced polymer composites—A review. *Nanomaterials*, 2021. **11**(9): p. 2186.

15. Rafiqah, S.A., A. Khalina, A.S. Harmaen, I.A. Tawakkal, K. Zaman, M. Asim, M. Nurrazi, and C.H. Lee, A review on properties and application of bio-based poly (butylene succinate). *Polymers*, 2021. **13**(9): p. 1436.

16. Merneedi, A., L. Natrayan, S. Kaliappan, D. Veeman, S. Angalaeswari, C. Srinivas, and P. Paramasivam, Experimental investigation on mechanical properties of carbon nanotube-reinforced epoxy composites for automobile application. *Journal of Nanomaterials*, 2021. **2021**: p. 1–7.

17. Shen, X., J. Jia, C. Chen, Y. Li, and J.-K. Kim, Enhancement of mechanical properties of natural fiber composites via carbon nanotube addition. *Journal of Materials Science*, 2014. **49**(8): p. 3225–3233.

18. Kohls, A., M. Maurer Ditty, F. Dehghandehnavi, and S.-Y. Zheng, Vertically aligned carbon nanotubes as a unique material for biomedical applications. *ACS Applied Materials and Interfaces*, 2022. **14**(5): p. 6287–6306.

19. Safdar, M., W. Kim, S. Park, Y. Gwon, Y.-O. Kim, and J. Kim, Engineering plants with carbon nanotubes: A sustainable agriculture approach. *Journal of Nanobiotechnology*, 2022. **20**(1): p. 1–30.

20. Hur, O., B.-H. Kang, and S.-H. Park, Optimization of electrical and mechanical properties of a single-walled carbon nanotube composite using a three-roll milling method. *Materials Chemistry and Physics*, 2023: p. 128354.

21. Kumar, A., K. Sharma, and A.R. Dixit, Carbon nanotube-and graphene-reinforced multiphase polymeric composites: Review on their properties and applications. *Journal of Materials Science*, 2020. **55**(7): p. 2682–2724.

22. Choudhary, V. and A. Gupta, Polymer/carbon nanotube nanocomposites. *Carbon Nanotubes-Polymer Nanocomposites*, 2011. **2011**: p. 65–90.

23. Hemath, M., S. Mavinkere Rangappa, V. Kushvaha, H.N. Dhakal, and S. Siengchin, A comprehensive review on mechanical, electromagnetic radiation shielding, and thermal conductivity of fibers/inorganic fillers reinforced hybrid polymer composites. *Polymer Composites*, 2020. **41**(10): p. 3940–3965.

24. Andrew, J.J., S.M. Srinivasan, A. Arockiarajan, and H.N. Dhakal, Parameters influencing the impact response of fiber-reinforced polymer matrix composite materials: A critical review. *Composite Structures*, 2019. **224**: p. 111007.

25. Mahmoud, M.E., A.M. El-Khatib, R.M. El-Sharkawy, A.R. Rashad, M.S. Badawi, and M.A. Gepreel, Design and testing of high-density polyethylene nanocomposites filled with lead oxide micro-and nano-particles: Mechanical, thermal, and morphological properties. *Journal of Applied Polymer Science*, 2019. **136**(31): p. 47812.

26. Hsissou, R., R. Seghiri, Z. Benzekri, M. Hilali, M. Rafik, and A. Elharfi, Polymer composite materials: A comprehensive review. *Composite Structures*, 2021. **262**: p. 113640.

27. Dodiuk, H., *Handbook of Thermoset Plastics*. 2021, William Andrew.

28. Asim, M., N. Saba, M. Jawaid, M. Nasir, M. Pervaiz, and O.Y. Alothman, A review on phenolic resin and its composites. *Current Analytical Chemistry*, 2018. **14**(3): p. 185–197.

29. Begum, S., S. Fawzia, and M. Hashmi, Polymer matrix composite with natural and synthetic fibres. *Advances in Materials and Processing Technologies*, 2020. **6**(3): p. 547–564.

30. Balakrishnan, P., M.J. John, L. Pothen, M. Sreekala, and S. Thomas, Natural fibre and polymer matrix composites and their applications in aerospace engineering. In *Advanced Composite Materials for Aerospace Engineering*. 2016, Elsevier. p. 365–383.

31. Semitekolos, D., P. Kainourgios, C. Jones, A. Rana, E.P. Koumoulos, and C.A. Charitidis, Advanced carbon fibre composites via poly methacrylic acid surface treatment; surface analysis and mechanical properties investigation. *Composites Part B: Engineering*, 2018. **155**: p. 237–243.

32. Fazeli, M., X. Liu, and C. Rudd, The effect of waterborne polyurethane coating on the mechanical properties of epoxy-based composite containing recycled carbon fibres. *Surfaces and Interfaces*, 2022. **29**: p. 101684.

33. Kausar, A., Performance of corrosion protective epoxy blend-based nanocomposite coatings: A review. *Polymer-Plastics Technology and Materials*, 2020. **59**(6): p. 658–673.

34. Kausar, A., Role of thermosetting polymer in structural composite. *American Journal of Polymer Science and Engineering*, 2017. **5**(1): p. 1–12.

35. Rubino, F., A. Nisticò, F. Tucci, and P. Carlone, Marine application of fiber reinforced composites: A review. *Journal of Marine Science and Engineering*, 2020. **8**(1): p. 26.

36. Kappenthuler, S. and S. Seeger, Assessing the long-term potential of fiber reinforced polymer composites for sustainable marine construction. *Journal of Ocean Engineering and Marine Energy*, 2021. **7**(2): p. 129–144.

37. Xu, Y., L. Guo, H. Zhang, H. Zhai, and H. Ren, Research status, industrial application demand and prospects of phenolic resin. *RSC Advances*, 2019. **9**(50): p. 28924–28935.

38. Jawaid, M. and M. Asim, *Phenolic Polymers Based Composite Materials*. 2021, Springer.

39. Shree, S. and J.S. Tate, *Mechanical Characterization of Core Shell Rubber Particles Modified Vinyl Ester and Glass Fiber Reinforced Composites*, SAMPE 2019- Charlotte, NC, May 2019.

40. Yang, G., M. Park, and S.-J. Park, Recent progresses of fabrication and characterization of fibers-reinforced composites: A review. *Composites Communications*, 2019. **14**: p. 34–42.

41. Patil, A.N., P.R. Kubade, and H.B. Kulkarni, Mechanical properties of hybrid glass micro balloons/fly ash cenosphere filled vinyl ester matrix syntactic foams. *Materials Today: Proceedings*, 2020. **22**: p. 1994–2000.

42. Nikolic, G., S. Zlatkovic, M. Cakic, S. Cakic, C. Lacnjevac, and Z. Rajic, Fast fourier transform IR characterization of epoxy GY systems crosslinked with aliphatic and cycloaliphatic EH polyamine adducts. *Sensors*, 2010. **10**(1): p. 684–696.

43. Islam, M., H. Ar-Rashid, F. Islam, N. Karmaker, F. Koly, J. Mahmud, K.N. Keya, and R. Khan, Fabrication and characterization of E-glass fiber reinforced unsaturated polyester resin based composite materials. 2019.

44. Arnold, J., *Environmental Effects on Crack Growth in Composites*. 2007.

45. Garland, C.A., *Effect of Manufacturing Process Conditions on the Durability of Pultruded Vinyl Ester/Glass Composites*. 2000, West Virginia University.

46. Lu, S., T. Yang, X. Xiao, X. Zhu, J. Wang, P. Zang, and J. Liu, Mechanical properties of the epoxy resin composites modified by nanofiller under different aging conditions. *Journal of Nanomaterials*, 2022. **2022**.

47. Rangaraj, R., S. Sathish, T. Mansadevi, R. Supriya, R. Surakasi, M. Aravindh, A. Karthick, V. Mohanavel, M. Ravichandran, and M. Muhibbullah, Investigation of weight fraction and alkaline treatment on catechu linnaeus/Hibiscus cannabinus/sansevieria ehrenbergii plant fibers-reinforced epoxy hybrid composites. *Advances in Materials Science and Engineering*, 2022. **2022**: p. 1–9.

48. Prasanna Venkatesh, R., K. Ramanathan, and V. Srinivasa Raman, Tensile, flexual, impact and water absorption properties of natural fibre reinforced polyester hybrid composites. *Fibres and Textiles in Eastern Europe*, 2016.

49. Ahmed, K.S., S. Vijayarangan, and C. Rajput, Mechanical behavior of isothalic polyester-based untreated woven jute and glass fabric hybrid composites. *Journal of Reinforced Plastics and Composites*, 2006. **25**(15): p. 1549–1569.

50. Mohd Nurazzi, N., A. Khalina, M. Chandrasekar, H. Aisyah, S. Ayu Rafiqah, R. Ilyas, and Z. Hanafee, Effect of fiber orientation and fiber loading on the mechanical and thermal properties of sugar palm yarn fiber reinforced unsaturated polyester resin composites. *Polimery*, 2020. **65**.

51. Keya, K.N., N.A. Kona, and R.A. Khan, Studies on the mechanical characterization of jute fiber reinforced unsaturated polyester resin based composites: Effect of weave structure and yarn density. *Advanced in Materials Research*, 2020. **1156**: p. 69–78.

52. Vu, C.M., D.D. Nguyen, L.H. Sinh, H.J. Choi, and T.D. Pham, Micro-fibril cellulose as a filler for glass fiber reinforced unsaturated polyester composites: Fabrication and mechanical characteristics. *Macromolecular Research*, 2018. **26**(1): p. 54–60.

53. Atiqah, A., M. Jawaid, S. Sapuan, and M. Ishak, Mechanical and thermal properties of sugar palm fiber reinforced thermoplastic polyurethane composites: Effect of silane treatment and fiber loading. *Journal of Renewable Materials*, 2018. **6**(5): p. 477.

54. Saha, A. and P. Kumari, Effect of alkaline treatment on physical, structural, mechanical and thermal properties of Bambusa tulda (Northeast Indian species) based sustainable green composites. *Polymer Composites*, 2023. **44**(4): p. 2449–2473.

55. Verleg, R. and S.K. Nagelvoort, *Enova Technology: Next Generation of Epoxy Bisphenol A Vinyl Ester Urethanes*.

56. Mbeche, S.M., P.M. Wambua, and D.N. Githinji, Mechanical properties of sisal/cattail hybrid-reinforced polyester composites. *Advances in Materials Science and Engineering*, 2020. **2020**: p. 1–9.

57. Athijayamani, A., B. Stalin, S. Sidhardhan, and A.B. Alavudeen, Mechanical properties of unidirectional aligned bagasse fibers/vinyl ester composite. *Journal of Polymer Engineering*, 2016. **36**(2): p. 157–163.

58. Memon, H., Y. Wei, and C. Zhu, Recyclable and reformable epoxy resins based on dynamic covalent bonds–Present, past, and future. *Polymer Testing*, 2022. **105**: p. 107420.

59. Sawicz, K., J. Ortyl, and R. Popielarz, Applicability of 7-hydroxy-4-methylcoumarin for cure monitoring and marking of epoxy resins. *Polimery*, 2010. **55**(7–8): p. 539–544.

60. Mishra, T., P. Mandal, A.K. Rout, and D. Sahoo, A state-of-the-art review on potential applications of natural fiber-reinforced polymer composite filled with inorganic nanoparticle. *Composites Part C: Open Access*, 2022: p. 100298.
61. Nassiopoulos, E. and J. Njuguna, Thermo-mechanical performance of poly (lactic acid)/flax fibre-reinforced biocomposites. *Materials and Design*, 2015. **66**: p. 473–485.
62. Jin, F.-L., X. Li, and S.-J. Park, Synthesis and application of epoxy resins: A review. *Journal of Industrial and Engineering Chemistry*, 2015. **29**: p. 1–11.
63. Prolongo, S.G., G. del Rosario, and A. Ureña, Comparative study on the adhesive properties of different epoxy resins. *International Journal of Adhesion and Adhesives*, 2006. **26**(3): p. 125–132.
64. Dall'Argine, C., A. Hochwallner, N. Klikovits, R. Liska, J. Stampfl, and M. Sangermano, Hot-lithography SLA-3D printing of epoxy resin. *Macromolecular Materials and Engineering*, 2020. **305**(10): p. 2000325.
65. Martulli, L.M., M. Alves, S. Pimenta, P.J. Hine, M. Kerschbaum, S.V. Lomov, and Y. Swolfs, Predictions of carbon fibre sheet moulding compound (CF-SMC) mechanical properties based on local fibre orientation. In *Proceedings ECCM 18, 18th European Conference on Composite Materials*. 2018.
66. Kumar, A., K. Sharma, and A.R. Dixit, A review on the mechanical properties of polymer composites reinforced by carbon nanotubes and graphene. *Carbon Letters*, 2021. **31**(2): p. 149–165.
67. Demir, E. and Ö. Güler, Production and properties of epoxy matrix composite reinforced with hollow silica nanospheres (HSN): Mechanical, thermal insulation, and sound insulation properties. *Journal of Polymer Research*, 2022. **29**(11): p. 477.
68. Wu, J., Y. Pan, Z. Ruan, Z. Zhao, J. Ai, J. Ban, and X. Jing, Carbon fiber-reinforced epoxy with 100% fiber recycling by transesterification reactions. *Frontiers in Materials*, 2022. **9**: p. 1045372.
69. Zhou, L., J. Yu, H. Wang, M. Chen, D. Fang, Z. Wang, and Z. Li, Dielectric and microwave absorption properties of resin-matrix composite coating filled with multi-wall carbon nanotubes and Ti 3 SiC 2 particles. *Journal of Materials Science: Materials in Electronics*, 2020. **31**: p. 15852–15858.
70. Nguyen, T.A., Q.T. Nguyen, and T.P. Bach, Mechanical properties and flame retardancy of epoxy resin/nanoclay/multiwalled carbon nanotube nanocomposites. *Journal of Chemistry*, 2019. **2019**.
71. Zhang, X., T. Cao, T. Zhao, C. Ma, and P. Liu, Preparation and properties of a new type of flexible epoxy resin based on a molecular network structure. *Express Polymer Letters*, 2021. **15**(11).
72. Jia, P., Y. Ma, Q. Kong, L. Xu, Q. Li, and Y. Zhou, Progress in development of epoxy resin systems based on biomass resources. *Green Materials*, 2019. **8**(1): p. 6–23.
73. Massoumi, B., M. Abbasian, R. Mohammad-Rezaei, A. Farnudiyan-Habibi, and M. Jaymand, Polystyrene-modified novolac epoxy resin/clay nanocomposite: Synthesis, and characterization. *Polymers for Advanced Technologies*, 2019. **30**(6): p. 1484–1492.
74. Thakur, T., S. Jaswal, B. Gaur, and A.S. Singha, Thermo-mechanical properties of rosin-modified o-cresol novolac epoxy thermosets comprising rosin-based imidoamine curing agents. *Polymer Engineering and Science*, 2021. **61**(1): p. 115–135.

75. Rudawska, A., 17 epoxy adhesives. *Handbook of Adhesive Technology*, 2017. **10**(11): p. 415.

76. Lu, M., Y. Liu, X. Du, S. Zhang, G. Chen, Q. Zhang, S. Yao, L. Liang, and M. Lu, Cure kinetics and properties of high performance cycloaliphatic epoxy resins cured with anhydride. *Industrial and Engineering Chemistry Research*, 2019. **58**(16): p. 6907–6918.

77. Langer, E., S. Waśkiewicz, and H. Kuczyńska, Application of new modified Schiff base epoxy resins as organic coatings. *Journal of Coatings Technology and Research*, 2019. **16**(4): p. 1109–1120.

78. Podkościelna, B., M. Wawrzkiewicz, and Ł. Klapiszewski, Synthesis, characterization and sorption ability of epoxy resin-based sorbents with amine groups. *Polymers*, 2021. **13**(23): p. 4139.

79. Venu, G., J.S. Jayan, A. Saritha, and K. Joseph, Thermal decomposition behavior and flame retardancy of bioepoxies, their blends and composites: A comprehensive review. *European Polymer Journal*, 2022. **162**: p. 110904.

80. Sukanto, H., W.W. Raharjo, D. Ariawan, J. Triyono, and M. Kaavesina, Epoxy resins thermosetting for mechanical engineering. *Open Engineering*, 2021. **11**(1): p. 797–814.

81. Souza, J.P. and J.M. Reis, Thermal behavior of DGEBA (diglycidyl ether of bisphenol A) adhesives and its influence on the strength of joints. *Applied Adhesion Science*, 2013. **1**(1): p. 1–10.

82. Woo, Y. and D. Kim, Cure and thermal decomposition kinetics of a DGEBA/amine system modified with epoxidized soybean oil. *Journal of Thermal Analysis and Calorimetry*, 2021. **144**(1): p. 119–126.

83. McCoy, J.D., W.B. Ancipink, S.R. Maestas, L.R. Draelos, D.B. Devries, and J.M. Kropka, Reactions of DGEBA epoxy cured with diethanolamine: Isoconversional kinetics and implications to network structure. *Thermochimica Acta*, 2019. **671**: p. 149–160.

84. Sindhu, P.S., N. Mitra, D. Ghindani, and S.S. Prabhu, Epoxy resin (DGEBA/TETA) exposed to water: A spectroscopic investigation to determine water-epoxy interactions. *Journal of Infrared, Millimeter and Terahertz Waves*, 2021. **42**(5): p. 558–571.

85. Wu, F., X. Zhou, and X. Yu, Reaction mechanism, cure behavior and properties of a multifunctional epoxy resin, TGDDM, with latent curing agent dicyandiamide. *RSC Advances*, 2018. **8**(15): p. 8248–8258.

86. Bard, S., M. Demleitner, R. Weber, R. Zeiler, and V. Altstädt, Effect of curing agent on the compressive behavior at elevated test temperature of carbon fiber-reinforced epoxy composites. *Polymers*, 2019. **11**(6): p. 943.

87. Boonlert-Uthai, T., K. Taki, and A. Somwangthanaroj, Curing behavior, rheological, and thermal properties of DGEBA modified with synthesized BPA/PEG hyperbranched epoxy after their photo-initiated cationic polymerization. *Polymers*, 2020. **12**(10): p. 2240.

88. Chen, C., B. Li, and M. Kanari, and D. Lu, Epoxy Adhesives. In *Adhesives and Adhesive Joints in Industry Applications*, 2019: p. 37–51.

89. Ali, A., S.S. Rahimian Koloor, A.H. Alshehri, and A. Arockiarajan, Carbon nanotube characteristics and enhancement effects on the mechanical features of polymer-based materials and structures – A review. *Journal of Materials Research and Technology*, 2023. **24**: p. 6495–6521.

90. Alhashmi Alamer, F. and G.A. Almalki, Fabrication of conductive fabrics based on SWCNTs, MWCNTs and graphene and their applications: A review. *Polymers*, 2022. **14**(24): p. 5376.

91. Ali, M.H. and R.I. Rubel, A comparative review of MG/CNTs and Al/CNTs composite to explore the prospect of bimetallic Mg-Al/CNTs composites. *AIMS Materials Science*, 2020. **7**(3): p. 217–243.

92. Bilalis, P., D. Katsigiannopoulos, A. Avgeropoulos, and G. Sakellariou, Non-covalent functionalization of carbon nanotubes with polymers. *RSC Advances*, 2014. **4**(6): p. 2911–2934.

93. Silvestro, L., A. Spat Ruviaro, P. Ricardo de Matos, F. Pelisser, D. Zambelli Mezalira, and P. Jean Paul Gleize, Functionalization of multi-walled carbon nanotubes with 3-aminopropyltriethoxysilane for application in cementitious matrix. *Construction and Building Materials*, 2021. **311**: p. 125358.

94. Park, J. and M. Yan, Covalent functionalization of graphene with reactive inter-mediates. *Accounts of Chemical Research*, 2013. **46**(1): p. 181–189.

95. Roy, S., R.S. Petrova, and S. Mitra, Effect of carbon nanotube (CNT) functional-ization in epoxy-CNT composites. *Nanotechnology Reviews*, 2018. **7**(6): p. 475–485.

96. Turan, F., M. Guclu, K. Gurkan, A. Durmus, and Y. Taskin, The effect of carbon nanotubes loading and processing parameters on the electrical, mechanical, and viscoelastic properties of epoxy-based composites. *Journal of the Brazilian Society of Mechanical Sciences and Engineering*, 2022. **44**(3): p. 93.

97. Chakraborty, A.K., T. Plyhm, M. Barbezat, A. Necola, and G.P. Terrasi, Carbon nanotube (CNT)–epoxy nanocomposites: A systematic investigation of CNT dispersion. *Journal of Nanoparticle Research*, 2011. **13**(12): p. 6493–6506.

98. Guan, L.-Z. and L.-C. Tang, Dispersion and alignment of Carbon nanotubes in polymer matrix. In *Handbook of Carbon Nanotubes*. 2021, Springer. p. 1–35.

99. Giliopoulos, D.J., K.S. Triantafyllidis, and D. Gournis, Chemical functionaliza-tion of carbon nanotubes for dispersion in epoxy matrices. In *Carbon Nanotube Enhanced Aerospace Composite Materials: A New Generation of Multifunctional Hybrid Structural Composites*, 2013: p. 155–183.

100. Zhang, Q., J. Wu, L. Gao, T. Liu, W. Zhong, G. Sui, G. Zheng, W. Fang, and X. Yang, Dispersion stability of functionalized MWCNT in the epoxy–amine system and its effects on mechanical and interfacial properties of carbon fiber composites. *Materials and Design*, 2016. **94**: p. 392–402.

101. Tang, L.-c., H. Zhang, J.-h. Han, X.-p. Wu, and Z. Zhang, Fracture mecha-nisms of epoxy filled with ozone functionalized multi-wall carbon nanotubes. *Composites Science and Technology*, 2011. **72**(1): p. 7–13.

102. Jiang, M., H. Zhou, and X. Cheng, Effect of rare earth surface modification of carbon nanotubes on enhancement of interfacial bonding of carbon nanotubes reinforced epoxy matrix composites. *Journal of Materials Science*, 2019. **54**(14): p. 10235–10248.

103. Wu, Q., H. Bai, X. Yang, and J. Zhu, Significantly increasing the interfacial adhe-sion of carbon fiber composites via constructing a synergistic hydrogen bond-ing network by vacuum filtration. *Composites Part B: Engineering*, 2021. **225**: p. 109300.

104. Alam, A., C. Wan, and T. McNally, Surface amination of carbon nanoparticles for modification of epoxy resins: Plasma-treatment vs. wet-chemistry approach. *European Polymer Journal*, 2017. **87**: p. 422–448.

105. Georgakilas, V., J.A. Perman, J. Tucek, and R. Zboril, Broad family of carbon nanoallotropes: Classification, chemistry, and applications of fullerenes, carbon dots, nanotubes, graphene, nanodiamonds, and combined superstructures. *Chemical Reviews*, 2015. **115**(11): p. 4744–4822.
106. Zabihi, O., M. Ahmadi, and M. Naebe, One-pot synthesis of aminated multiwalled carbon nanotube using thiol-ene click chemistry for improvement of epoxy nanocomposites properties. *RSC Advances*, 2015. **5**(119): p. 98692–98699.
107. Li, J., W. Zhu, S. Zhang, Q. Gao, J. Li, and W. Zhang, Amine-terminated hyperbranched polyamide covalent functionalized graphene oxide-reinforced epoxy nanocomposites with enhanced toughness and mechanical properties. *Polymer Testing*, 2019. **76**: p. 232–244.
108. Sakr, M.A., K. Sakthivel, T. Hossain, S.R. Shin, S. Siddiqua, J. Kim, and K. Kim, Recent trends in gelatin methacryloyl nanocomposite hydrogels for tissue engineering. *Journal of Biomedical Materials Research – Part A*, 2022. **110**(3): p. 708–724.
109. Liang, M., C. He, J. Dai, P. Ren, Y. Fu, F. Wang, X. Ge, T. Zhang, and Z. Lu, A high-strength double network polydopamine nanocomposite hydrogel for adhesion under seawater. *Journal of Materials Chemistry B*, 2020. **8**(36): p. 8232–8241.
110. Singh, N.P., V. Gupta, and A.P. Singh, Graphene and carbon nanotube reinforced epoxy nanocomposites: A review. *Polymer*, 2019. **180**: p. 121724.
111. Gonçalves, F.A., M. Santos, T. Cernadas, P. Alves, and P. Ferreira, Influence of fillers on epoxy resins properties: A review. *Journal of Materials Science*, 2022. **57**(32): p. 15183–15212.
112. Benra, J. and S. Forero, Epoxy resins reinforced with carbon nanotubes. *Lightweight Design Worldwide*, 2018. **11**(1): p. 6–11.
113. Farzanehfar, N., A. Nasr Esfahani, M. Sheikhi, and F. Rafiemanzelat, Imparting electrical conductivity in epoxy resins (chemistry and approaches). In *Multifunctional Epoxy Resins: Self-Healing, Thermally and Electrically Conductive Resins*. 2023, Springer. p. 365–413.
114. Abdulhameed, A., N.Z.A. Wahab, M.N. Mohtar, M.N. Hamidon, S. Shafie, and I.A. Halin, Methods and applications of electrical conductivity enhancement of materials using carbon nanotubes. *Journal of Electronic Materials*, 2021. **50**(6): p. 3207–3221.
115. Compton, B.G., J.K. Wilt, J.W. Kemp, N.S. Hmeidat, S.R. Maness, M. Edmond, S. Wilcenski, and J. Taylor, Mechanical and thermal properties of 3D-printed epoxy composites reinforced with boron nitride nanobarbs. *MRS Communications*, 2021. **11**(2): p. 100–105.
116. Vobornik, D., M. Chen, S. Zou, and G.P. Lopinski, Measuring the diameter of single-wall carbon nanotubes using AFM. *Nanomaterials*, 2023. **13**(3): p. 477.
117. Fraczek-Szczypta, A., E. Menaszek, T.B. Syeda, A. Misra, M. Alavijeh, J. Adu, and S. Blazewicz, Effect of MWCNT surface and chemical modification on in vitro cellular response. *Journal of Nanoparticle Research*, 2012. **14**(10): p. 1–14.
118. Wang, J. and T. Lei, Enrichment of high-purity large-diameter semiconducting single-walled carbon nanotubes. *Nanoscale*, 2022. **14**(4): p. 1096–1106.
119. Nanot, S., N.A. Thompson, J.-H. Kim, X. Wang, W.D. Rice, E.H. Hároz, Y. Ganesan, C.L. Pint, and J. Kono, *Single-Walled Carbon Nanotubes*. Springer Handbook of Nanomaterials, 2013: p. 105–146.

120. Salah, L.S., N. Ouslimani, D. Bousba, I. Huynen, Y. Danlée, and H. Aksas, Carbon nanotubes (CNTs) from synthesis to functionalized (CNTs) using conventional and new chemical approaches. *Journal of Nanomaterials*, 2021. **2021**: p. 1–31.

121. Zaghloul, M.M.Y., M.Y.M. Zaghloul, and M.M.Y. Zaghloul, Experimental and modeling analysis of mechanical-electrical behaviors of polypropylene composites filled with graphite and MWCNT fillers. *Polymer Testing*, 2017. **63**: p. 467–474.

122. Yang, Z., L. Yuan, Y. Gu, M. Li, Z. Sun, and Z. Zhang, Improvement in mechanical and thermal properties of phenolic foam reinforced with multiwalled carbon nanotubes. *Journal of Applied Polymer Science*, 2013. **130**(3): p. 1479–1488.

123. Loos, M., S. Pezzin, S. Amico, C. Bergmann, and L. Coelho, The matrix stiffness role on tensile and thermal properties of carbon nanotubes/epoxy composites. *Journal of Materials Science*, 2008. **43**(18): p. 6064–6069.

124. Yang, W., L. He, X. Tian, M. Yan, H. Yuan, X. Liao, J. Meng, Z. Hao, and L. Mai, Carbon-MEMS-based alternating stacked MoS2@ rGO-CNT microsupercapacitor with high capacitance and energy density. *Small*, 2017. **13**(26): p. 1700639.

125. Li, G.r., F. Wang, Q.w. Jiang, X.p. Gao, and P.w. Shen, Carbon nanotubes with titanium nitride as a low-cost counter-electrode material for dye-sensitized solar cells. *Angewandte Chemie International Edition*, 2010. **49**(21): p. 3653–3656.

126. Spinazzè, A., C. Zellino, F. Borghi, D. Campagnolo, S. Rovelli, M. Keller, G. Fanti, A. Cattaneo, and D.M. Cavallo, Carbon nanotubes: Probabilistic approach for occupational risk assessment. *Nanomaterials*, 2021. **11**(2): p. 409.

127. Ferreira, F.V., W. Franceschi, B.R. Menezes, A.F. Biagioni, A.R. Coutinho, and L.S. Cividanes, Synthesis, characterization, and applications of carbon nanotubes. In *Carbon-Based Nanofillers and their Rubber Nanocomposites*. 2019, Elsevier. p. 1–45.

128. Rathi, B.S., P.S. Kumar, and D.-V.N. Vo, Critical review on hazardous pollutants in water environment: Occurrence, monitoring, fate, removal technologies and risk assessment. *Science of the Total Environment*, 2021. **797**: p. 149134.

129. Kanwar, V.S., A. Sharma, A.L. Srivastav, and L. Rani, Phytoremediation of toxic metals present in soil and water environment: A critical review. *Environmental Science and Pollution Research*, 2020. **27**(36): p. 44835–44860.

130. Nabinejad, O., D. Sujan, M.E. Rahman, W.Y.H. Liew, and I.J. Davies, Hybrid composite using natural filler and multi-walled carbon nanotubes (MWCNTs). *Applied Composite Materials*, 2018. **25**(6): p. 1323–1337.

131. Rajmohan, T., K. Mohan, R. Prasath, and S. Vijayabhaskar, Tribological characterization of hybrid natural fiber MWCNT filled polymer composites In *Bio-Fiber Reinforced Composite Materials: Mechanical, Thermal and Tribological Properties*. 2022, Springer. p. 339–359.

132. Mirsalehi, S.A., A.A. Youzbashi, and A. Sazgar, Enhancement of out-of-plane mechanical properties of carbon fiber reinforced epoxy resin composite by incorporating the multi-walled carbon nanotubes. *SN Applied Sciences*, 2021. **3**(6): p. 630.

133. Mechanical properties of composite materials. In *Composites Science, Technology, and Engineering*, F.R. Jones, Editor. 2022, Cambridge University Press. p. 160–209.

134. Ramachandran, K., V. Boopalan, J.C. Bear, and R. Subramani, Multi-walled carbon nanotubes (MWCNTs)-reinforced ceramic nanocomposites for aerospace applications: A review. *Journal of Materials Science*, 2022. **57**(6): p. 3923–3953.

135. Madueke, C.I., R. Umunakwe, and O.M. Mbah, A review on the factors affecting the properties of natural fibre polymer composites. *Nigerian Journal of Technology*, 2022. **41**(1): p. 55–64.

136. Khieng, T.K., S. Debnath, E.T. Chaw Liang, M. Anwar, A. Pramanik, and A.K. Basak, A review on mechanical properties of natural fibre reinforced polymer composites under various strain rates. *Journal of Composites Science*, 2021. **5**(5): p. 130.

137. Ismail, N.H., M.H. Bin Mohamad, and M. Jaafar, Multi-walled carbon nanotubes/woven kenaf fabric-reinforced epoxy laminated composites. *Sains Malaysiana*, 2018. **47**(3): p. 563–569.

138. Peddini, S., C. Bosnyak, N. Henderson, C. Ellison, and D. Paul, Nanocomposites from styrene–butadiene rubber (SBR) and multiwall carbon nanotubes (MWCNT) part 2: Mechanical properties. *Polymer*, 2015. **56**: p. 443–451.

139. Amoroso, L., E.L. Heeley, S.N. Ramadas, and T. McNally, Crystallisation behaviour of composites of HDPE and MWCNTs: The effect of nanotube dispersion, orientation and polymer deformation. *Polymer*, 2020. **201**: p. 122587.

140. Subhani, T., M. Latif, I. Ahmad, S.A. Rakha, N. Ali, and A.A. Khurram, Mechanical performance of epoxy matrix hybrid nanocomposites containing carbon nanotubes and nanodiamonds. *Materials and Design*, 2015. **87**: p. 436–444.

141. Hadi, A.E., J.P. Siregar, T. Cionita, M.B. Norlaila, M.A.M. Badari, A.P. Irawan, J. Jaafar, T. Rihayat, R. Junid, and D.F. Fitriyana, Potentiality of utilizing woven pineapple leaf fibre for polymer composites. *Polymers*, 2022. **14**(13): p. 2744.

142. Hanan, F., M. Jawaid, M.T. Paridah, and J. Naveen, Characterization of hybrid oil palm empty fruit bunch/woven kenaf fabric-reinforced epoxy composites. *Polymers*, 2020. **12**(9): p. 2052.

143. Wulan, P.P.D.K. and Y. Yolanda, Mechanical property improvement of oil palm empty fruit bunch composites by hybridization using ramie fibers on epoxy–CNT matrices. *Science and Engineering of Composite Materials*, 2023. **30**(1).

144. Sathishkumar, T. and S. Ramakrishnan, Mechanical properties of nanococonut shell filler mixed jute mat-reinforced epoxy composites for structure application. In *Fiber-Reinforced Nanocomposites: Fundamentals and Applications*. 2020, Elsevier. p. 459–476.

145. Gogna, E., R. Kumar, Anurag, A.K. Sahoo, and A. Panda, A comprehensive review on jute fiber reinforced composites. *Advances in Industrial and Production Engineering: Select Proceedings of Flame*, 2018. **2019**: p. 459–467.

146. Sinha, A., A. Sinha, and R. Kumar, Comparative study of tensile behavior between epoxy/coir fiber and modified epoxy/coir fiber composite. In *Technology Innovation in Mechanical Engineering: Select Proceedings of Time 2021*. 2022, Springer. p. 325–333.

147. Nguyen, T.A. and T.H. Nguyen, Banana fiber-reinforced epoxy composites: Mechanical properties and fire retardancy. *International Journal of Chemical Engineering*, 2021. **2021**: p. 1–9.

148. Ahmad Nadzri, S.N.Z., M.T. Hameed Sultan, A.U.Md. Shah, S.N.A. Safri, and A.A. Basri, A review on the kenaf/glass hybrid composites with limitations on mechanical and low velocity impact properties. *Polymers*, 2020. **12**(6): p. 1285.
149. Ayrilmis, N., S. Jarusombuti, V. Fueangvivat, P. Bauchongkol, and R.H. White, Coir fiber reinforced polypropylene composite panel for automotive interior applications. *Fibers and Polymers*, 2011. **12**(7): p. 919–926.
150. Balda, S., A. Sharma, N. Capalash, and P. Sharma, Banana fibre: A natural and sustainable bioresource for eco-friendly applications. *Clean Technologies and Environmental Policy*, 2021. **23**(5): p. 1389–1401.
151. Goudenhooft, C., A. Bourmaud, and C. Baley, Flax (Linum usitatissimum L.) fibers for composite reinforcement: Exploring the link between plant growth, cell walls development, and fiber properties. *Frontiers in Plant Science*, 2019. **10**: p. 411.

7

Biodegradable Natural Fibre (Coconut Mesocarp) Filled Polypropylene Composites: Effects of the Compatibilizer and Coupling Agent on Tensile Properties

S. R. Ahmad, M. S. M. Sidik, M. F. Hamid, and S. S. M. Saleh

7.1 Introduction

Biodegradable natural fibre (Bio-NF) is made from natural fibre that can be obtained either from crops or from animals, such as hemp, kenaf, wool, and others. Those materials can be easily disposed of, which makes them commonly used as fillers in polymer composites. Today, waste materials are being transformed into engineering materials [1] suitable for the manufacturing, automotive [2–4], and aerospace industries [5]. The use of Bio-NF is cost-effective [6, 7]. Previously, the presence of bio-natural fibre was used to blend with thermosets. However, the widespread use of thermoplastics has caused the combination of bio-natural fibre and thermoplastic to attract interest from researchers in the development of novel materials [6–8]. Bio-NF thermoplastic composites offer several advantages: for instance, reducing the applications of host polymers that are hard to dispose of, benefiting the environment and sustainability; reducing the cost of host polymers from petroleum-based products; comparable properties; and suitability for specific applications [9].

Cocos nucifera (coconut palm) is a large palm, growing up to 30 m (100 ft) tall. A tall coconut palm tree can yield up to 75 fruits per year. Given proper care and growing conditions, coconut palms produce their first fruit in 6 to 10 years. Coconut palms are easy to plant, take care of, and grow with minimum effort. Coconut mesocarp (CM) is used in ropes, mats, doormats, brushes, and sacks, as well as boat caulking and mattress filling. It is used in potting compost, especially orchid mix, in horticulture. In most Asian countries, CM is used to produce brooms [9–11]. CM is well known as coir, and coconut fibrous husk is one of the bio-natural fibres. Furthermore, it is lightweight (low density, 1.1 g cm^{-3}) and has high ductility (15% to 40%) and good formability [10]. CM is made up of two components: pith particles that serve as a binder and coir fibres that provide structural strength. In nature, the only purpose of the CM is to protect the inside part of the coconut from harm when it falls from a tall coconut tree [11, 12]. Here, the CM itself is excellent at absorbing the impact.

The application of CM in polymers such as polypropylene has dramatically increased and is trending on the global market. Most researchers focus on investigating the properties of the CM polymer composites. This is because of the advantages of CM itself as a fibre with good properties in composites, especially increasing the stiffness [13, 14] and impact properties [15, 16]. Polypropylene composites are widely used as automotive components and parts [17], packaging [18, 19], and electrical insulators [20, 21, 22]. Their low density (light weight), the fact that they can be recycled, their non-toxic and non-hazardous properties, and their low cost make PP composites very significant engineering materials today. However, the addition of CM to PP reduces the mechanical properties, such as the strength of PP itself as a host polymer. To reduce these problems, a compatibilizer is applied to treat the host polymer [13, 14] and a coupling agent to treat the filler [23, 24]. Table 7.1 shows research papers that are related to this research using coir/coconut fibres in PP composites with and without treatment.

TABLE 7.1

Research Papers Related to This Research Using Coir/Coconut Fibres in Polypropylene Composites with and without Treatment

Authors	Polymer	Source of Nanocellulose	Description
Bettini et al. [23]	Polypropylene/ Coir fibre	Coir fibre	• The presence of PP-g-MA increased tensile properties.
Samia et al. [24]		Coir fibre	• The performance of coir fibre composites in industrial applications can be improved by chemical treatment.
Sabri et al. [25]	Polypropylene/ Coconut fibre/ MAPP	Coconut fibre	• By increasing filler loading from 10% to 40%, elastic modulus was increased while the tensile strength and elongation at break were decreased. The presence of MAPP increased tensile properties of the composites.
Sabri et al. [26]	Polypropylene/ Coir fibre/ 3-APE	Coconut fibre	• The addition of 3% 3-APE to treated composites enhanced their Young's modulus, but decreased their tensile strength and elongation at break with increased fibre content.

In this study, maleic anhydride grafted polypropylene (MAPP) is used as a compatibilizer and 3-aminopropyltriethoxysilane (3-APE) as a silane coupling agent that acts as a booster to the mechanical properties of the biodegradable natural fibre polymer composites. The comparison of both filler and matrix treatment compared with untreated PP/CM composites towards their tensile properties will be discussed. This chapter covers the tensile properties such as tensile strength, Young's modulus, and elongation at break of the composite.

7.2 Sample Preparation

7.2.1 Materials

The coconut fibre was supplied by Avasia Agro Sdn. Bhd, Simpang Pulai, Perak, Malaysia in the form of coconut husk. Polypropylene grade S12232 G112 was supplied by Polypropylene Malaysia Sdn. Bhd. Kuantan, Pahang in the form of pellets. MAPP with a grade of 426512 was supplied by ExxonMobil Chemical Asia Pacific, Singapore. The 3-APE was supplied by ZARM Scientific & Supplies Sdn Bhd, Bukit Mertajam, Pulau Pinang from Sigma Aldrich. Table 7.2 shows the formulation of PP/CM composites at different fibre loadings.

7.2.2 Preparation of Coconut Mesocarp Fibre

First, the coconut fibre was extracted from the long fibres of coconut. The CM was placed on the tray for the drying process. Then, it was cut into tiny pieces with an average length of 1 to 3 mm before being mixed with polypropylene. The preparation process of biodegradable natural fibre (coconut mesocarp short fibre) from the coconut tree to the PP/CM composites is shown in Figure 7.1.

TABLE 7.2

Formulation of Biodegradable Natural Fibre Polymer (PP/CM) Composites at Different Fibre Loadings

Material	PP/CM Composites (wt.%)	PP/CM/MAPP Composites (wt.%)	PP/CM/3-APE Composites (wt.%)
Polypropylene (PP)	100, 90, 80, 70, 60	100, 90, 80, 70, 60	100, 90, 80, 70, 60
Sansevieria fibre (SLF)	0, 10, 20, 30, 40	0, 10, 20, 30, 40	0, 10, 20, 30, 40
Maleic anhydride grafted polypropylene (MAPP)[a]	–	3	-
3-(Aminopropyl) triethoxysilane (3-APE)[a]	–	–	3

[a]*wt.% of filler.*

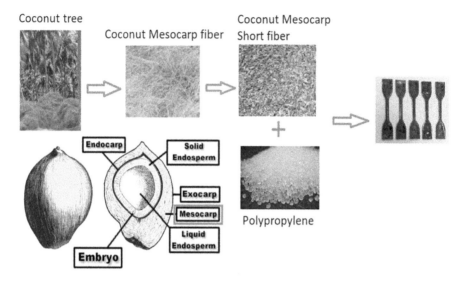

FIGURE 7.1
The preparation process of biodegradable natural fibre (coconut mesocarp short fibre) from the coconut tree to the PP/CM composites.

7.2.3 Fibre Treatment

Fibre treatment is not required for MAPP. It was added directly into the mixing chamber. However, for 3-APE, filler treatment is required. First, 3-APE was diluted in ethanol using a stirrer. The amount of 3-APE used was 3 wt.% of the filler weight. The filler was charged into the chamber mixer with the solution slowly to ensure uniform distribution of 3-APE. The filler was continuously mixed. After an hour, the filler was filtered out and then dried in a forced-convection oven at 80 C for 24 hours to allow complete evaporation of the ethanol.

7.2.4 Mixing Process

At a temperature of 180 C and a rotor speed of 50 rpm, composites were mixed in the Haake Polylab QC Mixer Machine (Thermo Fisher Scientific Inc., United States). Polypropylene was discharged into the mixing chamber. It melted completely into the mixer chamber after 5 minutes. Then, the fibre was inserted. The mixing process continued for another 5 minutes until the 10-minute mark was reached. It took 10 min for each mix to be completed. For compatibilized and treated composites, after 3 min of the fibre being inserted into the mixing chamber, the MAPP or 3-APE (after filler treatment) was added before the process continued to completion at 10 min. Based on Table 7.2, the composite mixtures are prepared with different loadings of CM fibre.

7.2.5 Compression Moulding

The composite mixtures were then compressed in a hydraulic hot press machine (GT 7014) (Dongguan Lixian Instrument Scientific Co., Ltd. Guangdong, China) to produce a ±1.0-mm sheet of composite using the moulding of a dumbbell shape to comply with the ASTM D638 standard. The force was set at 10 kN and the temperature at 180 C. The mould was placed on the machine to pre-heat it for 5 minutes. After that, the sample was pressed for 4 min. After completing the process, the sample was removed and blown with air to make it cool faster. The process was completed after it was removed from the mould. There were 10 samples per composition.

7.2.6 Measurement of Tensile Properties

The tensile test was carried out according to ASTM D412 using a Universal Testing Machine (UTM) (Instron Model 5569 from the United States). The testing was completed in a standard laboratory atmosphere of 25 C ± 3 C. The tensile testing machine was used with a crosshead speed of 50 mm/ minute, and the gauge length was set at 50 mm. The test was carried out on five samples for each formulation.

7.3 Results and Discussion

The tensile strength, Young's modulus, and elongation at break of PP/CM composites at 0, 10%, 20%, 30%, and 40% of CM fibre loading are discussed here. Figure 7.2 shows the tensile strength of the composites with and without treatment. Figure 7.2 indicates that the tensile strength of all composites has decreased because of an increasing CM fibre content. The size and shape of the fibres affected the decrease in the value of tensile strength [27]. The aspect ratio of fibres has a significant impact on tensile properties [28]. In addition, the CM fibres were randomly dispersed in the PP matrix when they were compounded using a hot compress machine.

At 10 wt.% CM fibre loading, the composites with 3-APE indicate the highest tensile strength (22.71 MPa), followed by the composites with MAPP (21.89 MPa). The lowest tensile strength is 20.13 MPa for composites without treatment. This happened due to the polarity difference between the phases. Therefore, the composites have poor fibre attachment to the matrix, which results in a loss in tensile strength [14, 26, 29].

MAPP was used to modify the host polymer PP. The tensile strength value of the composite with MAPP is higher than for the composites without any treatment in Figure 7.2. MAPP was used to enhance the interfacial compatibility between PP and CM [14, 25]. Vedat (2020) states that the addition of

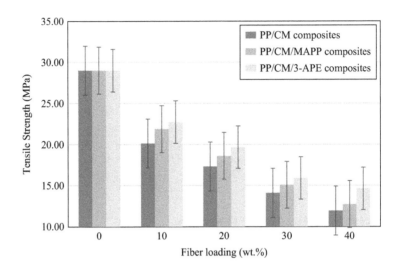

FIGURE 7.2
The tensile strength of PP/CM, PP/CM/MAPP, and PP/CM/3-APE composites at different fibre loading.

MAPP to PP and recycle PP (rPP) filled with mahogany wood fibre improved the tensile strength of the composites. However, after increasing the loading of wood fibre, these trends were reduced [30].

The CM fibre was treated with 3-APE. The tensile strength of composites with 3-APE was decreased [26]. However, the highest tensile strength is shown compared with the composite without treatment and the composite with MAPP at 10 wt.% CM fibre loading. This is due to better interfacial interaction between the matrix and the fibre. Other researchers reported that the mechanical properties improved when applying 3-APE in their composites [29, 31, 32, 33, 34, 35].

The Young's modulus of the composites increased steadily when the CM fibre content was increased, as shown in Figure 7.3. This occurred due to the increased stiffness of the composites [14, 25, 26]. At the maximum CM fibre loading (40 wt.%), in the presence of 3-APE as a silane coupling agent, the composites reached a Young's modulus near 2500 MPa. The composites with MAPP and without treatment show 2400 MPa and 2300 MPa at a maximum 40 wt.% fibre loading, respectively. At 40% fibre content, the stiffness of 3-APE composites is 7% higher than that of untreated composites. The Young's modulus of the PP/CM/3-APE composites increased 46% at 40 wt.% compared with 0% CM content. This happened due to strong interaction between the fibres and the matrices of the composites with the addition of 3-APE as a coupling agent.

The elongation of the composite gradually decreased as CM loading increased from 10% to 40% in Figure 7.4. This might be due to the hard

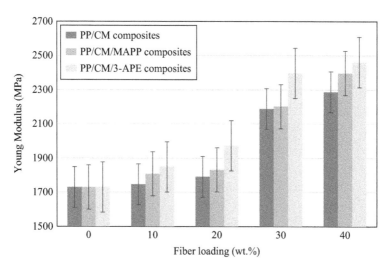

FIGURE 7.3
The Young's modulus of PP/CM, PP/CM/MAPP, and PP/CM/3-APE composites at different fibre loading.

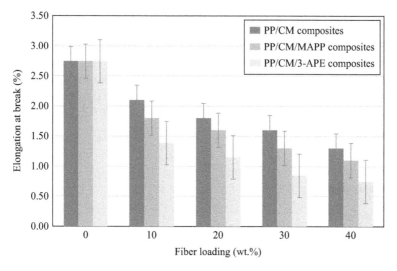

FIGURE 7.4
The elongation at break of PP/CM, PP/CM/MAPP, and PP/CM/3-APE composites at different fibre loading.

contact between the filler and the PP matrix, which has limited deformability. The drop is proportional to the amount of fibre present. The elongation at break is reduced because their brittleness increases due to the hardening of composites by fillers. [36]. The composite without treatment shows difficulty in fracturing, and its elongation at break is 1.3% at 40 wt.% CM fibre loading.

TABLE 7.3

Tensile Properties of Biodegradable Natural Polymer (PP/CM) Composites with and without Treatment

	Material/ Fibre Loading	PP/CM Composites (wt.%)[a]	PP/CM/MAPP Composites (wt.%)[a]	PP/CM/3-APE Composites (wt.%)[a]
Tensile strength (MPa)	0	29.00 ± 1.2	29.00 ± 1.2	29.00 ± 1.2
	10	20.13 ± 1.3	21.87 ± 1.1	22.71 ± 1.3
	20	17.31 ± 1.1	18.61 ± 1.3	19.66 ± 1.2
	30	14.09 ± 1.3	15.08 ± 1.2	15.91 ± 1.2
	40	11.93 ± 1.1	12.72 ± 1.2	14.62 ± 1.1
Young's modulus (MPa)	0	1730.19 ± 20	1730.19 ± 20	1730.19 ± 20
	10	1746.90 ± 23	1807.80 ± 25	1848.62 ± 28
	20	1790.57 ± 26	1831.30 ± 30	1971.96 ± 31
	30	2188.09 ± 25	2201.40 ± 20	2396.39 ± 26
	40	2286.91 ± 20	2396.60 ± 23	2460.33 ± 22
Elongation at break (%)	0	2.74 ± 0.03	2.74 ± 0.03	2.74 ± 0.03
	10	2.10 ± 0.06	1.80 ± 0.02	1.39 ± 0.05
	20	1.80 ± 0.04	1.60 ± 0.05	1.15 ± 0.04
	30	1.60 ± 0.03	1.30 ± 0.04	0.85 ± 0.03
	40	1.30 ± 0.05	1.10 ± 0.03	0.75 ± 0.05

[a] *Average values from 10 samples of each composition.*

The values for the composites with MAPP and 3-APE are 1.1% and 0.55%, respectively. Table 7.3 shows a summary of the tensile properties of PP/CM composites without treatment, with MAPP, and with 3-APE. The results are based on an average of 10 samples of each composition.

7.4 Conclusion

Polypropylene coconut mesocarp composites are low-cost composites that offer several advantages and have become versatile engineering materials. These composites are expected to be used in internal automotive component applications, packaging industries, and others according to their mechanical properties. The Young's modulus of the composites is enhanced by 24% with increasing CM fibre loading. Nevertheless, the tensile strength and elongation at break decreased as the CM loading increased by ±59% and ±60%, respectively. When MAPP was used as a compatibilizer in the composites, the tensile strength decreased by ±56% with increasing CM loading. However, the tensile strength increased by 8% after the addition of MAPP at a 10 wt.% fibre loading. This means that the presence of MAPP between PP and CM improved their interfacial interaction. The Young's modulus enhancement of ±28% resulted in better stiffness with increasing fibre loading. The elongation at break decreased by 60%, indicating that this composite became more brittle after adding more CM fibre. The coupling agent used in this composite was 3-APE. Tensile strength and elongation at break decreased by 50% and 73%, respectively. The significant improvement in this research is that the Young's modulus increased by 30% compared with composites without treatment. Therefore, the composites with MAPP showed enhanced tensile properties. The addition of 3-APE to PP/CM composites indicates an improvement in tensile properties compared with MAPP and without treatment (control) biodegradable natural polymer PP/CM composites.

Acknowledgements

The authors would like to thank Universiti Kuala Lumpur Malaysian Spanish Institute (UniKL MSI), Universiti Malaysia Perlis (UniMAP), and TAJ International College CITY COUNTRY for the collaboration towards the development of new materials for automotive applications.

References

1. K. S. Chun, C. M. Yeng, and S. Hussiensyah, "Green coupling agent for agro-waste based thermoplastic composites," *Polymer Composites*, vol. 39, no. 7, 2018, doi: 10.1002/pc.24228.
2. V. Naik, M. Kumar, and V. Kaup, "A review on natural fiber composite materials in automotive applications," *Engineered Science*, vol. 18, 2022, doi: 10.30919/es8d589.
3. J. Thomason and J. Rudeiros-Fernández, "Evaluation of injection molded natural fiber-polypropylenes for potential in automotive applications," in *11th Canadian – International Conference on Composites*, 2019.
4. J. L. Thomason and J. L. Rudeiros-Fernández, "Evaluation of injection moulded natural fiber-polyolefin for potential in automotive applications," in *Materials Today: Proceedings*, 2019, doi: 10.1016/j.matpr.2019.11.059.
5. D. Bhadra and N. R. Dhar, "Selection of the natural fiber for sustainable applications in aerospace cabin interior using fuzzy MCDM model," *Materialia (Oxf)*, vol. 21, 2022, doi: 10.1016/j.mtla.2021.101270.

6. R. Vijayan and A. Krishnamoorthy, "Review on natural fiber reinforced composites," in *Materials Today: Proceedings*, 2019, doi: 10.1016/j.matpr.2019. 05.175.

7. L. Kerni, S. Singh, A. Patnaik, and N. Kumar, "A review on natural fiber reinforced composites," in *Materials Today: Proceedings*, 2020, doi: 10.1016/j. matpr.2020.04.851.

8. P. Kaliappan, R. Kesavan, B. V. Ramnath, B. P. Kumar, and V. K. Rao, "A review on natural composites," *Advanced Science, Engineering and Medicine*, vol. 10, no. 3, 2018, doi: 10.1166/asem.2018.2162.

9. W. Stelte, N. Reddy, S. Barsberg, and A. R. Sanadi, "Coir from coconut processing waste as a raw material for applications beyond traditional uses," *BioResources*, vol. 18, no. 1, 2023, doi: 10.15376/biores.18.1.Stelte.

10. W. L. Bradley and S. Conroy, "Using agricultural waste to create more environmentally friendly and affordable products and help poor coconut farmers," in *E3S Web of Conferences*, 2019, doi: 10.1051/e3sconf/201913001034.

11. N. Reddy and Y. Yang, "Coconut husk fibers," in *Innovative Biofibers from Renewable Resources*, 2015, doi: 10.1007/978-3-662-45136-6_9.

12. P. H. Sankar, Y. V. M. Reddy, K. H. Reddy, M. A. Kumar, and A. Ramesh, "The effect of fiber length on tensile properties of polyester resin composites reinforced by the fibers of *Sansevieria trifasciata*," *International Letters of Natural Sciences*, vol. 8, 2014, doi: 10.18052/www.scipress.com/ilns.8.7.

13. A. G. Adeniyi, D. V. Onifade, J. O. Ighalo, and A. S. Adeoye, "A review of coir fiber reinforced polymer composites," *Composites Part B: Engineering*, vol. 176, 2019, doi: 10.1016/j.compositesb.2019.107305.

14. L. Galav, S. Mukhopadhyay, and B. L. Deopura, "Effect of compatibilizer on mechanical properties of chemically treated coir/polypropylene composite," *Indian Journal of Fibre and Textile Research*, vol. 44, no. 2, 2019.

15. S. Jayabal and U. Natarajan, "Influence of fiber parameters on tensile, flexural, and impact properties of nonwoven coir-polyester composites," *International Journal of Advanced Manufacturing Technology*, vol. 54, no. 5–8, 2011, doi: 10.1007/s00170-010-2969-8.

16. A. D. Shieddieque, R. Mardiyati, R. Suratman, B. Widyanto, "The effect of alkaline treatment and fiber orientation on impact resistant of bio-composites Sansevieria trifasciata fiber/polypropylene as automotive components material," in *AIP Conference Proceedings*, 2018, doi: 10.1063/1.5030263.

17. D. Shieddieque, R. Mardiyati, R. Suratman, B. Widyanto, "Preparation and characterization of sansevieria trifasciata fiber/high-impact polypropylene and Sansevieria trifasciata fiber/vinyl ester biocomposites for automotive applications," *International Journal of Technology*, vol. 12, no. 3, 2021, doi: 10.14716/ijtech.v12i3.2841.

18. T. Sharma and Dr. P. Alagh, "Process and factor analysis in the manufacturing of woven polypropylene packaging textiles," *International Journal of Home Science*, vol. 7, no. 2, 2021, doi: 10.22271/23957476.2021.v7.i2b.1155.

19. H. L. Nguyen et al., "Biorenewable, transparent, and oxygen/moisture barrier nanocellulose/nanochitin-based coating on polypropylene for food packaging applications," *Carbohydrate Polymers*, vol. 271, 2021, doi: 10.1016/j. carbpol.2021.118421.

20. M. Vetter, M. J. F. Healy, and D. W. Lane, "Investigating electric field induced molecular distortions in polypropylene using Raman spectroscopy," *Polymer Testing*, vol. 92, 2020, doi: 10.1016/j.polymertesting.2020.106851.

21. M. Weidenfeller Höfer and F. Schilling, "Thermal and electrical properties of magnetite filled polymers," *Composites Part A: Applied Science and Manufacturing*, vol. 33, no. 8, 2002, doi: 10.1016/S1359-835X(02)00085-4.

22. R. Rujiwarodom, N. Sunganun, W. Busayaporn, and A. Panyanuch, "Improvement of the electrical insulating property of polypropylene by electron beam irradiation," *Kasetsart Journal of - Natural Science*, vol. 48, no. 1, 2014.

23. S. H. P. Bettini et al., "Investigation on the use of coir fiber as alternative reinforcement in polypropylene," *Journal of Applied Polymer Science*, vol. 118, no. 5, 2010, doi: 10.1002/app.32418.

24. S. S. Mir, S. M. N. Hasan, M. J. Hossain, and M. Hasan, "Chemical modification effect on the mechanical properties of coir fiber," *Engineering Journal*, vol. 16, no. 2, 2012, doi: 10.4186/ej.2012.16.2.73.

25. M. Sabri, A. Mukhtar, K. Shahril, A. S. Rohana, and H. Salmah, "Effect of compatibilizer on mechanical properties and water absorption behaviour of coconut fiber filled polypropylene composite," in *Advanced in Materials Research*, 2013, doi: 10.4028/www.scientific.net/AMR.795.313.

26. M. Sabri, F. Hafiz, K. Shahril, S. S. Rohana, and H. Salmah, "Effects of silane coupling agent on mechanical properties and swelling behaviour of coconut fiber filled polypropylene composite," in *Advanced in Materials Research*, 2013, doi: 10.4028/www.scientific.net/AMR.626.657.

27. N. E. Zakaria, I. Ahmad, W. Z. Wan Mohamad, and A. Baharum, "Effects of fiber size on Sansevieria trifasciata/natural rubber/high density polyethylene biocomposites," *Malaysian Journal of Analytical Sciences*, vol. 22, no. 6, 2018, doi: 10.17576/mjas-2018-2206-16.

28. H. T. N. Kuan, M. Y. Tan, Y. Shen, and Y. Mohd, "Mechanical properties of particulate organic natural filler-reinforced polymer composite: A review," *Composites and Advanced Materials*, vol. 30, 2021, doi: 10.1177/26349833211007502.

29. N. E. Zakaria, A. Baharum, and I. Ahmad, "Mechanical properties of chemically modified Sansevieria trifasciata/natural rubber/high density polyethylene (STF/NR/HDPE) composites: Effect of silane coupling agent," in *AIP Conference Proceedings*, 2018, doi: 10.1063/1.5028011.

30. V. Çavuş, "1 Selected-properties of mahogany wood flour filled polypropylene composites: The effect of maleic anhydride-grafted polypropylene (MAPP)," *BioResources*, vol. 15, no. 2, 2020, doi: 10.15376/biores.15.2.2227-2236.

31. Z. Ahmad, H. A. Latif, H. A. H. Shaari, W. M. Aiman, N. I. Izwan, and N. M. A. Wahab, "The addition of silane coupling agent in coconut coir husk/PLA biocomposite: Mechanical and biodegradability studies," in *AIP Conference Proceedings*, 2018, doi: 10.1063/1.5066968.

32. Z. Demjén, B. Pukánszky, and J. Nagy, "Possible coupling reactions of functional silanes and polypropylene," *Polymer (Guildf)*, vol. 40, no. 7, 1999, doi: 10.1016/S0032-3861(98)00396-6.

33. S. Ragunathan et al., "Characterization and properties of PP/NBRv/Kenaf fibre composites with Silane treatment," in *Lecture Notes in Mechanical Engineering*, 2021, doi: 10.1007/978-981-16-0866-7_67.

34. M. N. M. Ansari and H. Ismail, "The effect of silane coupling agent on mechanical properties of feldspar filled polypropylene composites," *Journal of Reinforced Plastics and Composites*, vol. 28, no. 24, 2009, doi: 10.1177/0731684408095197.
35. Q. Wang, Y. Zhang, W. Liang, J. Wang, and Y. Chen, "Effect of silane treatment on mechanical properties and thermal behavior of bamboo fibers reinforced polypropylene composites," *Journal of Engineered Fibers and Fabrics*, vol. 15, 2020, doi: 10.1177/1558925020958195.
36. A. Supri and B. Lim, "Effect of treated and untreated filler loading on the mechanical, morphological, and water absorption properties of water hyacinth fibers-low density polyethylene," *Journal of Physical Science*, vol. 20, no. 2, 2009.

8

Tensile Properties of Polypropylene Composites Reinforced by Bio-Natural Fibre (Sansevieria Leaf Fibre): The Effect of Fibre Loading

S. R. Ahmad, M. F. Hamid, and S. S. M. Saleh

8.1 Introduction

Biodegradable-natural fibre (Bio-NF) can be used as a component of a composite material, where the orientation of fibres has a significant impact on the properties. Natural fibre–reinforced composites are an emerging area in polymer science, and the natural fibres are low-cost fibres with low density and high specific properties [1, 2]. Bio-NF is made from natural fibre that can be obtained from either crops or animals, such as hemp [3], kenaf [4], wool [5], jute [6], bamboo fibre [7], and others [8, 9]. Those materials can be easily disposed of, which leads to their being commonly used as reinforcement in polymer composites. After blending with polymers, waste materials are transformed into engineering materials suitable for the electronics, manufacturing [10–12], automotive [13–16], and aerospace [17, 18] industries today.

Previously, bio-natural fibre was used to blend with thermosets. In the new era of novel materials development, the widespread use of thermoplastics makes the combination of Bio-NF and thermoplastics one of the most interesting research topics [2, 19–21]. Bio-NF thermoplastic composites offer several advantages [22]. Although their low density, high specific strength to density ratio, low abrasiveness, biodegradability, and origin from a renewable resource have all been linked to their lower cost, natural fibres also have other advantages. Natural fibre composites can be blended and moulded with little loss of strength utilizing high-intensity, high-volume production equipment like extrusion and injection moulding. Due to the potential for lignocellulosic degradation, the processing temperature is normally kept below 200 C when manufacturing Bio-NF thermoplastics composites (Bio-NFTP). As a result, thermoplastics like polyethylene, polypropylene, and polystyrene are frequently utilized as the matrix in Bio-NFTP, since they may be treated below the 200 C threshold [23].

Sansevieria leaf is currently a popular or trending indoor plant in Malaysia. It is easy to plant, take care of, and grow with minimum effort. The leaves of Sansevieria are typically arranged in a rosette around the growing point, although some species are distichous. There is great variation in foliage within the genus. The SLF fibres contain hemicellulose (79.7%), cellulose (10.13%), lignin (3.8%), and wax (0.09%). Maximum cellulose content is necessary for effective reinforcing. Hemicellulose has high hydrophilicity. Lignin is very hydrophobic and amorphous [12]. Wax covers the surface of the leaf fibre. The thickness of SLF is 2.07×10^{-5} m, and its density is 0.9 g/cm^3 [24, 25] The mechanical properties of Sansevieria fibre are as follows. The tensile strength and Young's modulus are 348.6 MPa and 15.3 GPa, respectively. Elongation at break was reported to be 2.3% [26]. Sansevieria leaf fibre is one of the strongest plant fibres. The fibre is extracted from the leaves to make coarse fabrics, ropes, and tail-ropes [27]. Therefore, SLF can be very useful as reinforcement fibre by integrating it into thermoplastics.

Polypropylene (PP) is a petroleum-based material. In many aspects, it is similar to polyethylene, especially in solution behaviour and electrical properties. The methyl group improves mechanical properties and thermal resistance, while the chemical resistance decreases. PP composites are widely used as packaging, as insulators in electronic industries, and in automotive parts [16, 28, 29]. It is lightweight; the density of PP is between 0.895 and 0.92 g/cm^3. PP can be recycled and has non-toxic and non-hazardous properties. The relatively low cost of PP makes these composites very significant engineering materials today [30–34].

The addition of SLF to PP is expected to enhance the tensile properties of the polymer composites. It is expected to be applied to packaging, electronic sectors, automotive interior components, aerospace interior part applications, and others. In this study, the effect of fibre loading on the tensile properties of PP-filled SLF composites was investigated. Table 8.1 indicates the mechanical properties of Bio-NF (SLF)-filled PP composites.

8.2 Materials and Methods

8.2.1 Materials

A Sansevieria plant in Kulim (Kedah Darulaman, Malaysia) was used in this study. PP was obtained from Polypropylene Malaysia Sdn Bhd in the form of pallets with a grade of S12232 G112. Table 8.2 and Figure 8.1 show the formulation and sample preparation of PP/SLF composites at different SLF loadings.

TABLE 8.1

Mechanical Properties of Bio-NF (Sansevieria Leaf Fibre)–Filled Polypropylene Composites

Reference	Polymer Composites	Description
Abral and Kennedy [29]	*Sansevieria trifasciata/* Polypropylene	Addition of 2% ST content caused a slight increase of tensile strength in the PP/ST composite.
Shieddieque et al. [35]	*Sansevieria trifasciata/* Polypropylene	Untreated and unidirectional *Sansevieria trifasciata* fibre/polypropylene with a fibre volume fraction of 15% showed an impact resistance of 48.092 kJ/m².
Shieddieque et al. [28]	*Sansevieria trifasciata/* High-performance PP composite	The unidirectional orientation composites with alkali treatment and 15%vol of fibre had a tensile strength of 59.77 MPa and a stiffness of 1.97 GPa.
Aref et al. [36]	*Sansevieria trifasciata/*PP treated with fluorosilane	The addition of fluorosilane to the composites raised the flexural stress (35 MPa) and modulus (3.4 GPa) at 3% fibre content.

TABLE 8.2

Formulation of PP/SLF Composites at Different Fibre Loadings

Material	PP/SLF Composites (wt.%)
Polypropylene (PP)	100, 90, 80, 70, 60
Sansevieria fibre (SLF)	0, 10, 20, 30, 40

8.2.2 Preparation of Sansevieria Leaf Fibre

First, the leaves were dipped and rubbed by hand in clean water to remove the dirt and soil. After that, the leaves were placed on the tray for the drying process. After the leaves were completely dry, they were cut into small pieces by shortening the fibre. Then, the material was blended in a rough blender. After that, the product was again blended using a fine blender to get the tiniest particles. The product was filtered using a sieve to obtain 1-mm short fibres. The preparation process of SLF composites is shown in Figure 8.2.

8.2.3 Mixing Process

The mixing of composites was performed using a Thermo Scientific Haake Rheomix PolyLab QC Lab Mixer (Thermo Fisher Scientific Inc., United States) at a temperature of 180 C and a rotor speed of 50 rpm. PP was discharged into the mixing chamber. It melted completely into the mixer chamber after 5 min. Then, the SL fibre was inserted. The mixing continued for another 5 minutes until the 10-minute mark was reached. The mixing took 10 min for each mix to be completed. The previous steps were repeated five times for each composition. Based on Table 8.2, the composite mixtures were prepared with different loadings of SL fibre.

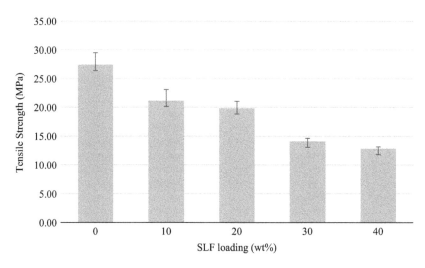

FIGURE 8.1
The tensile test of PP/SLF composites at different SLF loadings

8.2.4 Compression Moulding

The composite mixtures were then compressed in a hydraulic hot press machine (GT 7014) from Dongguan Lixian Instrument Scientific Co., Ltd. Guangdong, China to produce a 1.0-mm sheet of composite using the moulding of a dumb-bell shape to comply with the ASTM D-638 standard. The force was set at 10 kN, and the temperature was at 180 C. The mould was placed on the machine to pre-heat it for 5 minutes. After that, the sample was pressed for 4 min. After completing the process, the sample was removed and blown with air to make it cool faster. The process was completed after it was removed from the mould.

8.2.5 Measurement of Tensile Properties

The tensile test was carried out according to ASTM D412 using a Universal Testing Machine (UTM): Instron Model 5569 from the United States. The testing was done in a standard laboratory atmosphere of 25 C ± 3 C. The tensile testing machine was used with a crosshead speed of 50 mm/minute, and the gauge length was set at 50 mm. The test was carried out on five specimens for each formulation.

8.3 Results and Discussion

Figure 8.2 shows the tensile strength of SLF-filled PP at 0, 10%, 20%, 30%, and 40% of SLF loading. As reported in Zakaria et al. [37], the tensile strength decreased by about 53% due to the size and random orientation of the fibres. The tensile strength of Sansevieria fibre/composites was reduced in trends reported by Sankar et al. [38]. They used various sizes of Sansevieria fibre length in their research: 1 mm, 500 μm, 250 μm, and 125 μm. They state that a weak interface occurred between the hydrophobic matrix (PP with a hydrocarbon structure) and the hydrophilic fibre (containing OH groups) in their composites [39].

As mentioned in Sankar et al. [38], the presence of defects, such as voids, and poor interface bonding between matrix and reinforcement have been identified as the causes of the decreasing tendency in tensile strength as fibre length decreases from 8 to 2 mm. The tensile strength for their Sansevieria composites is in the range of 20 to 30 MPa for a fibre length of 1–3 mm at 20 wt.% of fibre. This is quite similar to the tensile strength near to 19.85 ± 1.2 MPa at 20 wt.% fibre loading in Priyadarsini et al. [14]. Besides, fibre orientation plays a significant role. The tensile strength of random dispersed fibre is lesser than unidirectional fibre orientation [35]. This depends on the compounding process using either hot compression or injection moulding and others.

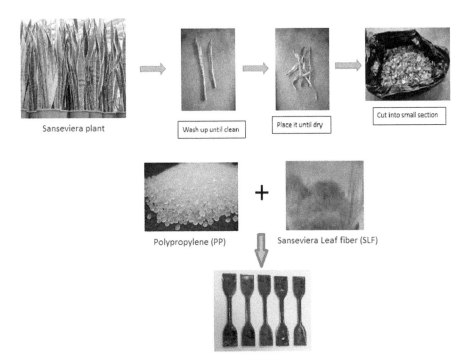

FIGURE 8.2
Biodegradable filler preparation from Sansevieria plant and polypropylene/Bio-NF(SLF) composites

Young's modulus of the composites at different SLF loading is shown in Figure 8.3. It increased gradually when the SLF content was increased (27%). This happened due to the increased stiffness of the composites [40, 41].The stiffness of both random and unidirectional SLF/Hi-Performance PP composites also increased as the fibre content added up to 15% [28]. The high amount of fibre provides a more rigid phase in the composites, leading to increased stiffness [42].

Figure 8.4 indicates the elongation at break of the composites up to 40 wt.% of SLF content. The elongation of the Sansevieria composite decreased as SLF loading increased from 10% to 40%. . This might be due to the weak interface

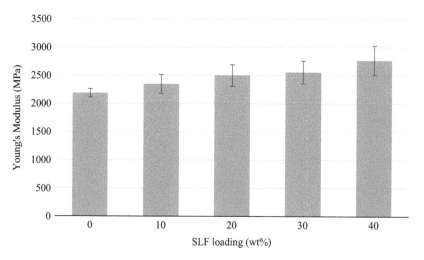

FIGURE 8.3
The Young's modulus of PP/SLF composites at different SLF loadings

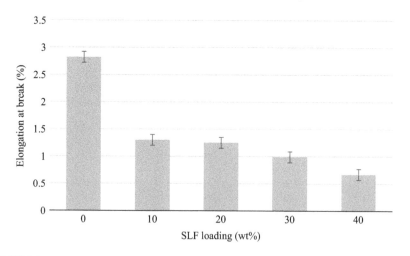

FIGURE 8.4
The elongation at break of PP/SLF composites at different SLF loadings

TABLE 8.3

Summary Results of the Tensile Properties of PP/SLF Composites

	Fibre Loading (wt.%)	PP/SLF Composites
Tensile strength (MPa)	0	27.40 ± 2.1
	10	21.13 ± 2.0
	20	19.85 ± 1.2
	30	14.09 ± 0.5
	40	12.83 ± 0.3
Young's modulus (MPa)	0	2182.95 ± 75
	10	2345.36 ± 169
	20	2498.50 ± 190
	30	2555.61 ± 200
	40	2770.39 ± 262
Elongation at break (%)	0	2.82 ± 0.033
	10	1.30 ± 0.030
	20	1.25 ± 0.025
	30	0.99 ± 0.045
	40	0.67 ± 0.042

between the filler and the PP matrix and its limited deformability. The drop is proportional to the amount of fibre present. According to Supri and Lim [43], the elongation at break is reduced because the fillers have hardened the composites and reduced their ductility. Table 8.3 shows the summary results of the tensile properties of the composites.

8.4 Conclusions

In conclusion, the addition of SLF enhanced the Young's modulus of PP/SLF composites. The maximum Young's modulus of the composites is 2770.39 ± 262 MPa at 40% fibre loading. The addition of SLF improves the stiffness of the composite. However, the tensile strength and elongation at break were decreased as the SLF loading increased. The composites experienced weak interfacial reaction due to fibre size, orientation, and shape. Hence, they were easy to fracture when the amount of SLF increased. The PP/SLF composites are expected to be used in packaging, electronics, and automotive applications.

Acknowledgements

The collaboration in the development of materials engineering for automotive use is acknowledged by the authors with gratitude to the Universiti

Kuala Lumpur Malaysian Spanish Institute, Kulim Hi-Tech Park, Kulim, the University of Malaysia Perlis, Perlis, and TAJ International College, Ipoh, Perak Malaysia.

References

1. X. Zhao et al., "Recycling of natural fiber composites: Challenges and opportunities," *Resources, Conservation and Recycling*, vol. 177. 2022. doi: 10.1016/j.resconrec.2021.105962.
2. A. Karimah et al., "A review on natural fibers for development of eco-friendly bio-composite: Characteristics, and utilizations," *Journal of Materials Research and Technology*, vol. 13. 2021. doi: 10.1016/j.jmrt.2021.06.014.
3. L. Stelea et al., "Characterisation of hemp fibres reinforced composites using thermoplastic polymers as matrices," *Polymers (Basel)*, vol. 14, no. 3, 2022. doi: 10.3390/polym14030481.
4. M. M. Owen, E. O. Achukwu, A. Z. Romli, and H. Md. Akil, "Recent advances on improving the mechanical and thermal properties of kenaf fibers/engineering thermoplastic composites using novel coating techniques: A review," *Composite Interfaces*, vol. 30, no. 8. 2023. doi: 10.1080/09276440.2023.2179238.
5. J. Tusnim, N. S. Jenifar, and M. Hasan, "Effect of chemical treatment of jute fiber on thermo-mechanical properties of jute and sheep wool fiber reinforced hybrid polypropylene composites," *Journal of Thermoplastic Composite Materials*, vol. 35, no. 11, 2022. doi: 10.1177/0892705720944220.
6. M. M. A. Sayeed et al., "Assessing mechanical properties of jute, kenaf, and pineapple leaf fiber-reinforced polypropylene composites: Experiment and modelling," *Polymers (Basel)*, vol. 15, no. 4, 2023. doi: 10.3390/polym15040830.
7. B. J. Wang and W. Bin Young, "The natural fiber reinforced thermoplastic composite made of woven bamboo fiber and polypropylene," *Fibers and Polymers*, vol. 23, no. 1, 2022. doi: 10.1007/s12221-021-0982-1.
8. H. Awais, Y. Nawab, A. Amjad, A. Anjang, H. Md. Akil, and M. S. Zainol Abidin, "Environmental benign natural fibre reinforced thermoplastic composites: A review," *Composites Part C: Open Access*, vol. 4, 2021. doi: 10.1016/j.jcomc.2020.100082.
9. A. Irshad and S. N. K., "Mechanical and thermal properties of hybrid composites reinforcing with natural fibres-a review," 2022.
10. J. Jaafar, J. P. Siregar, S. M. Salleh, M. H. Mohd. Hamdan, T. Cionita, and T. Rihayat, "Important considerations in manufacturing of natural fiber composites: A review," *International Journal of Precision Engineering and Manufacturing - Green Technology*, vol. 6, no. 3, 2019. doi: 10.1007/s40684-019-00097-2.
11. S. S. W. Kusuma, H. C. Saputra, and I. Widiastuti, "Manufacturing of natural fiber-reinforced recycled polymer-a systematic literature review," in *IOP Conference Series: Earth and Environmental Science*, 2021. doi: 10.1088/1742-6596/1808/1/012005.
12. S. Senthil Kumaran, K. Srinivasan, M. Ponmariappan, S. Yashwhanth, S. Akshay, and Y. C. Hu, "Study of raw and chemically treated Sansevieria ehrenbergii fibers for brake pad application," *Materials Research Express*, vol. 7, no. 5, 2020. doi: 10.1088/2053-1591/ab8f48.

13. K. Hariprasad, K. Ravichandran, V. Jayaseelan, and T. Muthuramalingam, "Acoustic and mechanical characterisation of polypropylene composites reinforced by natural fibres for automotive applications," *Journal of Materials Research and Technology*, vol. 9, no. 6, 2020. doi: 10.1016/j.jmrt.2020.09.112.

14. M. Priyadarsini, T. Biswal, and S. K. Acharya, "Study of mechanical properties of reinforced polypropylene (PP)/nettle fibers biocomposites and its application in automobile industry," *Materials Today: Proceedings*, vol. 74, 2023. doi: 10.1016/j.matpr.2022.11.349.

15. V. Naik, M. Kumar, and V. Kaup, "A review on natural fiber composite materials in automotive applications," *Engineered Science*, vol. 18, 2022. doi: 10.30919/es8d589.

16. H. Hadiji et al., "Damping analysis of nonwoven natural fibre-reinforced polypropylene composites used in automotive interior parts," *Polymer Testing*, vol. 89, 2020. doi: 10.1016/j.polymertesting.2020.106692.

17. S. Brischetto, "Analysis of natural fibre composites for aerospace structures," *Aircraft Engineering and Aerospace Technology*, vol. 90, no. 9, 2018. doi: 10.1108/AEAT-06-2017-0152.

18. M. R. Mansor, A. H. Nurfaizey, N. Tamaldin, and M. N. A. Nordin, "Natural fiber polymer composites: Utilization in aerospace engineering," in *Biomass, Biopolymer-Based Materials, and Bioenergy: Construction, Biomedical, and other Industrial Applications*, 2019. doi: 10.1016/B978-0-08-102426-3.00011-4.

19. M. Rouway et al., "Prediction of mechanical performance of natural fibers polypropylene composites: A comparison study," in *IOP Conference Series: Materials Science and Engineering*, 2020. doi: 10.1088/1757-899X/948/1/012031.

20. A. Mahesh, B. M. Rudresh, and H. N. Reddappa, "Potential of natural fibers in the modification of mechanical behavior of polypropylene hybrid composites," *Materials Today: Proceedings*, vol. 54, 2022. doi: 10.1016/j.matpr.2021.08.195.

21. Y. G. Thyavihalli Girijappa, S. Mavinkere Rangappa, J. Parameswaranpillai, and S. Siengchin, "Natural fibers as sustainable and renewable resource for development of eco-friendly composites: A comprehensive review," *Frontiers in Materials*, vol. 6, 2019. doi: 10.3389/fmats.2019.00226.

22. M. Li et al., "Recent advancements of plant-based natural fiber–reinforced composites and their applications," *Composites Part B: Engineering*, vol. 200, 2020. doi: 10.1016/j.compositesb.2020.108254.

23. Ankit and M. L. Rinawa, "Sustainable natural bio-composites and its applications," in *Lecture Notes in Mechanical Engineering*, 2021. doi: 10.1007/978-981-16-3033-0_41.

24. N. Balaji, S. Balasubramani, T. Ramakrishnan, and Y. Sureshbabu, "Experimental investigation of chemical and tensile properties of Sansevieria cylindrica fiber composites," in *Materials Science Forum*, 2020. doi: 10.4028/www.scientific.net/MSF.979.58.

25. B. Subramaniam et al., "Investigation of mechanical properties of Sansevieria cylindrica Fiber/polyester composites," *Advances in Materials Science and Engineering*, vol. 2022, 2022. doi: 10.1155/2022/2180614.

26. S. Rwawiire and B. Tomkova, "Morphological, thermal, and mechanical characterization of Sansevieria trifasciata Fibers," *Journal of Natural Fibers*, vol. 12, no. 3, 2015. doi: 10.1080/15440478.2014.914006.

27. A. G. Adeniyi, S. A. Adeoye, and J. O. Ighalo, "Sansevieria trifasciata fibre and composites: A review of recent developments," *International Polymer Processing*, vol. 35, no. 4, 2020. doi: 10.3139/217.3914.

28. D. Shieddieque, M. Mardiyati, R. Suratman, B. Widyanto, "Preparation and characterization of sansevieria trifasciata fiber/high-impact polypropylene and Sansevieria trifasciata fiber/vinyl ester biocomposites for automotive applications," *International Journal of Technology*, vol. 12, no. 3, 2021. doi: 10.14716/ijtech.v12i3.2841.

29. H. Abral and E. Kenedy, "Thermal degradation and tensile strength of sansevieria trifasciata-polypropylene composites," in *IOP Conference Series: Materials Science and Engineering*, 2015. doi: 10.1088/1757-899X/87/1/012011.

30. K. S. Chun, S. Husseinsyah, and C. M. Yeng, "Torque rheological properties of polypropylene/cocoa pod husk composites," *Journal of Thermoplastic Composite Materials*, vol. 30, no. 9, 2017. doi: 10.1177/0892705715618743.

31. K. Ben Hamou, H. Kaddami, F. Elisabete, and F. Erchiqui, "Synergistic association of wood /hemp fibers reinforcements on mechanical, physical and thermal properties of polypropylene-based hybrid composites," *Industrial Crops and Products*, vol. 192, 2023. doi: 10.1016/j.indcrop.2022.116052.

32. L. del Pilar Fajardo Cabrera, R. M. C. Santana de Lima, and C. D. C. Rodríguez, "Influence of coupling agent in mechanical, physical and thermal properties of polypropylene/bamboo fiber composites: Under natural outdoor aging," *Polymers (Basel)*, vol. 12, no. 4, 2020. doi: 10.3390/POLYM12040929.

33. K. S. Chun and S. Husseinsyah, "Agrowaste-based composites from cocoa pod husk and polypropylene: Effect of filler content and chemical treatment," *Journal of Thermoplastic Composite Materials*, vol. 29, no. 10, 2016. doi: 10.1177/0892705714563125.

34. A. K. Maurya and G. Manik, "Advances towards development of industrially relevant short natural fiber reinforced and hybridized polypropylene composites for various industrial applications: A review," *Journal of Polymer Research*, vol. 30, no. 1, 2023. doi: 10.1007/s10965-022-03413-8.

35. D. Shieddieque, M. Mardiyati, R. Suratman, B. Widyanto, "The effect of alkaline treatment and fiber orientation on impact resistant of bio-composites Sansevieria trifasciata fiber/polypropylene as automotive components material," in *AIP Conference Proceedings*, 2018. doi: 10.1063/1.5030263.

36. Y. M. Aref and A. Baharum, "Effect of fibre treatment using fluorosilane on Sansevieria trifasciata /Polypropylene composite," in *AIP Conference Proceedings*, 2018. doi: 10.1063/1.5028021.

37. N. E. Zakaria, A. Baharum, and I. Ahmad, "Mechanical properties of chemically modified Sansevieria trifasciata/natural rubber/high density polyethylene (STF/NR/HDPE) composites: Effect of silane coupling agent," in *AIP Conference Proceedings*, 2018. doi: 10.1063/1.5028011.

38. P. H. Sankar, Y. V. M. Reddy, K. H. Reddy, M. A. Kumar, and A. Ramesh, "The effect of fiber length on tensile properties of polyester resin composites reinforced by the fibers of *Sansevieria trifasciata*," *International Letters of Natural Sciences*, vol. 8, 2014. doi: 10.18052/www.scipress.com/ilns.8.7.

39. N. E. Zakaria, I. Ahmad, W. Z. Wan Mohamad, and A. Baharum, "Effects of fibre size on Sansevieria trifasciata/natural rubber/high density polyethylene biocomposites," *Malaysian Journal of Analytical Sciences*, vol. 22, no. 6, 2018. doi: 10.17576/mjas-2018-2206-16.

40. M. Sabri, F. Hafiz, K. Shahril, S. S. Rohana, and H. Salmah, "Effects of silane coupling agent on mechanical properties and swelling behaviour of coconut fiber filled polypropylene composite," in *Advanced in Materials Research*, 2013. doi: 10.4028/www.scientific.net/AMR.626.657.

41. M. Sabri, A. Mukhtar, K. Shahril, A. S. Rohana, and H. Salmah, "Effect of compatibilizer on mechanical properties and water absorption behaviour of coconut fiber filled polypropylene composite," in *Advanced in Materials Research*, 2013. doi: 10.4028/www.scientific.net/AMR.795.313.

42. M. A. I. Adnan, F. A. Zamri, M. S. Mohamad Sidik, and S. R. Ahmad, "Effect of maleic anhydride polypropylene on the properties of spear grass fiber in reinforced polypropylene and ethylene propylene diene monomer composites," in *Advanced Structured Materials*, 2020. doi: 10.1007/978-3-030-46036-5_19.

43. A. Supri and B. Lim, "Effect of treated and untreated filler loading on the mechanical, morphological, and water absorption properties of water hyacinth fibers-low density polyethylene," *Journal of Physical Science*, vol. 20, no. 2, 2009.

9

Sustainable Product Development Using Renewable Composites for Marine Engineering Applications

Yiow Ru Vern, Muhd Ridzuan bin Mansor, and
Mohd Adrinata bin Shaharuzaman

9.1 Introduction to Systematic Product Development

Product development is an essential process in the creation of a new product or in the improvement of an existing product already in the market. Product development is systematically approached in many applications, involving various stakeholders, including the customer, manufacturer, legislator, supplier, etc. Among the advantages of systematic product development are that it enables the product to be produced at a lower cost, at a faster time to market, and at a higher quality compared with competitors, which in turn, will capture a larger market share, gain high customer satisfaction and acceptance, and provide profit to the manufacturer (Sapuan, Salit, and Mansor 2014).

Generally, the overall product development process involves six main stages, which are identifying market need, creating product design specifications, developing the conceptual design of the product, producing a detailed design of the product, manufacturing the product to its targeted profile, and finally, selling the completed product to the targeted customer/putting the product into the market, as shown in Figure 9.1. Furthermore, the current execution of these stages is based on the concurrent engineering method, whereby all stakeholders (customer, manufacturer, designer, supplier, marketer, etc.) are jointly involved along the whole product development process. Synergetic collaboration is the key to the concurrent engineering concept, which ensures that all important requirements and constraints put forth by all stakeholders are identified and included during the development process. To help execute this product development process based on concurrent engineering methods in a systematic and scientific manner, many modern tools and techniques have been developed for each stage. Among them are

DOI: 10.1201/9781003408215-9

quality function deployment (QFD), Kano model, product design specification (PDS), Theory of Inventive Problem Solving (TRIZ), biomimetics, computer aided design (CAD), computer aided analysis (CAA), computer aided manufacturing (CAM), and additive manufacturing (AM), i.e., fused deposition modelling (FDM), selective laser sintering (SLS), and stereolithography (SLA). These stages are represented in Figure 9.2.

The start of the product development process involves first, the identification of market needs. This process ensures that the customer needs and

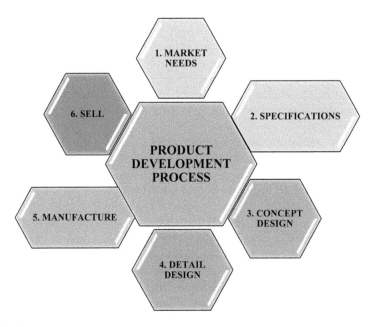

FIGURE 9.1
Overall product development process based on Pugh design core model.

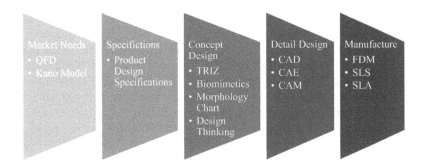

FIGURE 9.2
Example of modern tools and techniques in concurrent engineering for systematic product development process.

requirements are identified and incorporated into the product to be developed. The market analysis needs to be scrutinized to capture the true need of the customer, so that the final developed product fully satisfies the targeted market, thereby gaining complete customer acceptance. This important concept is also termed the user-centred approach (Kwon, Choi, and Hwang 2021). Information on the market gap will also be captured during the market need analysis, which will be used to help in the strategic positioning of the product to ensure a profitable outcome.

The second stage in the systematic product development process is creating a PDS. Kent (2016) essentially defined the PDS as what the product is required to provide, such that it is an answer to what the customer wants the product to achieve. The PDS serves as the main reference for the product to be developed with descriptions regarding technical details and constraints. Pugh (1991) listed 39 key parameters that cover the elements of the related PDS, including aesthetics, ergonomics, materials, standards, packaging, weight, maintenance, processing, size, product performance, and product cost.

The next stage of systematic product development is the third, which centres around the incarnation of the conceptual design. Compared with the other product development stages, this stage, where the conception of user/ customer requirements becomes its first realistically aesthetic workable form, demands the greatest amount of creative thinking (Chen and Terken 2023). While closely adhering to the PDS, designers need to painstakingly ensure that the idea generation meets customer requirements and all other design specifications. While this process requires significant creative thinking, one should not forget that the cost of execution must be kept to a minimum. In recent years, the element of sustainability has emerged as an important factor in product development; hence, it is now taken as critical that this be considered and integrated during the conceptual design stage of the product (Sapuan and Mansor 2021). There are two stages in conceptual design, i.e., idea generation and idea selection. The purpose of idea generation is to come out with as many design solutions as possible that meet the PDS, and the idea selection is performed to select the best design based on the design criteria identified at the earlier stage (Salwa et al. 2021). Some of the methods for performing the idea selection systematically and scientifically are the Technique for Order of Preference by Similarity to Ideal Solution (TOPSIS) (Yiow and Mansor 2022) and the Analytic Hierarchy Process (AHP) (Mastura et al. 2022). It is noted that on the one hand, material selection is regarded as part of the conceptual design stage, while on the other hand, this same material selection process is also interpreted as part of the embodiment stage in the product development process. Sapuan et al. (2017) concluded that the boundary between conceptual design and embodiment design is vague and cannot be differentiated easily due to the overlapping nature of both activities.

Next up in the product development process is the detail design. In this stage, dimensions, tolerances, and material information need to be defined in their entirety and precision in preparation for the determined manufacturing (Poli 2001). The outcome of the detail design includes, among others, product details, component drawings, and bill of materials (BOM). Furthermore, detailed engineering design analysis is also performed with the support of computer aided engineering (CAE) software to ensure conformance of the product with design specifications. Simulation studies around the extremes that the product is expected to endure can give insight into its full capabilities. Some of these studies include structural strength under static and dynamic loads, thermal variations, and vibration analysis. Information on the manufacturing and where necessary, joining processes leading to the final product is considered and finalized in agreement with the selected materials.

Then comes the fifth stage, where manufacturing takes place based on the best selected design and material. With a wide range of manufacturing methods, the best method would have already been singled out in the previous stage, taking into account the quality and quantity of the product. Cost and shortest lead time are the prime factors in determining the choice of the manufacturing method. One popular approach that has been used in recent times is AM. The more common term used among lay people to describe this technology is 3D printing, whereby a three-dimensional object is built for a CAD file through successive layerings of material, one on top of another, until a complete object is formed (Zhang and Liou, 2021). AM technology brings with it the freedom of fabricating complex geometry and the capability of mass customization. With modern methods, such as strip-cladding and multi-jet fusion, a speed of production comparable to conventional mass manufacturing is achievable (Khorasani et al. 2022). The product development process is complete once it is sold to the customer, otherwise phrased as putting the finished product out into the market.

9.2 Idea Generation Method in Systematic Product Development

9.2.1 TRIZ (Theory of Inventive Problem Solving)

TRIZ is the Russian acronym for "Teoriya Resheniya Izobretatelskikh Zadatch", which, when translated to English, would give us "Theory of Inventive Problem Solving". TRIZ is a concept of ideation, or problem solving, introduced by Genrikh Saulovich Altshuller, a Russian scientist who after studying numerous patents, noted common trends or patterns in which ideas are generated for designs of new or improvement of existing products (Lim et al. 2018; Mann 2001). This inspired him to introduce the various TRIZ tools, one of which is the 40 Inventive Principles, to the engineering community and society as a whole. The 40 Inventive Principles lists inventive methods used in approaching design "roadblocks" based on over 40,000 patents studied, which later were expanded to include over 200,000 in total (Childs 2019). In recent times, this database has increased to over two million patents studied.

To utilize the TRIZ method, a representation of a specific problem or idea is first generalized. This step aims to make it a more general problem so that the recommended solution in TRIZ can be looked up. From here, the recommended solution is applied to suit the specific problem at hand. A visual representation of the concept is shown in Figure 9.3.

When using the TRIZ method, it is commonly in reference to the matrix, more commonly known as the TRIZ Contradiction Matrix (Livotov 2018). This matrix is a 39 by 39 grid representing engineering parameters that have been identified as being involved in the problem-solving process. In any engineering problem, often the improvement of any parameter would take place at the expense, or sacrifice, of another. Consider the following statement: "to increase the strength of a wooden chair, the size of the legs and structure is increased, which then would make the chair heavier and bulkier". An everyday object like a wooden chair can be made stronger by simply making it out of thicker pieces of wood. However, this will also make it heavier and bulkier, making it harder to move around and taking up more space. Often, chairs are made to be moved around and stored easily. This example of the wooden chair shows the contradiction of sacrificing weight and size in the interest of gaining improved strength. In today's world, this matter is easily solved by using different alternative materials, as can be seen in plastic chairs or those made with hollow steel tubes. But if, specifically, it was in the interest of the builder to use only wood, then perhaps using the TRIZ Contradiction Matrix would provide some solution to the design of the chair. By determining the improving parameter, which would be inserted in the first column of the x-axis, and the worsening parameter, which would be inserted in the first row of the y-axis, the recommended general solutions will be shown in the intersecting grid. The recommended general solutions are actually from the TRIZ

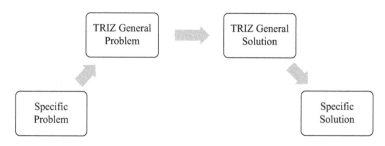

FIGURE 9.3
The generalized TRIZ process.

40 Inventive Principles, where the most commonly used methods from Altshuller's studies are recommended (Childs 2019). A small section of the Contradiction Matrix is depicted in Figure 9.4 for clarity.

As the TRIZ philosophy grew with time, other tools came to fruition. Among the notable concepts are the Substance Field, 76 Standard Inventive Solutions, and ARIZ (Algorithm of Inventive Problem Solving) (San, Jin, and Li 2009). With increased interest from researchers in various fields of expertise, other concepts have even been assimilated into the TRIZ method to develop other entities such as TRIZ+, CTRIZ, BioTRIZ, and so on (Sheu, Chiu, and Cayard 2020).

In innovation, many of the improvements of engineering systems do not simply come from random thoughts or ideas. Instead, there is a path of evolution that tends to be followed, and in TRIZ, this can be summed up by the S-Curve (Zhang, Liu, and Zhang 2006). The S-Curve is a graph plot of value/performance over time and tends to follow the shape of the letter "S". Depending on the specific point on the curve, the "maturity" of the technology can be estimated. It has been mentioned that the end of the curve signifies the full maturity of the technology, where no further improvement can be made. At the end of one S-Curve would be the beginning of a new S-Curve, which signifies the birth of a new technology. In one report, studies on the state of lubricant technology were carried out with the aid of

Which properties of the system change for the worse ?		1	2	3	4	5	6	7
Which properties are to be improved?		Weight of moving object	Weight of immobile object	Length of moving object	Length of immobile object	Area of moving object	Area of immobile object	Volume of moving object
1	Weight of moving object	+		15, 8, 29,34		29, 17, 38, 34		29, 2, 40, 28
2	Weight of immobile object		+		10, 1, 29, 35		35, 30, 13, 2	
3	Length of moving object	8, 15, 29, 34		+		15, 17, 4		7, 17, 4, 35
4	Length of immobile object		35, 28, 40, 29		+		17, 7, 10, 40	
5	Area of moving object	2, 17, 29, 4		14, 15, 18, 4		+		7, 14, 17, 4

FIGURE 9.4
A section of the TRIZ Contradiction Matrix showing the improving parameters on the left and the worsening parameters at the top. (From San, Y.T. et al., *TRIZ: Systematic Innovation in Manufacturing*, Firstfruits Publishing, Selangor, 2009.)

the S-Curve (Morina, Liskiewicz, and Neville 2006). It was mentioned that other TRIZ tools were also applied, among them the concepts of Ideality and Function/Attribute. The researchers mentioned that the state of tribology is approaching its maturity and suggested that new methods be explored, proposing that inspiration be taken from nature itself.

By employing the Su-Field (Substance-Field) method, new designs of hydraulic reciprocating seals were ideated (Zhang, Liu, and Zhang 2006). Through the exercise of identifying the key substances in the sealing mechanism together with its field, as well as the useful and harmful effects, a Su-Field diagram was envisaged. Improvements were then made based on the diagram. This method allows the engineer to "see" clearly the effects, be they useful or harmful, while finding the solutions from the 76 Standard Inventive Solutions (San, Jin, and Li 2009).

An exercise involving two key philosophies of TRIZ, namely, function analysis and resource analysis, was performed in the subversion analysis in identifying the causes of cracks in marine diesel engine injector fuel nozzles (Chybowski, Bejger, and Gawdzińska 2018). By drawing up a function diagram of the fuel injector, interrelations between key components were identified. Together with a list of potential sources of failure, or otherwise, stated as a cause-and-effect analysis, the factors that led to the cracking on the fuel injector nozzle were identified.

Many reports have demonstrated the use of the Contradiction Matrix. Effective use of the matrix involves first, the generation of the contradiction statement, which can be done quite easily using the key words "if", "then," and "but" (Ishak, Sivakumar, and Mansor 2018). Through the contradiction statement, references can be made to the matrix to find possible solutions. In some cases, the solutions themselves are easily applied to the problem at hand (Sapuan, Salit, and Mansor 2016). Others, Where more complex decision making is warranted, other applications are brought to play. One report on the design of a car handbrake involved the use of a morphological chart and AHP to help select the best option of the solutions suggested in the matrix. In another example, the solutions provided were then applied with biomimetics to find options in the design of a car side door impact beam (Shaharuzaman et al. 2020). Here, to determine the best design, the VIKOR (VIekriterijumsko KOmpromisno Rangiranje) method was employed as well.

The effectiveness of TRIZ was studied in a classroom setting where students were introduced to the problem-solving method and then given exercises to try out what had been learnt (Hernandez, Schmidt, and Okudan 2013). The students who were introduced to TRIZ tended to provide more ideas for the case studies given to them compared with a control group who were not exposed to TRIZ. In another study, the synergy of TRIZ with other classical forms of problem solving proved effective in increasing the number of solutions generated among participants (Schöfer et al. 2015).

9.3 Biomimetics Method

Biomimetics can generally be described as a design ideation method that involves searching for answers from the natural world. Fayemi et al. (2014) describes the process as a form of cohesion between biology and technology with the aim of problem solving through careful observation of nature's ways. In their research, it was noted that biomimetics is a subsection of a broader concept aptly termed bio-inspiration. In bio-inspiration, two other forms of design ideation are described, namely, biomimicry and bionics. Biomimicry is defined as observing nature more from an organizational perspective, relating to organisms and their interactions. Bionics, on the other hand, is focused on the technical aspect of replicating, modifying, or improving biological functions through electronic and/or mechanical means.

A brief search on the origins of biomimetics brings us to Otto H. Schmitt, who first coined the term "biomimetics" (Vincent et al. 2006; Lepora, Verschure, and Prescott 2013). In the early years, the use of the term "biomimetics" referred to the broad generalization of replication or referencing biology in engineering. Regardless of the variation in philosophy, the wonders of nature's intricate methods continue to inspire humans in various forms of technological development.

Biomimetics has been the basis for some very clever ideas over the years, one of which is the Velcro strap. George de Mestral was walking his dog one day when he noticed the dried burdock seeds stuck to his clothes (Stephens 2007; Bhushan 2009). Upon further scrutiny, he found that the seeds had fine hairs, which functioned as hooks that allowed them to easily attach to the coats of passing creatures, transporting them to other places for germination. Using this same principle, the Velcro strip can be found in fasteners everywhere, from clothing to equipment.

The Shinkansen bullet train in Japan has a nose inspired by the profile of the kingfisher's beak (Hood 2006; Fayemi et al. 2014). Noticing how minor was the splash created in the water as the kingfisher dove in to catch its prey, Eiji Nakatsu, the Director of the Technical Development and Test Operation Department of JR-West, found inspiration for the design of the bullet train nose. By incorporating such a design, the greatly improved aerodynamics eliminated the loud noise generated when the train exited tunnels at speeds exceeding 300 km/h.

Needles used in syringes are an example of biomimetics in the medical field. The design of the needle is taken from the profile of the mosquito's proboscis (Izumi et al. 2011; Kim, Park, and Prausnitz 2012). The needle is made from biodegradable polymer, making the experience of receiving injections more pleasant for patients as well as making it more biodegradable compared with the older design of needles, which were made from steel.

Modern paints applied to the underside of ships are made to be hydrophobic, making it difficult for marine organisms to attach to the ship's hull (Han et al. 2021). This technology was inspired by the lotus leaf and is a more environmentally friendly alternative to the biocide-based paints used in the past.

Janine M Benyus created the website AskNature.org as a common site for gathering discoveries of nature's strategies (Hwang et al. 2015). The information on the site continues to grow with contributions from discoveries across the globe, and it has been dubbed the best cataloguing of natural solutions known to mankind (Lim et al. 2018). The referencing for this project was also done using resources from this website.

9.4 Other Methods

Besides TRIZ and biomimetics, other methods have also been developed with the purpose of stimulating growth in idea generation for product development. One of a wide variety is the classical method known as the Morphology Chart. The Morphological Chart is a visually based idea generation method, whereby ideas are generated by combining different functional alternatives for the product. Mansor et al. (2014) implemented the Morphology Chart in developing a new conceptual design for a manually operated parking brake lever. Azammi et al. (2018) also applied a similar Morphological Chart method during the conceptual design generation phase of developing a new automotive engine rubber mounting using natural fibre composite materials. An example of the Morphological Chart is shown in Figure 9.5.

TRIZ solution principles and design strategy	Design features	Solutions				
		1	2	3	4	5
#1. Segmentation i) Product the component in different sections and the sections should be asymmetric as possible to simplify and ease the design and manufacturing process	Cross-section profile	Double-T	Square	U-type		
	No. of sections	2	3	4		
ii) Joining the sections using pin-and-boss method for easy assembly and disassembly process	Sectioning type	Symmetrical	Non-symmetrical			
	Part section assembly method	Screw-boss	Circular pin-boss	Square pin-boss	Adhesive	
#3. Local quality i) Vary the thickness of the component according to the stress concentration value. Thicker component at higher stress location points.	Body type	Solid	Shell			
ii) Brake lever body casing designed with ribs to reinforce and strengthened the structure as well as with pin-and-boss features to provide quick and easy assembly method. Joint different functioning feature together to the same component	Ribbing pattern	I	V	X	I+V	I+X

Note: *Example for Concept Design 1 = Square + 2-sections + symmetrical + circular pin-boss + shell + 1-rib*

FIGURE 9.5
Example of Morphological Chart for developing new conceptual design of automotive parking brake lever. (From Mansor, M.R. et al., *Materials and Design*, 54, 473–482, 2014.)

FIGURE 9.6
Design Thinking methodology (Balakrishnan 2022)

Another exciting method for performing product development is called Design Thinking. This method consists of a five-step process in product development (Balakrishnan 2022). This method is also widely applied for problem solving in various applications, both technical and non-technical (Boyle et al. 2022). The Design Thinking method has some similarity with the conventional product development process flow, but its uniqueness lies at its forefront step/process, which emphasizes empathy. In this first step, the problem to be solved is based on the actual understanding of the user's needs, so that the actual user requirements are incorporated into the development process. In this way, the final solution developed will fully satisfy the customer requirements, eliminating the rejection of the solution by the targeted customer/end user, and hence allowing full user acceptance for the solution provided in the form of the final product developed. The sequential processes are defining the actual problem to solve, ideating the potential solutions to solve the problem, creating a prototype of the solution, and finally, testing the prototype to gauge its effectiveness in solving the problem and user acceptance. The Design Thinking method is also often called User-Centered Design (Parizi et al. 2022). Figure 9.6 shows the overall Design Thinking methodology applied in a product development process.

9.5 Systematic Product Development Using Integrated TRIZ and Biomimetics Methods

9.5.1 Introduction

The motivation behind this product development project is to come up with a new design for the crosshead bearing of a two-stroke marine diesel engine. The crosshead bearing is a journal bearing comprising a steel-backed piece with tin-based white metal overlay (Hutchings and Shipway 2017). The crosshead operates as an oscillating assembly similar to revolute joints and is supported by a hydrodynamic oil film while in operation (Wakuri, Hamatake, and Soejima 1982). The crosshead bearing is tasked with providing support for the entire piston assembly as well as the firing forces during the running of the engine (Li et al. 2021). At rest, the bearing supports the piston, piston rod, and components, which weigh around 1460 kg for a 60-cm bore piston as found on a MAN B&W S60MC ("Volume II: Maintenance" 2003). With ever-increasing pressure from environmental groups, alternatives to conventional manufacturing materials such as steel are being considered (Thakur and Thakur 2014). One such alternative is the use of plant fibres with polymer resins, more categorically known as natural fibre composites (NFCs) (Kim 2012).

9.5.2 Initial Ideation with TRIZ

In this project, the steps involved in the ideation of a new marine diesel engine crosshead bearing design are represented by the flow chart in Figure 9.7. The first step is to define the problem based on the design intent. This helps to clarify that the focus is to rethink the geometry of the journal bearing. Next is to decide the improving and worsening parameters, respectively. These

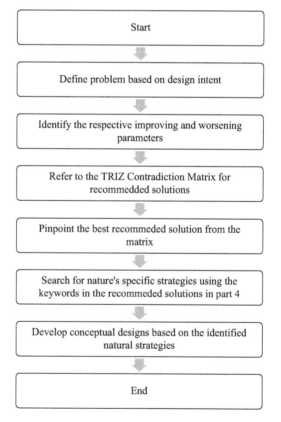

FIGURE 9.7
Design ideation flow process.

two parameters are then used to find the recommendations of the TRIZ 40 Inventive Principles from the TRIZ Contradiction Matrix. Upon evaluation of the best-suited solution, the second step is where biomimetics is put forth by referring to the AskNature.org database. Keywords related to the selected solution in the first step are used to search for suitable strategies employed by nature in solving similar challenges. From here, new geometrical designs of the crosshead bearing are envisaged, taking inspiration from the strategies implemented by nature.

To recreate the crosshead bearing design with the intention of substituting steel with NFC, careful considerations have to be made. NFC possesses lower mechanical properties compared with steel, and this is a key factor (Mastura et al. 2017). To rethink the design of a journal bearing such as the crosshead bearing using NFC requires the employment of the TRIZ method. The contradiction in this case can be seen in the fact that steel is replaced with NFC. The crosshead bearing, in this case, is viewed as a system, with parameters that can be improved or worsened depending on the material selection. Deliberating on the idea allowed the narrowing of parameters to just two that could be used on the TRIZ Contradiction Matrix. The improving and worsening parameters were "durability of moving object" and "strength", respectively. To elaborate, if NFC was used as the material for the crosshead bearing, it would be of a lower "strength". However, it is necessary that the bearing be able to perform its duty of withstanding deformation during loading, hence the necessity to improve its "durability". The statement used to describe this contradiction would be: "By using material of a lower mechanical strength, the durability of the bearing must still be maintained".

By entering the improving and worsening parameters into the TRIZ Contradiction Matrix as "Durability of moving object" and "Strength", respectively, the suggested solutions of "#3 Local quality", "#10 Preliminary action", and "#27 Cheap short living objects" were recommended (Table 9.1). In "#3 Local quality", the description given involves the change of structure, either internal or external, to suit the operation that is required. This means that for the bearing to operate, structural changes have to be made to accommodate the use of NFC. "#10 Preliminary action" involves an action to take place just before it is needed. This suggested solution would mean that there is an additional step in the operations that must be

TABLE 9.1

Contradicting Parameters and Solutions from Contradiction Matrix

	Engineering Parameters		Solutions from Matrix
+	15: Durability of moving object	#3:	Local quality
-	14: Strength	#10:	Preliminary action
		#27:	Cheap short living objects

performed just before the crosshead bearing is loaded. Finally, the solution "#27 Cheap short living objects" suggests that cheap disposable material or components be used for the crosshead bearing. The details of the descriptions of the suggested solutions from the TRIZ 40 Inventive Principles are listed in Table 9.2.

The selection of the best from the three suggested solutions had to be made based on the main purpose of the project, that is, the use of NFC, because it is a more environmentally friendly resource compared with steel. Hence, the choice of "#27 Cheap short living objects" is not desirable, as it suggests that the crosshead bearing material be made cheaper and easily disposable. This option would increase material consumption over time and would also lead to increased wastage. Additionally, the replacement of crosshead bearings in two-stroke marine diesel engines is a tedious and time-consuming job, something that one of the authors had the opportunity to experience in the past. "#10 Preliminary action" is also not suitable in this context, as any additional action in the crosshead would complicate the intricate workings of the engine itself. For a change of bearing material, altering the internal engine mechanics would be unnecessary and might even add complication to the mechanisms. Moving to "#3 Local quality", the idea of a new bearing structure seemed most plausible compared with the previous two suggested solutions; however, clear visualization could not be materialized. At this juncture in the project, reference was then made to the natural world for readily available solutions.

TABLE 9.2

Brief Descriptions of the Meaning of the Suggested Solution from Contradiction Matrix

Solution		Description
#3:	Local quality	i. Change an object's structure from uniform to non-uniform, change an external environment (or external influence) from uniform to non-uniform.
		ii. Make each part of an object function in conditions most suitable for its operation.
		iii. Make each part of an object fulfil a different and useful function.
#10:	Preliminary action	i. Perform, before it is needed, the required change of an object (either fully or partially).
		ii. Pre-arrange objects such that they can come into action from the most convenient place and without losing time for their delivery.
#27:	Cheap short living objects	i. Replace an inexpensive object with a multiple of inexpensive objects, comprising certain qualities (such as service life, for instance).

9.6 Idea Development with Biomimetics

The next step in the process was to refer to AskNature.org to find solutions that have been used by nature to solve its own problems. AskNature.org is a database containing various strategies used in the natural world, contributed by various researchers over time (Hwang et al. 2015). Through millennia, nature has evolved to what it is today, and in the process, developed the best solutions for the numerous challenges along the way (Bhushan 2009). After trying out several keywords in the search box provided on the AskNature.org website, where many biological strategies were displayed, the best strategies were obtained when "structure" was used. By using "structure" in the search box, it represented the core attribute that had to be improved in the crosshead bearing design to accommodate NFC material. In an analogy, "structure" was to biomimetics what "local quality" was to the TRIZ Inventive Principles.

From the search, three biological strategies were found to be most suitable, namely, the Amazon water lily, the macadamia nut shell, and the hedgehog spine. These strategies offered insight into the way that local structural strength is improved through clever use of fortification at areas where increased stress is experienced. Other strategies, like the wasp nest and beehive, were also included in the list of suggested strategies. However, these strategies were actually meant to tackle temperature control rather than enhance structural robustness. From the three mentioned strategies, ideas were generated for geometrical designs of the crosshead bearing.

Each of the chosen three biological strategies showed how these respective organisms evolved to ensure survivability within their local surroundings. The hedgehog is a small mammal equipped with an armour of spines to protect itself from potential predators These spines also have been reported to provide a damping effect, cushioning the impact from falls of up to 10 metres above ground (Vincent and Owers 1986). Research on the thin-walled structure of the spines has shown that complex inner construction supported by longitudinal stringers is to be credited for these impact absorbing capabilities (Drol et al. 2019; Swift et al. 2016). Inspirations have been drawn from this unique structure in design ideation, as seen in the design of new American football helmets, as reported by Shaw (2019), and car side door impact beams, by Shaharuzaman et al. (2020).

The leaves of the water lily native to the Amazon basin in the South American jungles can grow up to 6 feet in diameter (BBC Studios 2022). The leaves are supported by rib structures spreading out from the centre as they stretch out to fully gather sunlight for photosynthesis. In the crowded waters of the basin, these leaves also provide a means of passage for passing animals and have been reported to be able to support up to 70 pounds (31.75 kg). These abilities of spreading wide across the water and supporting weight show how the ribs help to distribute the load across the entire surface of the leaf, avoiding any stress concentrations that may cause damage. It has been

TABLE 9.3

Crosshead Bearing Design Ideation from TRIZ and Biomimetics

Biological Strategy	CAD Drawing (Side View)	Label
Hedgehog spine (Swift et al. 2016)		Design A1
		Design A2
Amazon water lily (Asknature.org, 2021)		Design B
Macadamia nut shell (Kaupp and Kaupp 2007)		Design C

reported that this unique plant spurred ideas in several architectural marvels, such as the Olympic Games Arena in Rome (Makram and Ouf 2019) and the ancient Crystal Palace Building in London (Srisuwan 2019).

The macadamia nut is encased by a hard shell that has attracted the interest of researchers (Jennings and Macmillan 1986; Kaupp and Kaupp 2007). A series of tests confirmed that the fracture hardness of this shell can be likened to that of glass and other ceramics. Investigations into the microstructure of the shell showed an advanced fullerene network, which is the reason for the high hardness and strength. This advanced structure of the enclosure protects the macadamia nut from damage when it hits the ground, falling from great heights.

Based on the three biological strategies described, new geometrical designs for the two-stroke marine diesel engine crosshead bearing were envisaged. These new designs took inspiration from the novel and intricate internal builds employed by these creations of nature and can be seen in Table 9.3.

9.7 Conclusion

Systematic product development process is described in this chapter, which involves six main stages, starting from identifying customer needs until the final stage, which is putting the product on the market. In addition, conceptual design processes for new products using modern and innovative methods, namely, TRIZ and biomimetics, are also included. To showcase how both idea generation methods may be implemented, a case study on marine engineering application is demonstrated, where a total of four different designs were ideated for the two-stroke marine diesel engine crosshead bearing design. Both methods were used in synergy to visualize geometrical characteristics of the NFC-based journal bearing. The inspirations were taken from the hedgehog spine, macadamia nutshell, and Amazon water lily. In conclusion, the use of both TRIZ and biomimetic methods in systematic product development offers many benefits, especially in spurring new and innovative ideas during the conceptual design stage of the product, as well as breaking the trade-off faced in conventional problem-solving situations towards achieving an ideal solution with nature as the inspiration.

Acknowledgement

The authors would like to thank Universiti Teknikal Malaysia Melaka (UTeM) and Akademi Laut Malaysia (ALAM) for the opportunity and support given in this research and writing.

References

AskNature. 2021. Ribbed Structure Provides Support — Biological Strategy — AskNature [Online]. Available at: https://asknature.org/strategy/ribbed-structure-provides-support/ [Accessed 11 December 2021].

Azammi, A. M. Noor, Salit Mohd Sapuan, Mohamad Ridzwan Ishak, and Mohamed T. H. Sultan. 2018. "Conceptual Design of Automobile Engine Rubber Mounting Composite Using TRIZ-Morphological Chart-Analytic Network Process Technique." *Defence Technology* 14 (4): 268–77. doi:10.1016/j.dt.2018.05.009.

BBC Studios. 2022. Amazing! Giant Waterlillies in the Amazon - The Private Life of Plants - David Attenborough - BBC wildlife [Online Video]. 9 February 2007. Available at: https://www.youtube.com/watch?v=igkjcuw_n_U [Accessed: 11 June 2022].

Balakrishnan, Balamuralithara. 2022. "Exploring the Impact of Design Thinking Tool Among Design Undergraduates: A Study on Creative Skills and Motivation to Think Creatively." *International Journal of Technology and Design Education.* doi:10.1007/s10798-021-09652-y.

Bhushan, Bharat. 2009. "Biomimetics: Lessons from Nature - An Overview." *Philosophical Transactions of the Royal Society A: Mathematical, Physical and Engineering Sciences* 367 (1893): 1445–86. doi:10.1098/rsta.2009.0011.

Boyle, Fiona, Joseph Walsh, Daniel Riordan, Cathal Geary, Padraig Kelly, and Eilish Broderick. 2022. "REEdI Design Thinking for Developing Engineering Curricula." *Education Sciences* 12 (3). doi:10.3390/educsci12030206.

Chen, Fang, and Jacques Terken. 2023. "Design Process." In *Automotive Interaction Design*, Springer Tracts in Mechanical Engineering, edited by Chen, Fang, and Jacques Terken, 165–179. Singapore: Springer.

Childs, Peter R. N. 2019. "Ideation." In *Mechanical Design Engineering Handbook*, 2nd ed., 75–144. doi:10.1016/b978-0-08-102367-9.00003-2.

Chybowski, Leszek, Artur Bejger, and Katarzyna Gawdzińska. 2018. "Application of Subversion Analysis in the Search for the Causes of Cracking in a Marine Engine Injector Nozzle." *World Academy of Science, Engineering and Technology International Journal of Industrial and Manufacturing Engineering* 12 (4): 302–8. doi:doi.org/10.5281/zenodo.1316087.

Drol, Christopher J., Emily B. Kennedy, Bor Kai Hsiung, Nathan B. Swift, and Kwek Tze Tan. 2019. "Bioinspirational Understanding of Flexural Performance in Hedgehog Spines." *Acta Biomaterialia* 94: 553–64. doi:10.1016/j.actbio.2019.04.036.

Fayemi, Pierre Emmanuel Ifeolohoum, Nicolas Maranzana, Améziane Aoussat, and Giacomo Bersano. 2014. "Bio-Inspired Design Characterisation and Its Links with Problem Solving Tools." *Proceedings of International Design Conference (Design 2014)* 1: 173–182.

Han, Xu, Jianhua Wu, Xianhui Zhang, Junyou Shi, Jiaxin Wei, Yang Yang, Bo Wu, and Yonghui Feng. 2021. "The Progress on Antifouling Organic Coating: From Biocide to Biomimetic Surface." *Journal of Materials Science and Technology* 61 (September 2008): 46–62.

Hernandez, Noe Vargas, Linda C. Schmidt, and Gül E. Okudan. 2013. "Systematic Ideation Effectiveness Study of TRIZ." *Journal of Mechanical Design, Transactions of the ASME* 135 (10): 1–10. doi:10.1115/1.4024976.

Hood, Christopher. 2006. *Shinkansen From Buller Train to Symbol of Modern Japan*, Vol. 1., 1st ed. London: Routledge.

Hutchings, Ian, and Philip Shipway. 2017. "Applications and Case Studies." In *Tribology: Friction and Wear of Engineering Materials*, 303–52. Oxford: Butterworth-Heinemann. doi:10.1016/b978-0-08-100910-9.00009-x.

Hwang, Jangsun, Yoon Jeong, Jeong Min Park, Kwan Hong Lee, Jong Wook Hong, and Jonghoon Choi. 2015. "Biomimetics: Forecasting the Future of Science, Engineering, and Medicine." *International Journal of Nanomedicine* 10: 5701–13. doi:10.2147/IJN.S83642.

Ishak, Noordiana Mohd, D. Sivakumar, and Muhd Ridzuan Mansor. 2018. "The Application of TRIZ on Natural Fibre Metal Laminate to Reduce the Weight of the Car Front Hood." *Journal of the Brazilian Society of Mechanical Sciences and Engineering* 40 (2): 1–12. doi:10.1007/s40430-018-1039-2.

Izumi, Hayato, Masato Suzuki, Seiji Aoyagi, and Tsutomu Kanzaki. 2011. "Realistic Imitation of Mosquito's Proboscis: Electrochemically Etched Sharp and Jagged Needles and Their Cooperative Inserting Motion." *Sensors and Actuators, A: Physical* 165 (1): 115–23. doi:10.1016/j.sna.2010.02.010.

Jennings, J. S., and Norman H. Macmillan. 1986. "A Tough Nut to Crack." *Journal of Materials Science* 21 (5): 1517–24. doi:10.1007/BF01114704.

Kaupp, Gerd, and Maria Kaupp. 2007. "Advanced Fullerene-Type Texture and Further Features of the Macadamia Nutshell as Revealed by Optical 3D Microscopy." *Scientific Research and Essays* 2 (5): 150–58.

Kent, Robin. 2016. *Quality Management in Plastics Processing.* Elsevier. doi:10.1016/C2016-0-03226-6

Khorasani, Mahyar, Jennifer Loy, Amir Hossein Ghasemi, Elmira Sharabian, Martin Leary, Hamed Mirafzal, Peter Cochrane, Bernard Rolfe, and Ian Gibson. 2022. "A Review of Industry 4.0 and Additive Manufacturing Synergy." *Rapid Prototyping Journal* 28 (8): 1462–75. doi:10.1108/RPJ-08-2021-0194.

Kim, Y. K. 2012. "Natural Fibre Composites (NFCs) for Construction and Automotive Industries." In *Handbook of Natural Fibres*, 254–79. Woodhead Publishing Limited. doi:10.1533/9780857095510.2.254.

Kim, Yeu Chun, Jung Hwan Park, and Mark R. Prausnitz. 2012. "Microneedles for Drug and Vaccine Delivery." *Advanced Drug Delivery Reviews* 64 (14): 1547–68. doi:10.1016/j.addr.2012.04.005.

Kwon, Jieun, Younghyun Choi, and Yura Hwang. 2021. "Enterprise Design Thinking: An Investigation on User-Centered Design Processes in Large Corporations." *Designs* 5 (3). doi:10.3390/designs5030043.

Lepora, Nathan F., Paul Verschure, and Tony J. Prescott. 2013. "The State of the Art in Biomimetics." *Bioinspiration and Biomimetics* 8 (1). doi:10.1088/1748-3182/8/1/013001.

Li, Rui, Xianghui Meng, Jingjin Dong, and Wenda Li. 2021. "Transient Tribo-Dynamic Analysis of Crosshead Slipper in Low-Speed Marine Diesel Engines During Engine Startup." *Friction* 9 (6): 1504–27. doi:10.1007/s40544-020-0433-9.

Lim, Chaeguk, Dooseob Yun, Inchae Park, and Byungun Yoon. 2018. "A Systematic Approach for New Technology Development by Using a Biomimicry-Based TRIZ Contradiction Matrix." *Creativity and Innovation Management* 27 (4): 414–30. doi:10.1111/caim.12273.

Livotov, Pavel. 2018. "Altshuller's Contradiction Matrix. A Critical View and Best-Practice Recommendations." In *TRIZ Innovation Technology: Product Development and Inventive Problem Solving*, edited by Vladimir Petrov, 237–46. Berlin: TriS Europe.

Makram, Abeer, and Tarek Abou Ouf. 2019. "Biomimetic and Biophilic Design as an Approach to the Innovative Sustainable Architectural Design." *AR-UP 2019: Third International Conference of Architecture and Urban Planning*, no. October: 509–18.

Mann, Darrell. 2001. "An Introduction to TRIZ: The Theory of Inventive Problem Solving." *Creativity and Innovation Management* 10 (2): 123–25. doi:10.1111/1467-8691.00212.

Mansor, Muhd Ridzuan, Salit Mohd Sapuan, E. S. Zainudin, A. A. Nuraini, and Arep A. Hambali. 2014. "Conceptual Design of Kenaf Fiber Polymer Composite Automotive Parking Brake Lever Using Integrated TRIZ-Morphological Chart-Analytic Hierarchy Process Method." *Materials and Design* 54: 473–82. doi:10.1016/j.matdes.2013.08.064.

Mastura, Mohammad Taha, Razali Nadlene, Ridhwan Jumaidin, Syahibudil Ikhwan Abdul Kudus, Muhd Ridzuan Mansor, and Hussin Mohamed Saiful Firdaus. 2022. "Concurrent Material Selection of Natural Fibre Filament for Fused

Deposition Modeling Using Integration of Analytic Hierarchy Process/Analytic Network Process." *Journal of Renewable Materials* 10 (5): 1221–38. doi:10.32604/jrm.2022.018082.

Mastura, Mohammad Taha, Salit Mohd Sapuan, Muhd Ridzuan Mansor, and A. A. Nuraini. 2017. "Environmentally Conscious Hybrid Bio-Composite Material Selection for Automotive Anti-Roll Bar." *International Journal of Advanced Manufacturing Technology* 89 (5–8): 2203–19. doi:10.1007/s00170-016-9217-9.

Morina, Ardian, Tomasz Liskiewicz, and Anne Neville. 2006. "Designing New Lubricant Additives Using Biomimetics." *WIT Transactions on Ecology and the Environment* 87: 157–66. doi:10.2495/DN060151.

Parizi, Rafael, Matheus Prestes, Sabrina Marczak, and Tayana Conte. 2022. "How Has Design Thinking Being Used and Integrated into Software Development Activities? A Systematic Mapping." *Journal of Systems and Software* 187: 111217. doi:10.1016/j.jss.2022.111217.

Poli, Corrado. 2001. 'Chapter 1 – Introduction." In *Design for Manufacturing*, edited by C. Poli, 1–12. Butterworth-Heinemann. doi: 10.1016/B978-0-7506-7341-9.50005-7.

Pugh, Stuart. 1991. *Total Design: Integrated Methods for Successful Product Engineering.* Boston: Addison-Wesley Pub. Co.

Salwa, H N, Mohd Sapuan, Salit, Mohammad Taha Mastura, and Mohd Yusoff Moh Zuhri. 2021. "Conceptual Design and Selection of Natural Fibre Reinforced Biopolymer Composite (NFBC) Takeout Food Container." doi:10.32604/jrm.2021.013977.

San, Yeoh Teong, Yeoh Tay Jin, and Song Chia Li. 2009. *TRIZ: Systematic Innovation in Manufacturing.* Firstfruits Publishing.

Sapuan, Salit Mohd, and Muhd Ridzuan Mansor. 2014. "Concurrent Engineering Approach in the Development of Composite Products: A Review." *Materials and Design* 58: 161–67. doi:10.1016/j.matdes.2014.01.059.

Sapuan, Salit Mohd, and Muhd Ridzuan Mansor. 2016. "Design of Natural Fiber-Reinforced Composite Structures." In *Natural Fiber Composites*, edited by R. D. S. G. Campilho, 255–75. Boca Raton: Taylor and Francis Group.

Sapuan, Salit Mohd, and Muhd Ridzuan Mansor. (eds.). 2021. *Design for Sustainability: Green Materials and Processes.* Elsevier.

Sapuan, Salit Mohd. 2017. *Composite Materials: Concurrent Engineering Approach.* Butterworth-Heinemann.

Schöfer, Malte, Nicolas Maranzana, Améziane Aoussat, Claude Gazo, and Giacomo Bersano. 2015. "The Value of TRIZ and Its Derivatives for Interdisciplinary Group Problem Solving." *Procedia Engineering* 131: 672–81. doi:10.1016/j.proeng.2015.12.353.

Shaharuzaman, Mohd Adrinata, Salit Mohd Sapuan, Muhd Ridzuan Mansor, and Mohd Yusoff Moh Zuhri. 2020. "Conceptual Design of Natural Fiber Composites as a Side-Door Impact Beam Using Hybrid Approach." *Journal of Renewable Materials* 8 (5): 549–63. doi:10.32604/jrm.2020.08769.

Shaw, Meg. 2019. "Doctors at the University of Akron Create New of Football Helmet That Could Reduce Concuss." Akron. February 4. Available at: https://www.news5cleveland.com/news/local-news/akron-canton-news/doctors-at-the-university-of-akron-create-new-style-of-football-helmet-that-could-reduce-concussions.

Sheu, D. Daniel, Ming Chuan Chiu, and Dimitri Cayard. 2020. "The 7 Pillars of TRIZ Philosophies." *Computers and Industrial Engineering* 146 (101): 106572. doi:10.1016/j.cie.2020.106572.

Srisuwan, Touchaphong. 2019. "Biomimetic in Lightweight Structures : Solution for Sustainable Design - A Review Article." *Built* 14: 7–16. doi:10.14456/built.2019.8.

Stephens, Thomas. 2007. "How a Swiss Invention Hooked the World." *Swiss Info.Ch.* Available at: http://www.swissinfo.ch/eng/archive/How_a_Swiss_invention _hooked_the_world.html?cid=5653568.

Swift, Nathan B., Bor Kai Hsiung, Emily B. Kennedy, and Kwek Tze Tan. 2016. "Dynamic Impact Testing of Hedgehog Spines Using a Dual-Arm Crash Pendulum." *Journal of the Mechanical Behavior of Biomedical Materials* 61 (December 2017): 271–82. doi:10.1016/j.jmbbm.2016.03.019.

Thakur, Vijay Kumar, and Manju Kumari Thakur. 2014. "Processing and Characterization of Natural Cellulose Fibers/Thermoset Polymer Composites." *Carbohydrate Polymers* 109: 102–17. doi:10.1016/j.carbpol.2014.03.039.

Vincent, Julian F. V., Olga A. Bogatyreva, Nikolaj R. Bogatyrev, Adrian Bowyer, and Anja Karina Pahl. 2006. "Biomimetics: Its Practice and Theory." *Journal of the Royal Society Interface* 3 (9): 471–82. doi:10.1098/rsif.2006.0127.

Vincent, Julian F. V., and P. Owers. 1986. "Mechanical Design of Hedgehog Spines and Porcupine Quills." *The Zoological Society of London* 210 (0): 55–75.

"Volume II: Maintenance." 2003. In *Electronic Instruction Manual for Engine Type: 6S60MC, Data* 1 (1). Kyungnam: HSD Engine Co., Ltd.

Wakuri, Yutaro, Toshiro Hamatake, and Mitsuhiro Soejima. 1982. "Studies on the Load Carrying Capability of a Crosshead-Pin Bearing in a Two-Stroke Cycle Diesel Engine." *Journal of the Marine Engineering Society in Japan* 25: 1457–64.

Yiow, Ru Vern, and Muhd Ridzuan Mansor. 2022. "Natural Fibre Selection for Sustainable Two-Stroke Marine Diesel Engine Crosshead Bearing." *Journal of Natural Fibre Polymer Composites (JNFPC)* 1 (1): 1–11.

Zhang, Fuying, Hui Liu, and Linjing Zhang. 2006. "The Innovative Design of Reciprocating Seal for Hydraulic Cylinder Based on TRIZ Evolution Theory." In *International Federation for Information Processing (IFIP)*, edited by K. Wang, G. Kovacs, M. Wozny, and M. Fang, 207: 440–49. Boston: Springer. doi:10.1007/0-387-34403-9_60.

Zhang, Xinchang, and Frank Liou. 2021. "Introduction to Additive Manufacturing." In *Additive Manufacturing: Handbooks in Advanced Manufacturing*, edited by J. Pou, A. Riveiro, and J. P. Davim, Chapter 1: 1—31. Amsterdam: Elsevier.

10

Seed/Fruit Fibre–Reinforced Composites

Madeha Jabbar, Mariam Jabbar, and Khubab Shaker

10.1 Seed Fibres

The development of seed/fruit fibre–reinforced composites has gained significant interest in recent years. The textile industry is always in a phase of adapting to new technologies and generating new types of fibre, which are innovative and beneficial to the sector from the perspective of social and environmental sustainability. The upcoming phase focuses on the development and use of more environment friendly fibers. Cotton, hemp, and linen are some of the natural fibres that are used as raw materials in the textile industry [1, 2].

In recent decades, there has been a steadily rising demand from the world community for products that are less harmful to the environment. Natural fibres have found use as reinforcement in a variety of applications, including concrete, acoustic, absorbent materials, buildings, aerospace, electronics, bridges, and indoor applications in the automotive, sports equipment, and home furnishings industries, as well as portable structures. Natural fibres are gaining popularity as a result of their several advantages, including their recyclability, biocompatibility, low abrasiveness, light weight, and low cost, in addition to their contribution to the reduction of carbon dioxide emissions. USA, China, India, Pakistan, Indonesia, Bangladesh, Belgium, Belarus, France, etc. have a diverse collection of natural resources that may be used to produce natural fibres. These resources include bark fibre, leaf fibre, seed/fruit fibre, grass fibre, stalk fibre, and wood fibre [2].

Seed fibres are an interesting and underutilized material for reinforcing composites. They are typically stronger and stiffer than the fruit fibres typically used in composite manufacturing, making them an ideal choice for high-strength applications. In addition, seed fibres are usually more durable and resistant to degradation than fruit fibres, making them a good choice for long-term performance in harsh environments. Seed fibre composites have the potential to be lighter and more environmentally friendly than traditional composites, making them an attractive option for a variety of applications [3].

DOI: 10.1201/9781003408215-10

The biggest advantage of using seed fibres in composites is their ability to absorb a wide range of loads. This means that they can be effectively used in applications where other types of fibre reinforcements would not be suitable, such as in wind turbines and car bodies. Additionally, because seed fibres are so small, they can also be easily inserted into gaps or holes in the fibre matrix, offering an improved structural performance [4].

10.1.1 Properties of Seed Fibre

The properties of fibre–reinforced composites are determined by the type and arrangement of the fibres. The most common type of fibre–reinforced composite is a biodegradable matrix reinforced with a comparatively long, strong fibre. The strength and toughness of these composites are dramatically improved by incorporating high concentrations of reinforcing fibres into the matrix. For example, in carbon/epoxy composites, up to 95% of the mass may be composed of reinforcing fibres [5, 6].

Seed fibres have unique characteristics that make them well suited for use in fibre-reinforced composites. First, seed fibres are highly aligned within the matrix and possess good bonding properties. This helps to create a strong and durable composite material [7].

Second, seed fibres have a small diameter that offers higher interfacial area to bond with the matrix. As a result, they can effectively reinforce the composite material and improve its stiffness, strength, and durability. Seed fibres are relatively easy to use and distribute within a composite material. This makes them a good choice for applications that require excellent mechanical properties [8].

10.1.2 Types of Seed Fibres

Due to their distinctive structural and compositional characteristics, seed fibres play a vital historical and contemporary role in the manufacture of textiles. Numerous plant-based fibres, such as cotton, coir, kapok, and milkweed, are produced from seeds and have played a significant role in the development of human societies. These fibres, which are frequently thought of as the basic building blocks of material culture, have a wide range of physical properties, including length, diameter, surface morphology, and tensile strength. Cotton, the best-known example of seed fibres, has had an enduring impact on the world's textile industry thanks to its brilliant whiteness and fine structure. Kapok fibers, which are derived from the silky floss around the kapok tree's seeds, exhibit distinctive profiles distinguished by their tenacity and buoyancy, respectively. The intrinsic hollow nature of milkweed seed fibres also gives them outstanding buoyancy and insulating abilities. In the area of natural fibre–reinforced materials and sustainable resource utilization, the many facets of seed fibres and their botanical origins form an intriguing nexus for scholarly research and technical study [9].

10.2 Cotton Fibre

Cotton represents the most common type of natural fibre, and it comes from the seed hair of plants of the genus *Gossypium* (see Figure 10.1). Cotton also contains the purest form of cellulose that can be found in nature. Cotton's strength depends on its highly fibrillar and crystalline structure; also, cotton's strength increases by 25% when it is wet. Cotton is a good insulator of heat, but it may be damaged by mildew. It also becomes yellow when exposed to the sun over an extended period, and it loses its strength [10].

Additionally, it is a key cash crop for a lot of countries, including China, India, and the United States. From the sixteenth century, when colonists first started growing cotton in the United States, it has been an essential source of fibre. Cotton has subsequently spread to more than 80 different countries and accounts for 56% of all fibres that are used in the United States to produce garments and home furnishings. Cotton is grown at a rate of 3.8 million tons per year in the United States, whereas consumption is only 1.7 million tons per year [11, 12].

10.2.1 Production and Processing

Cotton seed is typically planted in the spring, and subsequent thinning of the young plants into rows occurs in the summer. After some time has passed, a great number of blooms with a creamy white exterior and a pink inside will begin to emerge [13].

After three days, the bloom will fade and die, leaving behind a little green seed pod, often known as a "boll". Cotton fibres begin their development on the plant as long hairs that become linked to the seeds that are enclosed within the boll. The boll becomes increasingly densely packed

FIGURE 10.1
Cotton plant.

with fibres as the plant continues to develop. When it has reached its full potential, the boll will rupture, revealing the cotton within as a fluffy ball of very fine fibres [14].

Cotton fibres are made up of seed hairs, each of which is a single-celled structure. It develops from the epidermis, also known as the seed coat, of the cotton plant. A single cotton boll can contain upto 150,000 fibres, whereas a single cotton seed can develop as many as 20,000 fibres on the surface of its hull. Cotton plants produce bolls, which are really fruits that develop after the flowers fall off the plant. The remaining immature fruit will continue to grow in size for around seven weeks, at which point it will form the ripe boll. The boll will then open, exposing the mass of cotton fibres, which will subsequently expand and dry into a light, fluffy mass [15].

During the time that the bolls are ripening, the cotton fibres continue to expand and become more developed. One or two days before the onset of blooming, very fine fibres can be observed on the surface of the embryo seed of some kinds of cotton. Other types don't start producing fibre until a day or two after they've finished blooming.

10.2.2 Structure and Appearance

Cotton fibres have a complex, multilayered structure that has been the subject of research for almost a century. The structure of the main cell wall of the cotton fibre, and more specifically, the outer surface layer (the cuticle), has a significant impact on the fibre's qualities as well as the processing and end application. The cotton fibre has a fibrillar shape that is made up of a primary wall, a secondary wall, and a lumen in between them [13].

10.2.3 Physical and Chemical Properties

Cotton fibre is a natural product that has a large degree of variation. The physical characteristics of cotton fibres are closely connected to the economic and industrial worth of the cotton they produce. When it comes to selling cotton, the length of the fibre, as well as its strength, colour grade, and micronaire, is of the utmost significance [10].

Researchers need to get an understanding not only of the effects that fibre qualities have on the market, but also of the effects that these variables have on performance, along with the effects of other properties like fineness, maturity, neps, length distribution, and fibre cohesion.

In addition to this, it is essential to have a solid understanding of the degree to which each of these qualities might vary within a cotton sample. Moreover, the way the fibre characteristics are distributed plays a significant part in the performance of cotton. Some of the properties of cotton fibres are listed in Table 10.1 and Table 10.2 [16]. Length is the most important attribute of cotton fibre. Cotton fibre length is determined as the upper half mean length (average length of the longest 50% of fibres) to a precision

TABLE 10.1

Physical Properties of Cotton Fibre [16]

Property	Value
Length (inches)	0.5–2.5
Density (g/cm³)	1.54–1.56
Tenacity (g/den)	3–5
Breaking elongation (%)	5–10
Diameter (microns)	11–22
Moisture at 65% RH (%)	7–10 (standard 8.5)

TABLE 10.2

Chemical Properties of Cotton Fibre

Property	Value (%)
Lignin	0
Cellulose	96.0
Holocellulose	96.3
Ash	1.3

of 1/100th of an inch. Fibre strength is the amount of force required to break the strands of fibres that are gripped in two sets of jaws that are 1/8 inch apart from one another. The gram-per-denier scale is used to quantify the fibre's strength [17].

The degree of reflectance (Rd) and the yellowness (+b) are the two characteristics that define the colour of cotton samples. Brightness may be determined by the sample's degree of reflection, while yellowness reveals the amount of cotton pigmentation present.

The temperature, the presence of pests and fungus, and other factors can all have an influence on the colour of the fibres. White, grey, speckled, tinted, and yellow stained are the five categories of colour that are commonly recognized. Cotton loses its capacity to be processed into finished goods as its colour becomes more discoloured [17].

10.2.3.1 Physical Properties of Cotton

The micronaire scale is used to quantify the fineness and maturity of fibres. Cotton fibres with a consistent mass of 2.34 grams are packed into a cylinder with a known volume, and then, the air permeability of the compressed sample is measured.

10.2.3.2 Chemical Properties of Cotton Fibres

Cotton is a natural cellulosic fibre that swells when it is exposed to high levels of humidity, when it is submerged in water, and when it is exposed to strong concentrations of certain acids, salts, and bases. The adsorption of highly hydrated ions is typically thought to be the cause of the swelling effect. Cotton has a moisture regain of around 7.1–8.5%, whereas its moisture absorption ranges from 7% to 8% [18]. The chemical composition of cotton fibre is presented in Table 10.2.

10.2.4 Cotton Fibre–Reinforced Composites

In recent years, natural fibres have attracted a lot of attention as a potential substitute for man-made fibre in fibre–reinforced composites. They offer several benefits such as biodegradability, decreased energy consumption, low cost, light weight, and competitive specific mechanical properties, and as a result, they are quickly becoming a viable option instead of inorganic fibres. This is happening because they are evolving quickly as a possible alternative. Because of their qualities, natural fibres have the potential to be used in textile composite material in place of glass or carbon fibres [19].

The leftover fibres from cotton production were used to make composite panels by Wafa Baccouch and her team. Cotton waste fibres are utilized as a reinforcement to the epoxy resin due to the benefits that they offer, which include being favourable to the environment and being cost-effective. The waste from cotton fibres is first shredded into fibres, and then, the carding and needle punching processes are used to turn the fibres into nonwoven mats. The production of the panels is accomplished by a process known as vacuum-assisted resin transfer moulding [20].

T. Alomayri created geopolymer composites reinforced with short cotton fibres and evaluated their hardness, compressive strength, and impact strength to determine their mechanical characteristics. Investigations were done into how fibre contents (0.3, 0.5, 0.7, and 1 wt.%) and their dispersion affected mechanical qualities. For these composites to have the best mechanical characteristics, a fibre content of 0.5 wt.% was found to be necessary [20].

10.3 Coconut Fibre/Coir Fibre

Coconut fibre is among the natural fibres that are readily accessible in tropical locations. This fibre is collected from the shell of the coconut fruit. Coir fibre is another name for this substance. The husk or shell of a coconut is where the fibre of the coconut is obtained (see Figure 10.2). Five million metric tons of coconut fibres are produced every year throughout the globe, with most of this output coming from Sri Lanka and India. It is believed that the overall worth is 100 million dollars. Following India and Sri Lanka as the most important exporters are Thailand, Vietnam, Philippines, and Indonesia. Roughly half of the coconut fibres that are generated are sent out of the country in their natural state [21].

FIGURE 10.2
Coconut husk for fibre extraction.

10.3.1 Production and Processing

Coconut fibre is a natural fibre that is derived from the shell of unripe coconuts. It may also be obtained from mature coconuts. The coconut is soaked in hot saltwater, and then, the fibres are extracted from the shell by combing and crushing the fibres. Each fibre cell is about 1 millimetre in length and 10–20 micrometres in diameter, and the fibre cells themselves are thin and hollow with strong cellulosic walls [22].

Traditional methods of extracting coir fibre include separating the husks from the coconuts and soaking them in ponds or salt water for as long as 10 months. This procedure is quite tough and time-consuming. The husks go through anaerobic process while they are immersed, which enables them to loosen and split from one another. After that, the fibres go through a hand washing, drying, and cleaning process. This time-consuming method of retting results in the production of fibre of the greatest possible quality, which may then be spun into yarn or woven into fabric.

To reduce the amount of time needed for soaking, several mechanical approaches may be used. After a five-day soaking in water, the husks are then crushed, which causes the fibres inside them to become fragmented. Drums are used in the process of separating the coarse fibres from the shorter, more woody sections, which are then combed, rinsed, dried, and cleaned. As manufacturers shift their focus to producing goods of uniformly high quality, these methods are gradually becoming less hazardous to the environment [23].

10.3.2 Structure and Appearance

The surface of coconut fibres is exceedingly heterogeneous, with noticeable breaks, micro-pores, uneven wax-like deposits due to the presence of wax, and fatty and spherical bulges. More abnormalities and contaminants can be seen on the surface of backwater retted fibre compared with raw fibre, which may be the result of the creation of an extra salt coating during backwater retting and subsequent drying without rinsing when compared with raw and untreated retted coconut fibres.

These pollutants cannot be removed by washing with simple water, but alkaline treatments can clean the surface and disclose pit-like interface pores that were hidden on the top of raw and backwater retted fibres [24].

10.3.3 Physical and Chemical Properties

10.3.3.1 Physical Properties of Coconut Fibre

The length of the fibres ranges anywhere from 50 to 150 millimetres on average, and they have the ability to expand beyond their elastic region without snapping. The durability of coir in the face of deterioration and exposure to sea water is exceptional [25]. There are two distinct types of fibres found in coir. The first kind of fibre is a brown one, and this is the one that is used most frequently. It is obtained from fully matured coconuts. The second kind consists of a white fibre that is much finer in texture and originates from unripe green coconuts. It is difficult to colour brown coco fibre because it includes significant quantities of lignin. Coconuts that are between 10 and 12 months old produce fibres of the greatest possible quality. These coconuts produce a fine fibre that may be spun into yarn that is of a light hue and very fine texture. The thickness of jute and sisal fibres is comparable to that of coconut fibre. The diameter of coconut fibre may vary anywhere from 0.1 to 0.75 mm, while the average diameter is 0.18 mm. Table 10.3 displays the mechanical and physical characteristics of coconut fibre, which is contrasted with two prominent materials [21, 26, 27].

10.3.3.2 Chemical Composition of Coconut Fibre

There is a significant amount of lignin found in coconut fibres. Cellulose, hemicellulose, and lignin are the three primary components that make up coconut fibre, as shown in Table 10.4 [28].

TABLE 10.3

Physical Properties of Coconut Fibre

Property	Value
Length (in.)	6–8
Density (g/cc)	1.40
Tenacity (g/tex)	10
Breaking elongation (%)	30
Diameter (mm)	0.1–1.5
Rigidity of modulus (dyne/cm^2)	1.8924
Moisture at 65% RH (%)	10.50

TABLE 10.4

Chemical Composition of Coconut Fibre

Property	Value (%)
Lignin	45.84
Cellulose	43.44
Hemicellulose	0.225
Pectin and related compounds	3
Water soluble	5.25
Ash	2.22

10.3.4 Coconut Fibre–Reinforced Composites

As a result of the fact that plastic has emerged as the preferred material in the modern world, a significant amount of focus has been placed on the creation of bio-based plastics and polymers, as well as their composites. Coir fibre is a kind of biocompatible material that is often used, and it is derived from plants. In the production of biocomposites, coir fibres are being employed as reinforcement in polymers, in either their untreated or their treated form. Additionally, coir fibres are finding widespread use as a reinforcing component in ceramic-based composites, most notably in concrete-based materials [29, 30].

It is possible to achieve the desired level of resilience or elastic modulus in a composite by considering the bond that forms between the fibre surface and the matrix. The stability of natural fibres with a matrix is a significant factor in the determination of the properties of the composite. In most cases, coir fibre is burned or used for fertilization; nevertheless, some research has concentrated on the production of bio-insulating materials utilizing coir fibres [31].

10.4 Milkweed Fibre

Milkweed fibres are obtained from the seeds of milkweed plants, which belong to species of the genus *Asclepias*. The seed pods generate a fuzz that is known as silk or floss and has a smooth, lightweight texture (see Figure 10.3). The glossy and tender strands have a hue that is somewhere between yellow and white. Milkweed fibres are too fragile to be spun; hence, they are typically used for cushioning in upholstery. They also have high stability, which allows them to be utilized in place of kapok [32].

Milkweed fibres might be taken from either the seed pod or the stem of the plant, based on individual preference. Additionally, an extract of the white milk sap that can be produced from the plant's stem has several medical

FIGURE 10.3
Milkweed plant and fibres.

applications, particularly for the treatment of pain, asthma, bronchitis, dyspepsia, leprosy, tumours, and some gastro-intestinal illnesses. Among the several acceptable types of milkweed fibres, those that are removed from the seed pod are typically used in technical applications, particularly as filler materials [33].

10.4.1 Structure and Appearance

The fabric made from milkweed is yellowish white in colour and has a silky, glossy texture. It is composed of individual fibres that range in length from approximately 1 to 3 centimetres and in diameter from around 20 to 50 micrometres. Milkweed fibres consist of a single cell, just like cotton fibres, yet there is no twisting along the length of the fibre. Because of their hollow form, the fibres are lightweight and offer outstanding insulating qualities. The seed pods, which carry the fibre, are subjected to mechanical processing. The fibres of milkweed include oily material as well as lignin, which is a component found in woody plants. Because of this substance, milkweed fibre cannot be spun, because it is too fragile. There are several challenges while spinning the fibres because of the smooth surface of the milkweed fibres and their limited durability against external stresses [34–36].

10.4.2 Physical Properties of Milkweed Fibres

An essential requirement for using milkweed ligno-cellulosic fibres is a thorough understanding of their shape and chemical composition as these factors may have a significant impact on the end use of these fibres. Among the several acceptable types of milkweed fibres, those that are removed from the seed pod are typically used in technical applications, particularly as filler materials. The lightweight and effective insulating qualities of milkweed fibres are due to their hollow structure [33]. The physical properties of milkweed fibre are presented in Table 10.5.

TABLE 10.5

Physical Properties of Milkweed Fibre

Property	Value
Mean length (mm)	21.2
Strength (g/tex)	23.0
Fineness (micronaire)	2.4
Elongation (%)	6
Diameter (µm)	10–18
Density (g/cc)	0.97
Mean length (mm)	21.3

TABLE 10.6

Chemical Composition of Milkweed Fibre

Property	Value (%)
Lignin	18
Cellulose	50–55
Hemicellulose	24
Sugar	2–3
Wax	1–2
Ash	0.5–1

10.4.3 Chemical Composition of Milkweed Fibre

Ligno-cellulosic fibres are generally made up of a variety of materials, primarily a combination of both cellulose and hemicellulose, lignin, and other organic components such as ash and wax. Table 10.6 presents the primary chemical composition of the fibres obtained from milkweed seeds [36].

10.4.4 Milkweed Fibre–Reinforced Composites

Technical applications of milkweed fibres might also be attributed to the composite manufacturing sectors, particularly for the purpose of strengthening cement composite constructions. The low density of milkweed fibres allows the inclusion of a greater number of fibres per unit weight during the production of composite materials, which ultimately results in a lighter-weight composite structure with improved properties.

Moghgan Sayanjali Jasbi et al. investigated milkweed fibre's influence on the tensile and flexural characteristics of composites after being subjected to an alkaline treatment. The composite reinforcement consisted of fibres that were coated with a polyvinyl acetate (PVAc) resin. These fibres were placed in the form of needle-punched nonwoven sheets. The findings of the statistical investigation showed that treatment with alkali had a major impact on the mechanical properties of the composite specimens.

Narendra Reddy et al. studied the properties of milkweed and kenaf fibres as a reinforcement in composites for automotive applications. When compared with similar polypropylene (PP) composites reinforced using kenaf fibres, milkweed-reinforced lightweight PP composites have much improved tensile and flexural characteristics. As compared with kenaf fibres, milkweed floss contains a lower proportion of crystallinity, a higher concentration of lignin, and lower levels of cellulose. In contrast to other natural cellulose fibres, milkweed floss threads are entirely hollow despite having short lengths and low elongation. Milkweed fibres have a low density due to the hollow core and perhaps poor crystallinity. Due to their low density, milkweed fibres may be used in higher concentrations per unit weight of a composite, which results in fewer voids and improved flexural and tensile capabilities [37–39].

10.5 Kapok Fibre

Kapok, also known as capok, is a cellulosic natural fibre. The seed coats of kapok trees are the source of its production. There are several other names for kapok, including silk cotton and java cotton. It has a hollow body. This is spherical in form, is extremely buoyant, and has a twistless fibre. Pure kapok fibre is incapable of being spun into yarn like cotton fibre due to its brittleness, poor cohesivity, and low strength; however, kapok fibre may be effectively combined with cotton fibre to make yarns (see Figure 10.4). In addition to a loss in the overall cost of manufacturing for the yarns, a rise in the kapok content results in a reduction in the regularity and tenacity of the yarn while simultaneously resulting in an increase in its extensibility [40, 41].

FIGURE 10.4
Kapok plant and fibres.

The big lumen and waxy surface of kapok fibre, on the other hand, make it difficult for hydrophilic colouring agents or dyes to penetrate the fibre. As a result, kapok fibre has a poorer dyeing efficiency than other fibres. Despite this, the mixture of these characteristics gives kapok fibre improved hydrophobic and emulsifying capabilities, which in turn makes it possible for this fibre to be used as a potential buoyant material as well as an oil-absorbing material [42, 43].

10.5.1 Production and Processing

Kapok fibre is obtained from the seed of the plant. Knocking the mature pods that have not yet opened off the tree is the typical method of harvesting them. However, it is also possible to cut them off the tree itself or collect them after they have fallen to the ground. The kapok fibre is exposed to the sun so that the hulls may dry off. Hand labour is required to remove the seed and fibres from the pods, as well as to hull the fruit. The seeds are dispersed throughout the floss, and they are able to make their way to the base of the container, where they may be readily extracted. To finish the drying process, the kapok fibre is exposed to the sun for three to five hours. After that, the kapok fibres are tied up into bundles and sent out to undergo further processing [41, 44].

10.5.2 Structure and Appearance

Kapok fibres have a circular cross section that is uniformly packed with air and have a large lumen. The wall thickness of an air-filled lumen is around 1–2 micrometres. Because of this, its cell wall is very thin, and it has a thick film of wax covering it. Microscopic observation reveals that kapok fibres are transparent and contain air bubbles that are distinctive to the lumen. Kapok fibres have an oval to circular cross section when its cross section is observed [45].

10.5.3 Physical Properties of Kapok Fibres

A detailed overview of the geometry and chemical composition of kapok fibres is a prerequisite for employing them, since these characteristics may have a major influence on the applications for which they are used. The physical properties of kapok fibres are listed in Table 10.7 [46, 47].

10.5.4 Chemical Composition of Kapok Fibre

The main ingredients of ligno-cellulosic fibres are a mixture of cellulose and hemicellulose, lignin, and other organic elements including ash and wax. The chemical composition of fibres generated from kapok is shown in Table 10.8 [48–50].

TABLE 10.7

Physical Properties of Kapok Fibre

Property	Value
Mean length (mm)	20
Strength (MPa)	45–64
Fineness (denier)	0.4–0.7
Elongation (%)	1.8–4.23
Diameter (μm)	20.5
Average linear density (tex)	0.064
Tenacity (gram/denier)	1.4–1.74

TABLE 10.8

Chemical Composition of Kapok Fibre

Property	Value (%)
Lignin	14.10
Holocellulose	83.73
Moisture content	11.23
Wax	2.34
Ash	1.05

10.5.5 Kapok Fibre–Reinforced Composites

Because of its thin air-filled lumen and significant void content, kapok fibre is traditionally used as a filling for insulating against sound and heat, as well as for bedding, life preservers, and similar aquatic safety equipment. Kapok fibre–based products have opened a range of potential in a variety of applications, since they possess several distinctive characteristics, such as absorption. Jiaming Sun et al. made supercapacitors by using 3D microtubular kapok fibre composites. Murilo J. P. Macedo et al. used recycled polyethylene and kapok fibres to make composites [51, 52].

Yian Zheng et al. reviewed kapok fibres applications as an absorbent material. Lihua Lyu et al. investigated the effect of kapok fibre on flame-retardant and absorption properties. The utilization rate of kapok fibre was improved by preparing flame-retardant and sound-absorption composites using the hot-pressing technique. Kapok fibre was used as the reinforcement material, poly-caprolactone was used as the matrix material, and magnesium hydroxide was used as the fire retardant. Then, using the mean sound-absorption coefficient as the index, the impacts of hot-pressing temperature, hot-pressing duration, density of composites, mass percentage of kapok fibre, thickness of composites, and air layer thickness on the sound-absorption capabilities of composites were examined [53].

10.6 Coffee Fibre

Coffee is the second most popular beverage after water in terms of consumption all across the globe, making it one of the most significant agricultural commodities in the world. It is planted in around 80 nations, and the world's total output in 2019 was approximately 10.2 million tons. More than 90% of the coffee that is produced globally each year, which includes the husk, is considered trash. On the other hand, the usage of plastic is growing at an alarming rate, making the need for eco-friendly substitutes more pressing. Singtex Industries developed a method to use coffee grounds, which were previously regarded as trash. This is then used in the production of a technological composite fibre, which may be woven or knitted into garment items. Just one cup of coffee is all that is needed to produce two T-shirts. It is feasible to combine the coffee fibre with other materials, such as polyester, and materials that are more conventional, such as nylon. The advantages that may be gained from using these items are rather extensive [54–56].

10.6.1 Production and Processing

The production process of coffee ground fibre involves several stages. It begins with the preparation of coffee residual material, followed by the filtration of coffee ashes to isolate specific components. Subsequently, the elimination of organic components takes place, ensuring the refinement of the material. The process advances with the preparation of carbonized particles, contributing to the desired characteristics. The final stage entails the meticulous mixing of both carbonized and uncarbonized materials, culminating in the creation of coffee ground fibre with its distinctive properties and applications [57, 58].

A microencapsulation process is performed on the leftover coffee residue once the essential oil is taken from the beans. In this process, the waste from the coffee beans is washed in clean water and then dried; the result is ground particles that range in size from 20 to 100 microns. After that, the powdered mixture is strained through a screen. The resulting mixture has the capability of being sieved into various fine particle sizes ranging from 80 to 100 μm [56].

The powdered mixture is then treated with certain solvents in order to eliminate the organic components that were present in the mixture. Ethyl ether is used in an extractor of the Soxhlet type to carry out the process of extracting the fat. After the fatty acids have been removed, the water-soluble elements of the aqueous solution that contains those ingredients are evaporated to lower the pressure, and then, the glycerol is removed using absolute alcohol extraction [59].

The mixture created by these three processes is collected and then subjected to carbonization. For instance, pyrolysis is the process in which a combination of coffee is heated, degraded, and subsequently transformed into the product that is wanted while air is present in the environment. Carbonization, the processing of charcoal, gasification, and the processing of activated carbon are all included in pyrolysis. In the presence of chemicals such as zinc chloride, magnesium chloride, calcium chloride, or phosphoric acid, the first stages of the carbonization process of coffee's basic materials are carried out. In this method, ground particles and polymer chips (such as polypropylene (PP), nylon, or polyethylene terephthalate (PET) are mixed in a weight ratio of 1:9 so that the masterbatch may be produced. Alternatively, 75% of carbonized particles and 25% of the substance containing coffee scent are mixed with polymer chips (such as PP, nylon, or PET) to produce the masterbatch [59].

10.6.2 Properties of Coffee Fibre

The coffee ground fibre has a higher capacity for rapid drying, which indicates that it draws moisture away from the skin and toward the outside of the fabric in a continuous manner. The moisture from the body is drawn away from the body and into the atmosphere by the wicking action of the S.Café® fabric. This characteristic, which is permanent, will never be removed by washing, since it is not a form of finish in the cloth that is transient. The fibre contains nano-sized coffee granules, which are permanently bonded. These coffee particles can absorb odours [59].

The ground coffee from S.Café® is designed with an enormous number of tiny pores that provide a natural and chemical-free protection for fibre, yarn, or cloth. This shield may persist for a long time, deflect UV rays, and provide a comfortable outdoor experience. As S.Café® technologies make use of recycled coffee grounds that otherwise would have been dumped in a landfill, increasing the amount of coffee that is recycled provides the waste with a value [56].

In addition to this S.Café® yarn, there is another sustainable yarn that is able to reduce the temperature of our bodies by about 1 to 2 degrees Celsius when compared with traditional materials. In point of fact, the cloth itself is what imparts the feeling of coolness [60].

10.6.3 Coffee Fibre–Reinforced Composites

Coffee grounds are the primary component in the production of the technological composite fibre known as S.Café® . The process of turning coffee grounds into fabric is quite like the process of turning bamboo into a material that is comparable to viscose. Barbara Maria Mateus Goncalves researched how to effectively use coffee husk fibre waste in polymer biocomposites [61].

The possible utilization of coffee hull (CH) in reinforced polyethylene (PE) matrix composites was investigated for the very first time by Zhihao Wang. Experiments with composites that had CH added to them were investigated for their effects on mechanical characteristics, hygroscopicity, thermogravimetric analysis, fibre treatment, and microstructures [62, 63].

Kelen Cristina Reis investigated the potential of cellulosic particles derived from byproducts of the coffee industry to be converted into polyhydroxy butyrate (PHB) and used as a reinforcing filler in biocomposites. The level of crystallinity was reduced when coffee parchment was included, whereas the thermal properties of the composites were increased by using coffee dregs at varying concentrations [64].

10.7 Conclusion

In conclusion, research into natural fibres has shown that they have tremendous potential as competitive substitutes in a variety of industries. These fibres have unique properties that range from toughness and buoyancy to thermal insulation and environmental friendliness. By utilizing these natural resources' inherent advantages, composite materials are being developed that not only perform admirably but also meet the urgent demand for sustainability in the modern world.

These natural fibres' wide range of uses demonstrates how adaptable they are. These materials have proven their capacity to adapt to a variety of functional purposes, from milkweed's use in cushioning and sound insulation to the use of coconut fibre in reinforcing composites. The buoyant qualities of kapok and its potential as a sound insulator, as well as the moisture-wicking and odour-absorbing qualities of coffee fibre, serve as excellent illustrations of how novel uses can result from unlikely origins.

The creation of composite materials utilizing these natural fibres becomes more important as companies continue to adopt eco-friendly solutions. A critical step towards a more sustainable future is the development of materials that both improve product performance and reduce environmental impact. In essence, the study of coconut, milkweed, kapok, and coffee fibres demonstrates the potential for nature-inspired innovation to help create a world that is greener and more resource-efficient.

References

1. C. Felgueiras, N. G. Azoia, C. Gonçalves, M. Gama, and F. Dourado, "Trends on the cellulose-based textiles: Raw materials and technologies," *Frontiers in Bioengineering and Biotechnology*, vol. 9, p. 608826, Mar. 29, 2021, doi: 10.3389/fbioe.2021.608826.
2. Md. I. H. Mondal, *Fundamentals of Natural Fibres and Textiles*. Woodhead Publishing, 2021.

3. K. Shaker, Y. Nawab, "Lignocellulosic fibers: Sustainable biomaterials for green composites", *Springer*, 2022.

4. C. Felgueiras, N. G. Azoia, C. Gonçalves, M. Gama, and F. Dourado, "Trends on the cellulose-based textiles: Raw materials and technologies," *Frontiers in Bioengineering and Biotechnology*, vol. 9, p. 608826, Mar. 2021, doi: 10.3389/FBIOE.2021.608826.

5. G. Lufi et al., "Influence of seed color and hull proportion on quality properties of seeds in Brassica napus L.," *Wiley Online Library*, vol. 87, no. 6, pp. 235–237, 1985, doi: 10.1002/lipi.19850870605.

6. J. Turner, H. H. Ramey Jr, and S. Worley Jr, "Relationship of yield, seed quality, and fiber properties in upland cotton1," *Wiley Online Library*, vol. 16, no. 4, pp. 578–580, Jul. 1976, doi: 10.2135/cropsci1976.0011183X001600040038x.

7. R. Egala, G. V. Jagadeesh, and S. G. Setti, "Experimental investigation and prediction of tribological behavior of unidirectional short castor oil fiber reinforced epoxy composites," *Friction*, vol. 9, no. 2, pp. 250–272, Apr. 2021, doi: 10.1007/S40544-019-0332-0.

8. E. Mukhametshina, R. Muradov, I. Abbazov, and A. Usmankulov, "Improving fiber quality by reducing seed damage in the gin machine," *E3S Web of Conferences*, vol. 304, p. 03018, Sep. 2021, doi: 10.1051/E3SCONF/202130403018.

9. S. K. Ramamoorthy, M. Skrifvars, A. Persson, and S. K. Ramamoorthy, "A review of natural fibers used in biocomposites: Plant, animal and regenerated cellulose fibers," *Polymer Reviews*, vol. 55, no. 1, pp. 107–162, Jan. 2015, Taylor & Francis, doi: 10.1080/15583724.2014.971124.

10. R. Kozłowski, and M. Mackiewicz-Talarczyk, "Introduction to natural textile fibres," in *Handbook of Natural Fibres*. Elsevier, Accessed: Aug. 23, 2023 [Online]. Available: https://www.sciencedirect.com/science/article/pii/B9780128183984000013.

11. M. Negm, and S. Sanad, "Cotton fibres, picking, ginning, spinning and weaving," in *Handbook of Natural Fibres*. Elsevier, 2020. Accessed: Aug. 23, 2023 [Online]. Available: https://www.sciencedirect.com/science/article/pii/B9780128187821000018.

12. A. R. Bunsell, "Handbook of tensile properties of textile and technical fibres," 2009. Accessed: Aug. 23, 2023 [Online]. Available: https://books.google.com.pk/books?hl=en&lr=&id=51KkAgAAQBAJ&oi=fnd&pg=PP1&dq=Cotton+fibres.+Woodhead+Publishing&ots=9opospLNmf&sig=XUyiccxmwXFKA3Wmebq RrIZ0Wi0.

13. M. Ansell, and L. Y. Mwaikambo, "The structure of cotton and other plant fibres," in *Handbook of Textile Fibre Structure*. Elsevier, 2009, doi: 10.1533/9781845697310.1.62.

14. S. Ahmad, and M. Hasanuzzaman, "Cotton production and uses," in *Agronomy, Crop Protection, and Postharvest Technologies*. Springer, 2020. Accessed: Aug. 23, 2023 [Online]. Available: https://link.springer.com/content/pdf/10.1007/978-981-15-1472-2.pdf.

15. P. J. Wakelyn, N. R. Bertoniere, and A. D. French, "Cotton fiber chemistry and technology," 2006. Accessed: Aug. 23, 2023 [Online]. Available: https://books.google.com.pk/books?hl=en&lr=&id=3FUVqE3zczwC&oi=fnd&pg=PP1&dq=Production+and+processing+of+cotton+fibers&ots=psiMddoIjM&sig=1JtBFNVDna3WrzDD_vOCbHSLJoI.

16. Y. Nawab, and K. Shaker, "Textile raw materials" in *Textile Engineering: An introduction,* 2nd edition, De Gruyter, 2023. doi: 10.1515/9783110799415.

17. O. Harzallah, H. Benzina, and J. Y. Drean, "Physical and mechanical properties of cotton fibers: single-fiber failure," *Textile Research Journal,* vol. 80, no. 11, pp. 1093–1102, 2010, doi: 10.1177/0040517509352525.

18. N. Abidi, "Chemical properties of cotton fiber and chemical modification," *Cotton Fiber: Physics, Chemistry and Biology,* pp. 95–115, Jan. 2018, doi: 10.1007/978-3-030-00871-0_5.

19. B. Wankhede, H. Bisaria, S. Ojha, and V. S. Dakre, "A review on cotton fibre-reinforced polymer composites and their applications," *Proceedings of the Institution of Mechanical Engineers, Part L: Journal of Materials: Design and Applications,* 2022, doi: 10.1177/14644207221143876.

20. T. Alomayri, and I. M. Low, "Synthesis and characterization of mechanical properties in cotton fiber-reinforced geopolymer composites," *Journal of Asian Ceramic Societies.* Taylor & Francis, vol. 1, no. 1, pp. 30–34, 2013, doi: 10.1016/j.jascer.2013.01.002.

21. I. O. Oladele, S. O. Adelani, B. A. Makinde-Isola, and T. F. Omotosho, "Coconut/coir fibers, their composites and applications," *Plant Fibers, their Composites, and Applications.* Elsevier, 2022 [Online]. Available: https://www.sciencedirect.com/science/article/pii/B9780128245286000047. Accessed on: Aug. 23, 2023.

22. K. F. Hasan, P. G. Horváth, M. Bak, and T. Alpár, "A state-of-the-art review on coir fiber-reinforced biocomposites," *Rsc Advances.* pubs.rsc.org, 2021 [Online]. Available: https://pubs.rsc.org/en/content/articlehtml/2021/ra/d1ra00231g. Accessed on: Aug. 23, 2023.

23. J. A. Delarue, "Tensile Strength of coconut fiber waste as an organic fiber on concrete," *Civil and Environmental Research,* vol. 9, no. 11, 2017 [Online]. Available: https://core.ac.uk/download/pdf/234678621.pdf. Accessed on: Aug. 23, 2023.

24. R. R. Singh, and E. S. Mittal, "Improvement of local subgrade soil for road construction by the use of coconut coir fiber," *International Journal of Research in Engineering and Technology,* 2014 [Online]. Available: https://www.academia.edu/download/34193475/IMPROVEMENT_OF_LOCAL_SUBGRADE_SOIL_FOR_ROAD_CONSTUCTION_BY_THE_USE_OF_COCONUT_COIR_FIBER.pdf. Accessed on: Aug. 23, 2023.

25. W. Ahmad et al., "Effect of coconut fiber length and content on properties of high strength concrete," *Materials,* vol. 13, no. 5, p. 1075, Feb. 2020, doi: 10.3390/MA13051075.

26. M. Ali, A. Liu, H. Sou, and N. Chouw, "Mechanical and dynamic properties of coconut fibre reinforced concrete," *Construction and Building Materials.* Elsevier, 2012 [Online]. Available: https://www.sciencedirect.com/science/article/pii/S0950061811007586. Accessed on: Aug. 23, 2023.

27. M. Ali, "Coconut fibre: A versatile material and its applications in engineering," *Journal of Civil Engineering and Construction Technology,* vol. 2, no. 9, pp. 189–197, 2011 [Online]. Available: https://academicjournals.org/journal/JCECT/article-full-text-pdf/D540A213064. Accessed on: Aug. 23, 2023.

28. M. Arsyad and I. Wardana, "The morphology of coconut fiber surface under chemical treatment," *SciELO Brasil* [Online]. Available: https://www.scielo.br/j/rmat/a/YYnkfFw6mDFkHS5XTPZkNCk/?lang=en. Accessed on: Aug. 23, 2023.

29. S. Konduru, M. R. Evans, and R. H. Stamps, "Coconut husk and processing effects on chemical and physical properties of coconut coir dust," *HortScience,* vol. 34, no. 1, 1999 [Online]. Available: https://journals.ashs.org/hortsci/view/journals/hortsci/34/1/article-p88.xml. Accessed on: Aug. 23, 2023.

30. C. Y. Lai, S. M. Sapuan, M. Ahmad, N. Yahya, and K. Z. H. M. Dahlan, "Mechanical and electrical properties of coconut coir fiber-reinforced polypropylene composites," *Polymer - Plastics Technology and Engineering,* vol. 44, no. 4, pp. 619–632, 2005, doi: 10.1081/PTE-200057787.

31. G. S. Babu, "Strength and stiffness response of coir fiber-reinforced tropical soil," *Journal of Materials in Civil Engineering,* vol. 20, no. 9, pp. 571–579, Sep. 2008 [Online]. Available: https://ascelibrary.org/doi/abs/10.1061/(ASCE)0899-1561(2008)20:9(571). Accessed on: Aug. 23, 2023.

32. S. Hassanzadeh, and H. Hasani, "A review on milkweed fiber properties as a high-potential raw material in textile applications," *Journal of Industrial Textiles,* vol. 46, no. 6, pp. 1412–1436, Feb. 2017, doi: 10.1177/1528083715620398.

33. S. Hassanzadeh, and H. Hasani, "A review on milkweed fiber properties as a high-potential raw material in textile applications," *Journal of Industrial Textiles,* vol. 46, no. 6, pp. 1412–1436, Feb. 2017, doi: 10.1177/1528083715620398.

34. G. L. Louis, and B. A. K. Andrews, "Cotton/milkweed blends: A novel textile product," *Textile Research Journal,* vol. 57, no. 6, pp. 339–345, 1987, doi: 10.1177/004051758705700604.

35. M. Mula et al., "Sustainable milkweed fiber composites for medical textile application," *ACS Sustainable Chemistry & Engineering,* Aug. 2023, doi: 10.1021/ACSSUSCHEMENG.3C01508.

36. J. Y. F. Dréan et al., "Mechanical characterization and behavior in spinning processing of milkweed fibers," *Textile Research Journal,* vol. 63, no. 8, pp. 443–450, 1993, doi: 10.1177/004051759306300803.

37. A. Zarehshi, "Study of the water vapor permeability of multiple layer fabrics containing the milkweed fibers as the middle layer," *Journal of Textile Institute,* vol. 113, no. 5, pp. 765–776, 2022 [Online]. Available: https://www.tandfonline.com/doi/abs/10.1080/00405000.2021.1936387. Accessed on: Aug. 23, 2023.

38. R. Rengasamy, and D. Das, "Study of oil sorption behavior of filled and structured fiber assemblies made from polypropylene, kapok and milkweed fibers," Elsevier, Accessed: Aug. 23, 2023 [Online]. Available: https://www.sciencedirect.com/science/article/pii/S030438941001455X.

39. A. Sabziparvar, and L. Boulos, "Effect of fiber length and treatments on the hygroscopic properties of milkweed fibers for superabsorbent applications," Elsevier, Accessed: Aug. 23, 2023 [Online]. Available: https://www.sciencedirect.com/science/article/pii/S2352186422003534

40. S. Meiwu, X. Hong, and Y. W. Journal, "The fine structure of the kapok fiber," *Textile Research Journal,* vol. 78, no. 15, pp. 1386–1397, Oct. 2008, doi: 10.1177/0040517508095594.

41. R. H. Sangalang, "Kapok fiber-structure, characteristics and applications: (A review)," *Journal of Chemistry,* vol. 37, no. 3, pp. 2018, pdfs.semanticscholar.org, doi: 10.13005/ojc/370301.

42. Y. Zheng, and A. Wang, "Kapok fiber: Applications," *Biomass and Bioenergy: Applications.* Springer, Accessed: Aug. 23, 2023 [Online]. Available: https://link.springer.com/chapter/10.1007/978-3-319-07578-5_13

43. Z. Yang, and J. Yan, "Pore structure of kapok fiber," *Cellulose*. Springer, Accessed: Aug. 23, 2023 [Online]. Available: https://idp.springer.com/authorize /casa?redirect_uri=https://link.springer.com/article/10.1007/s10570-018-1767-6 &casa_token=226p-ng18b4AAAAA:nALsumV6Fv5go5Bus3uBmeB1QKJVwyV Rd2nNafkwvEJfjFjOKldneCcMCnDNmwN8KoeiOt3zU5Xts4gCGQ

44. T. Dong, G. Xu, and F. Wang, "Adsorption and adhesiveness of kapok fiber to different oils," *Journal of Hazardous Materials*. Elsevier, Accessed: Aug. 23, 2023 [Online]. Available: https://www.sciencedirect.com/science/article/pii/ S0304389415002459

45. Y. Zheng, J. Wang, Y. Zhu, and A. Wang, "Research and application of kapok fiber as an absorbing material: A mini review," *Journal of Environmental Sciences*. Elsevier, Accessed: Aug. 23, 2023 [Online]. Available: https://www.sciencedi-rect.com/science/article/pii/S1001074214002393

46. Y. Zheng, J. Wang, and A. Wang, "Recent advances in the potential applica-tions of hollow kapok fiber-based functional materials," *Cellulose*. Springer, Accessed: Aug. 23, 2023 [Online]. Available: https://idp.springer.com/authorize /casa?redirect_uri=https://link.springer.com/article/10.1007/s10570-021-03834 -6&casa_token=GUkFvKPkfcEAAAAA:IU4E0b2-8RK2KEXXornE2Vclsxk PmvlFuaTL8iuqYjeAFchtTLfIBlcMXl2AFOqTWSnaTvT7lBiZrF_9eA

47. Y. Zheng, and A. Wang, "Recent advances in the potential applications of hol-low kapok fiber-based functional materials," *Cellulose*, vol. 28, no. 9, pp. 5269–5292, Jun. 2021, doi: 10.1007/S10570-021-03834-6.

48. J. Wang, and A. Wang, "Acetylated modification of kapok fiber and application for oil absorption," *Fibers and Polymers*, vol. 14, no. 11, pp. 1834–1840, Nov. 2013, Springer, doi: 10.1007/s12221-013-1834-4.

49. K. Hori, M. Flavier, S. Kuga, T. Lam, and K. Iiyama, "Excellent oil absorbent kapok fiber: Fiber structure, chemical characteristics, and application," *Journal of Wood Science*. Springer, 2000, Accessed: Aug. 23, 2023 [Online]. Available: https://link.springer.com/article/10.1007/BF00776404

50. J. Wang, Y. Zheng, and A. Wand, "Effect of kapok fiber treated with various solvents on oil absorbency," *Industrial Crops and Products*. Elsevier, vol. 40, pp. 178–184, 2012, doi: 10.1016/j.indcrop.2012.03.002.

51. L. Mwaikambo and E. B.-P. Testing, "The performance of cotton–kapok fab-ric–polyester composites," Elsevier, vol. 18, pp. 181–198, 1999, doi: 10.1016/ S0142-9418(98)00017-8.

52. J. Sun et al., "Fabrication of three-dimensional microtubular kapok fiber car-bon aerogel/RuO2 composites for supercapacitors," Elsevier, Accessed: Aug. 23, 2023 [Online]. Available: https://www.sciencedirect.com/science/article/ pii/S0013468619301203.

53. B. Tang et al., "Fabrication of kapok fibers and natural rubber composites for pres-sure sensor applications," Springer, Accessed: Aug. 23, 2023 [Online]. Available: https://idp.springer.com/authorize/casa?redirect_uri=https://link.springer .com/article/10.1007/s10570-020-03647-z&casa_token=O3nf37zU8SAAAAAA:- -pC7wvjl0lR3Kjv2znFpc__dV3KLD7FDLviR0t9soTOyIn_obu8vVdEQNCXE8c h7kfX1SOJlnbA0cLpgw.

54. J. Li, Y. Zhu, L. Chang, N. Herlina, and A. Santika Kurniati, "Coffee bean skin waste extraction for silk dyeing," iopscience.iop.org, doi: 10.1088/1757-899X/801/1/012075.

55. L. Wang, Q. Cao, and Y. Cao, "Study on the properties of coffee carbon filament yarns," *Advanced Materials Research*. Trans Tech Publ, Accessed: Aug. 23, 2023 [Online]. Available: https://www.scientific.net/AMR.821-822.64.

56. G. Karunakaran, A. Periyasamy, and A. Tehrani, "Extraction of micro, nano-crystalline cellulose and textile fibers from coffee waste," *Journal of Testing and Evaluation*, astm.org, Accessed: Aug. 23, 2023 [Online]. Available: https://www.astm.org/jte20220487.html.

57. C. Zhang, L. Zhao, and X. Gu, "Effect of blending ratio on the hollow coffee carbon polyester/cotton blended yarn," *Textile Research Journal*, vol. 92, no. 5–6, pp. 906–918, Mar. 2022, doi: 10.1177/00405175211032085.

58. C. Zhang, L. Zhao, and X. G.-T. R. Journal, "Effect of blending ratio on the hollow coffee carbon polyester/cotton blended yarn," journals.sagepub.com, Accessed: Aug. 23, 2023 [Online]. Available: https://journals.sagepub.com/doi/abs/10.1177/00405175211032085.

59. M. Y. Tan, H. T. Nicholas Kuan, and M. C. Lee, "Characterization of alkaline treatment and fiber content on the physical, thermal, and mechanical properties of ground coffee waste/oxobiodegradable HDPE," *International Journal of Polymer Science*. hindawi.com, Accessed: Aug. 23, 2023 [Online]. Available: https://www.hindawi.com/journals/ijps/2017/6258151/abs/.

60. N. A. Kalebek, "Fastness and antibacterial properties of polypropylene surgical face masks dyed with coffee grounds," *The Journal of The Textile Institute*, vol. 113, no. 7, pp. 1309–1315, 2022.

61. L. Huang, B. Mu, X. Yi, S. Li, and Q. Wang, "Sustainable Use of Coffee Husks For Reinforcing Polyethylene Composites," *Journal of Polymers and the Environment*, vol. 26, no. 1, pp. 48–58, Jan. 2018, doi: 10.1007/S10924-016-0917-X.

62. L. Huang, B. Mu, X. Yi, S. Li, and Q. W. Wang, "Sustainable use of coffee husks for reinforcing polyethylene composites," *Journal of Polymers and the Environment*, vol. 26, no. 1, pp. 48–58, Jan. 2018 [Online]. Available: https://link.springer.com/article/10.1007/s10924-016-0917-x. Accessed on Aug. 23, 2023.

63. N. Gama, A. Ferreira, and D. V. Evtuguin, "New poly(lactic acid) composites produced from coffee beverage wastes," *Journal of Applied Polymer Science*, vol. 139, no. 1, Jan. 2022, doi: 10.1002/APP.51434.

64. N. Gama, A. Ferreira, and D. E. V. Tuguin, "New poly(lactic acid) composites produced from coffee beverage wastes," *Journal of Applied Polymer Science*, vol. 139, no. 1, Jan. 2022 [Online]. Available: https://onlinelibrary.wiley.com/doi/abs/10.1002/app.51434. Accessed on Aug. 23, 2023.

11

Biocomposites from Durian Biomass Wastes: Properties, Characterisation, and Applications

E. S. Zainudin, H. A. Aisyah, N. M. Nurazzi, and R. A. Ilyas

11.1 Introduction

Durian, scientific name *Durio zibethinus*, is one of the best-known tropical fruits in Southeast Asia, a member of the Bombaceae family. A total of 29 species are reported to be present in the genus Durio, 23 of them having been discovered in Malaysia (Idris, 2011). At least 9 species have been identified in Malaysia that generate edible fruits, out of a total of 23 species. The Malay Peninsula is the original area where this fruit was first farmed, even though it has lately been developed mostly within the Southeast Asian region (Zhan et al., 2021).

Although the durian is known for its terrible scent, it is a popular fruit throughout the Southeast Asian region. Whereas durian trees are considered to have been introduced first in Peninsular Malaysia and Borneo rather than any other region, their growth has spread throughout tropical Asia, from Sri Lanka to Papua New Guinea, with certain varieties also being grown in Hawaii, the United States, and Northern Australia (Muhammad et al., 2021). According to Tan et al., (2019), it was dubbed the "King of Fruits" by South Asian people, attributed to its heavenly flavour and mouth-watering aroma. Furthermore, according to Tan et al.'s (2019) study, durian has a variety of health benefits, including the availability of numerous bioactive and antioxidative chemicals and being an excellent source of protein.

Durian fruit also has its own harvest season, usually occurring in mid-May to July. The standard weight of a durian fruit is somewhere around 1.5 kg, and the nutritious component of the fruit contributes around 15%–30% of the total weight (Ngabura et al., 2019). About 80% of durians planted in Malaysia are seeds of original species, while the remaining 20% are cultivated utilizing cloned seeds (Mohd Ali et al., 2020). Approximately 50%

DOI: 10.1201/9781003408215-11

to 65% of a durian fruit consists of creamy yellow pulp, which is usually eaten, while the balance of the content, which is estimated to be around 45% to 55%, is categorized as waste, including seed and skin (Penjumras et al., 2014).

In general, according to the research by Manshor et al., (2014), around 40% of durian skin fibre (DSF) may be extracted from 1 kg of durian skin waste. Figure 11.1 shows the consumable and non-consumable parts of durian fruit. In recent years, durian waste, always one of the biggest contributors to agricultural waste due to being the king of fruit, has become a favourite among fruits not only in Malaysia but also in the Southeast Asian region. Basically, agricultural waste increases drastically in the durian season, which is of real concern for society as it can lead to environmental problems. According to Xu et al. (2017), durian waste can be considered the major contributor to agricultural waste, with roughly 480,000 metric tons of durian waste generated annually. A chart in Figure 11.2 illustrates that in 2022, Malaysia achieved a noteworthy uptick in durian production, reaching around 455,000 metric tons. This marked a notable increase compared to the previous year when production stood at approximately 448,000 metric tons (Statista, 2024). Figure 11.2 illustrates that in 2020, the greatest amount of waste was produced, 273,448 metric tons. As a result, landfills are experiencing an increase in the rate at which they are being filled, mainly due to the enormous volume and quantity of durian waste (Payus et al., 2021). According to *The Sun Daily* (2020), by 2030, Malaysia's national durian production is predicted to reach 443,000 metric tons, equivalent to 265,800–310,100 metric tons of the total waste production.

The expanding volume of durian waste in recent years requires serious monitoring, since it would cause environmental pollution if not burned or disposed of in a landfill. Essentially, as the quantity of durian waste

FIGURE 11.1
Consumable and non-consumable part from durian fruit (Adapted from Koay et al., 2018).

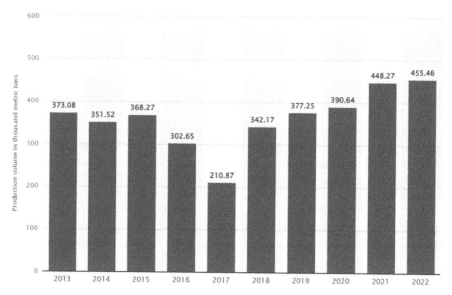

FIGURE 11.2
The production of durian waste in Malaysia (Adopted from Statista, 2024).

grows, it requires a larger area of land for disposal. To make ends meet, many durian sellers disregard hygiene by improperly disposing of the waste. As a result, the area has become a dirty and desolate place. As was reported by Mamat (2020), Solid Waste Management and Public Cleansing Corporation (SWCorp) are responsible for arresting individuals for illegally disposing of durian skins. The act is in contravention of Section 71 (1) of the Solid Waste Management and Public Cleansing Act 2007 (Act 672), which carries a fine of between RM10,000 and RM100,000 or a six-month to a five-year prison sentence. The adverse impacts of burning on a wider scale will harm the environment, while there is insufficient space for disposal. A new method of utilizing durian waste, as opposed to the old method of disposing of or burning it, must be developed. One alternative method in order to control the situation effectively is to convert durian waste into useful products such as composite, board, water, pulp and adsorbent material.

In light of the preceding discussion, this review aims to provide an overview on durian waste properties, including its physical, mechanical, and chemical properties. In addition, the utilization of durian waste fibre in composite fields is also described and discussed, as well as the main properties of the products. Finally, the challenges of the potential applications, especially applicability towards sustainable applications, will also be discussed. This comprehensive review is expected to pave the way for innovative approaches to the creation of novel durian fibre–based materials for a wide range of applications.

11.2 Properties of Durian Skin Fibre

The durian skin is composed of three unique layer structures: exocarp (firm thorny exterior layer), mesocarp (thick white middle layer), and endocarp (the thin inner layer that is in contact with the flesh) (San Ha et al., 2020). The durian skin acts as a shielding layer for the flesh when the durian drops to the ground. Exocarp, known as thorn, and mesocarp fibres are embedded in soft foam. When compressive force is applied, the fibres and the foam delaminate and commonly detach from the foam. Durian fibre is classified as short fibre with a diameter (D) and length (L) between 0.170 and 0.447 mm and between 0.84 and 2.38 mm, respectively (Manshor et al., 2012). Meanwhile, it has an average aspect ratio (L/D) between 4.94 and 5.32 mm, density (ρ) between 1.152 and 1.311 g/cm^3, and moisture content (MC) about 7.92–9.10%. Table 11.1 shows the data on the diameter, length, and aspect ratio of DSF.

Being one of the lignocellulosic materials, durian fibre mainly consists of cellulose, hemicelluloses, lignin, extractives, and inorganic matter. Durian fibre has a complex structure composed of 60.7% cellulose, 22.1% hemicellulose, and 17.2% lignin (Anuar et al., 2018). In other research by Anuar et al. (2015), cellulose (60.7%), hemicellulose (22.1%), and lignin (17.2%) are the three main components in durian fibre. The higher cellulose content in durian fibre has a significant influence on the mechanical properties of the fibre, such as high stiffness, high crystallinity, and high tensile and impact strength; thus, it has increased rigidity and does not dissolve easily in organic solvents (Lubis et al., 2018; Nur et al., 2014). The high cellulose and hemicellulose contents in durian fibre both contribute to the brittleness reduction of the fibre (Nur et al., 2014). The lignin content in the durian fibre influences the fibre's structure, properties, morphology, flexibility, and hydrolysis. Lignin is a polyphonic, amorphous, three-dimensional branching network polymer that provides vital mechanical support for plants. Lignin serves as a binder and stiffening agent, binding the fibre together (Mossello et al., 2010). Furthermore, the lignin content of the DSF affects the structure, characteristics, morphology, flexibility, and hydrolysis of the fibre (Nur Aimi et al., 2014). Table 11.2 shows the chemical composition of durian fibre.

TABLE 11.1

The Diameter, Length, and Aspect Ratio of DSF

	Diameter, D (mm)	Length (mm)	Aspect Ratio, L/D	Density, ρ (g/cm^3)	Moisture Content, MC (%)
Min	0.170	0.84	4.94	1.152	7.92
Max	0.447	2.38	5.32	1.311	9.10
Average	0.309	1.61	5.21	1.243	8.44

Source: Manshor, M.R. et al., *Materials and Design*, 59, 279–286, 2012.

TABLE 11.2

Chemical Composition of Durian Fibre

Chemical Composition	Dried Durian Peel (%)	Dried Durian Peel Fibre (%)	Durian Peel (%)
Ash Content	5.5	4.3	4.35
Lignin	10.9	10.7	15.45
Holocellulose	47.1	54.2	73.54
Alpha-cellulose	31.6	35.6	60.45
Hemi-cellulose	15.5	18.6	13.09

Source: Khedari, J. et al., *Building and Environment*, 39(1), 59–65, 2004.

In terms of mechanical properties, durian fibre has a low elongation break and absorbs less impact energy (Manshor et al., 2014). The tensile strength of durian fibre ranges from 60.50 to 298.80 MPa, the strain ranges from 1.1 to 4.9% 0.011 to 0.049, and Young's modulus ranges from 5707 to 5988 MPa (Lubis et al., 2018).

11.3 Applications

Being a natural fibre, durian fibre is a renewable source that is being used to create a generation of reinforcements for composite and polymer-based industries. Because of rising environmental concern, the development of green composite materials has been a popular issue recently. Durian fibres are being used in a wide range of interests, including building materials, boards, absorbent materials, bioenergy, furniture, biopolymers, pulp and paper, and as a reinforcement material, as shown in Figure 11.3. This is primarily because they have benefits over synthetic fibres, such as low cost, light weight, energy saving applications, enhanced surface smoothness of composite components, good relative mechanical qualities, and renewable resources. Durian fibre–reinforced composites are being used extensively to replace conventional synthetic fibre, as well as to fully utilize the fibre and reduce environmental problems. This section will focus on the applications of durian fibre in composite sectors, mainly on engineered board and polymer composite. Some of the applications of durian fibre are listed in Table 11.3.

11.3.1 Engineered Board

Engineered board is often created from lumber-producing timber that has been blended with adhesives. In order to make wood that satisfies size standards that are challenging to satisfy in nature, this form of board frequently

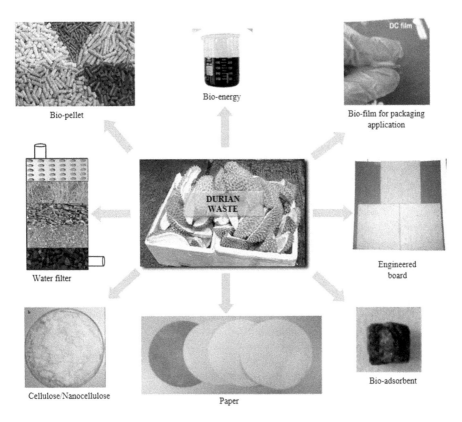

FIGURE 11.3
Applications of durian fibres.

uses waste from sawmills, or other natural fibre, which is then treated by chemical or heating procedures. Engineered board panels are utilized to overcome the limitations of natural wood in construction and industry to avoid catastrophe. Engineered board panels include particleboard, medium-density fibreboard (MDF), hardboard, oriented strand board (OSB), and cement board. Particleboard is one of the best-known engineered boards because it is cheaper, easy to process, and denser, and it is utilized when aesthetics and durability are less important than cost. MDF is a composite product made by decomposing wood into fibres, frequently in a defibrator, and mixing them with a synthetic resin binder, followed by applying extreme heat and pressure. Hardboard is produced from compacted shattered wood fibres bonded with a resin, followed by a hot-pressing process. The hardboard is dimensionally stable, scratch- and dent-resistant, and water-resistant; however, the thickness is inconsistent due to uneven MC and surface interaction. Cement board is made from wood strands, particles, or fibres bonded with Portland cement as a mineral binder and is used in numerous

TABLE 11.3

Applications of Durian Fibers in the Composite Industry

Application	Fibre Type and Matrix	Reference
Paper making	Durian rind pulp	Masrol et al. (2015), Masrol et al. (2017)
Food packaging film	DSFs + Polylactic acid (PLA)	Nur E'zzati et al. (2017)
Carbon aerogels for adsorption of organic liquids and oils	Durian shell pyrolysis	Wang et al. (2017)
Paper making	Durian rind soda-anthraquinone pulp	Masrol et al. (2018)
Food packaging film	Polylactic acid (PLA) + epoxidized palm oil (EPO) + cinnamon essential oil	Anuar et al. (2018)
Brake canvas	Durian fibres and teak leaf fibres + polyester resin + magnesium oxide	Matsuri et al. (2018)
Water filter	Durian husk and palm fibres	Marzuki and Daud (2018)
Packaging film	Durian-rind cellulose	Zhao et al. (2019)
Carbon-rich biochar (CRB)	Cassava rhizome, durian peel, pineapple peel, and corncob	Aup-Ngoen and Noipitak (2020)
Cellulose fibre casted film	Cellulose fibre from durian rind	Lubis et al. (2020)
Bioenergy for industrial pyrolytic applications	Durian shell pyrolysis	Liu et al. (2021)
Bio-pellets	Empty fruit bunch and durian rinds + cornstarch adhesive	Selvarajoo et al. (2021)
Methylene blue (MB) and water hardness removal	Durian and passion fruit peels	Gan and Cheok (2021)
Adsorbent for removal of physical pollutants	Durian husk	Payus et al. (2021)
Sulfur carrier for lithium-sulfur batteries	Durian shell derived hierarchical porous carbon (DPC)	Xue et al. (2021)
Water hardness removal	Jackfruit, passion fruit, and durian peel	Tan et al. (2021)
Bio-energy	Anaerobic digestion of durian shell and jackfruit peel	Wang et al. (2022)

construction applications. Cement board is substantially more resistant to both decay and combustion than resin-bonded or solid wood boards.

Khedari et al. (2003) prepared particleboard from durian and coconut coir fibres with low thermal conductivity for construction panels to apply in building insulation, such as walls and ceilings. In this study, three types of resins were used, namely, urea formaldehyde (UF), phenol formaldehyde (PF), and isocyanate (IC), via the hot press method. Mechanical and thermal conductivity tests were conducted to observe the behaviours of the composites employing different types of resin. According to the authors, the superior PF and IC resins are advised for external exposure, while certain UF boards performed better than PF and IC boards when bending strength and internal bonding were taken into account. In terms of thermal conductivity, the values range from 0.054 to 0.1854 W/mK, which is a rather low value. Having such low values will undoubtedly help such boards gain more acceptance.

In the next study, Khedari et al. (2004) developed a mixture of durian fibre and coconut coir insulation particleboard at different mixture ratios and board densities. The mixing ratios of the fibres were 10:90, 25:75, 75:25, and 90:10 durian fibre and coconut coir by weight, respectively. The resin used in this study was 12% UF. The qualities of insulation particleboards were found to be significantly influenced by the mixture ratio and board density. With a board density of 856 kg/m³, the ideal mixing ratio of durian fibre and coconut coir was 10:90. The high ratio of coconut coir created a strong cross-linking network between the fibres, which eventually reduced the void and open spaces between the particles in the particleboard structure. Furthermore, the chemical component make-up of this ratio also revealed that the board included a significant amount of lignin, which has a strong water resistance and internal bonding strength. Due to their low thermal conductivity in any mixing ratio, particleboards made from durian fibres and coconut coir can be utilized as a component of construction materials, to help save energy, and also assist in managing and using agricultural waste.

In a separate study, durian husk from two varieties, namely, D24 and D163, was analysed for chemical composition, gross energy (GE) value, and fibre strength for use as an alternative material for the particleboard industry (Zddin and Risby, 2010). The chemical composition and GE of durian husks for both varieties were found to be suitable and within the normal values for other natural fibres, and it was possible to use them as a material for particleboard production. The micrograph images of the durian husk in Figure 11.4 also show a microfibrillar angle that is an important aspect of fibre resistance to deformation when load is applied. Thus, according to the analysis, the tested fibre can be used as a replacement material in the particleboard sector.

Charoenvai (2013) applied durian peel powder as an adhesive to reduce the formaldehyde resin usage of particleboard production. The main objective of this study was to reduce the dependency on utilizing hazardous formaldehyde resin as well as to produce green insulating particleboard. The results show that durian powder was able to act as a natural binder, with good thermal conductivity properties. It was also concluded that the ideal ratio of durian fibres: durian powder: water is 2:1:1.5, dried at a temperature of 100 C for 24 h. A study on using durian fibres and polyvinyl alcohol to produce formaldehyde-free compressed fibreboard was performed by Watanapa and Wiyaratn (2013). The outcome revealed that a blend of durian fibres and polyvinyl alcohol at a weight ratio of 1:4 and 4% for alkyl ketene produces the best compacted durian fibreboard. The outcomes of the experiment demonstrated that adding polyvinyl alcohol may greatly enhance physical qualities. Therefore, it is feasible and more intriguing to create compressed durian fibreboard utilizing polyvinyl alcohol as the binder.

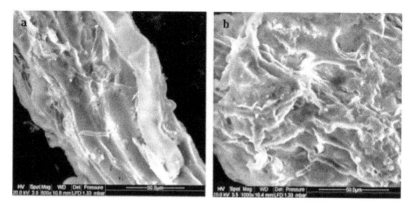

FIGURE 11.4
Micrograph images of (a) D24 durian fibre and (b) D163 durian fibre at 20,000x magnification (Adapted from Zddin and Risby, 2007).

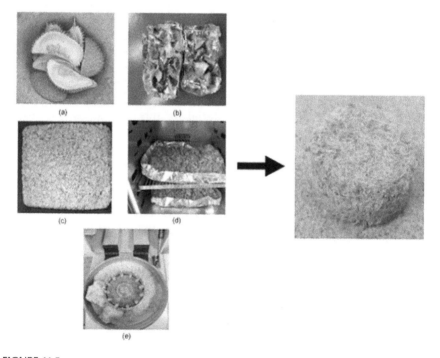

FIGURE 11.5
The preparation process of durian fibres. (a) durian waste, (b) cutting process, (c) soaking process, (d) drying process, (e) milling process (Adapted from Putra et al., 2022).

Most recently, Putra et al. (2022) investigated the absorption performance of durian fibres according to the manufacturing process of the samples, such as the alkali treatment shown in Figure 11.5. The acoustical properties were calculated for different thickness and density samples.

The sample with 20 mm thickness shows an absorption coefficient above 1.3 kHz, while the sample with 30 mm thickness and density of 214 kg/m^3 shows an absorption coefficient above 1 kHz. This performance meets the requirement for commercial rockwool with a thicker sample of 45 mm. Durian husk fibres can thus be used as an alternative natural sound absorber.

11.3.2 Polymer Composite

Durian fibre composites have been applied as a reinforcement material in many types of applications. It is claimed that durian fibre–reinforced polymer composites have been developed during the past several decades, and many industries have reported using these composites to manufacture a wide range of products. Table 11.4 shows the summary of the durian fibre–reinforced thermoset and thermoplastic polymer composites.

The influence of fibre content and pre-treatment of durian fibre using 4% sodium hydroxide (NaOH) on the impact and thermal properties of polylactic Acid (PLA) composites has been identified. Manshor et al. (2014) engineered durian fibre–reinforced PLA composites, where the fibre content was varied between 10% and 40%, and the effect of the fibre content of the composite on the impact, flexural, and thermal performance was also investigated. The results show that when compared with an untreated composite, treated durian fibres significantly improved the characteristics of PLA composites. The impact strength and flexural modulus of composites with 30% pre-treated durian fibres were found to be comparable to those of unreinforced PLA. The micrograph analysis of the impact sample reveals that the impact strength decreases with fibre content due to the weakness of interfacial bonding, which creates microscopic voids between the matrix and fibres and subsequently makes it simple for fractures to spread. Thermogravimetry analysis (TGA) values revealed that pre-treated fibres increased the thermal stability of PLA composite, while differential scanning calorimetry (DSC) revealed that the presence of pre-treated fibres only slightly increased the glass transition temperature (T_g), crystallization temperature (T_c), and melting temperature (T_m) compared with untreated DSF, suggesting that NaOH pre-treatment improves the fibre's properties.

In a separate study, a blend polymer of polypropylene (PP) and maleic anhydride polypropylene (MAPP) was used to fabricate a durian fibre composite (Nur Aimi et al., 2014). The main objective of this study was to study the effect of fibre content, fibre size, and different amounts of MAPP added, namely, 0, 2.5, and 5.0 wt.%, on the composite properties. Response surface methodology (RSM) was used to determine the effect of independent factors, either individually or in combination, and to assess the connection between the responses and the independent variables. It was found that the strength of the composites was mainly influenced by the fibre content, followed by

TABLE 11.4

Studies on DSFs-Reinforced Polymer Composites

Type of Fibre	Matrix	Processing Technique	Reference
Durian skin fibre (10, 20, and 30% weight)	Polylactic acid (PLA)	Twin screw extruder and injection moulding	Kaiser et al. (2012)
NaOH treated DSFs (10, 20, 30, and 40% weight)	PLA	Extrusion and injection moulding	Manshor et al. (2014)
NaOH treated DSFs (10, 20, 30, and 40% weight)	PLA and maleic anhydride	Twin screw extruder and injection moulding	Nur Aimi et al. (2012)
Durian skin fibre (10, 30, and 50% weight)	Polypropylene (PP) and maleic anhydride polypropylene (MAPP)	Twin screw extruder and injection moulding	Nur Aimi et al. (2014)
Durian peel fibre (5, 10, 15, and 20 %)	High density polyethylene (HDPE)	Twin screw extruder and mat forming	Charoenvai (2014)
Durian skin fibre (10, 20, 30, 40, 50, 60, and 70%)	Low linear density polyethylene (LLDPE)	Internal mixer and compression moulding	Wan Nazri et al. (2014)
30% durian skin fibre	PP and MAPP	Extrusion and injection moulding	Anuar et al. (2015)
Durian skin fibre	Polylactic acid (PLA) + epoxidized palm oil (EPO)	Solution casting	Anuar et al. (2016)
Durian skin fibre	PLA + EPO + cinnamon essential oil (CEO)	Solution casting	Anuar et al. (2018)
Durian skin fibre	PLA	Solution casting	Nur E'zzati et al. (2017)
Durian skin nanofibre	PP	Melt blending	Nur E'zzati et al. (2018a)
Durian skin nanofibre	PP	Internal mixer and compression moulding	Nur E'zzati et al. (2018b)
Durian skin fibre	High density polyethylene (HDPE) + maleic anhydride grafted polyethylene (MAPE)	Melt blending and compression moulding	Abdul Rashid et al. (2018)
Durian skin fibre and Kevlar ply	Phenolic resin	Vacuum infusion	Salman et al. (2018)
Durian fibres and teak leaf fibres	Polyester resin + magnesium oxide	Hot press	Matsuri et al. (2018)
Durian husk fibres	Recycled polystyrene foam	Melt compound and compression moulding	Koay et al. (2018)
Durian husk fibres	Recycled polystyrene foam + Ultra-Plast	Melt compound and compression moulding	Chun et al. (2019)
NaOH treated DSFs (0.1, 0.2, 0.3, and 0.4 g)	Polyester resin	Hot press	Sembiring et al. (2018)
NaOH treated DSFs (15, 25, 30, and 40 wt.%)	PLA	Melt compound and compression moulding	Pang et al. (2019)

the fibre size and the MAPP content. At 50 wt.% fibre content, 250–500 m fibre size, and without MAPP content, the optimum values for impact and stiffness behaviour were obtained. The sample without MAPP showed fibre pullout, fibres debonding, and fibre fracture, indicating that the energy needed to debond, pull out, and break the fibres was used, which enhanced the toughness of the composite. The sample with MAPP showed that the contact between the fibres and the matrix was still there, as were the energy-absorbing processes brought about by fibre pullout and fibre breakage.

The effect of PLA composites was studied by Nur Aimi et al. (2015), who used durian skin nanofibres (DSNF) with the fibre size in the range of 49–81 nm. Five types of composite with different DSNF formulations were developed, namely. 1, 2, 3, 4, and 5% of DSNF weight. The tensile, impact, and thermal properties of the PLA-DSNF composites were analysed through a series of tests, and they found that composites with low DSNF loading exhibited notable tensile and impact strength due to their higher aspect ratio, which created a higher possibility to arrange fibres in the direction of tensile and impact deformation. For thermal analysis, it was noted that the presence of 1% DSNF improves the composite's thermal stability, as it was indicated that the T_g, T_c, and T_m of the PLA biocomposite were likewise raised by the addition of DSNF. The inclusion of DSNF suggested that the PLA-DSNF composite had a lower storage modulus and loss modulus.

The influence of fibre composition on the mechanical performance and interaction of durian skin and teak leaf powder composites developed for brake canvas has been identified. Matsuri et al. (2018) engineered durian skin and teak leaf powder–reinforced polyester composites, where the fibre composition was varied between 20 and 60% (Figure 11.6), and the effect of the fibre composition of the composite on the hardness, wear resistance, and absorbency performance of the brake canvas was also investigated. The results show that with a composition of 40% durian fibres and 40% of teak leaf fibres, the composite is similar to the standard brake canvas that is widely used and meets the standard. Due to the durian fruit's coarse texture

FIGURE 11.6
The different composition of durian skin and teal leaves powder for brake canvas appalication (Adapted from Masturi et al., 2018).

and highly permeable fibres, the amount of durian fibres had a substantial impact on all of the tested properties.

Alternative methods of improving the mechanical performance of durian fibre composites were investigated by researchers. Cellulose or nanosize fillers are being added to resin to boost matrix strength in addition to minimizing void formation and improving water resistance. This method was adopted by Mohd Nordin et al. (2021) in their study by enhancing the properties of PLA composite using DSNF via a freeze drying (FD) method. In this study, cinnamon essential oil was added as a plasticizer, and the effect of the FD process on DSF and its effect on PLA composite tensile strength was studied. The addition of DSF and DSNF to PLA significantly changed the tensile strength of these composites. The addition of nanofiller was found to reduce the tensile strength because of the agglomeration of DSNF and hydrophobicity with PLA. However, in a separate study by Lubis et al. (2020), the incorporation of cellulose from durian fibres and 2-[acryloyloxyl] ethyl-trimethylammonium chloride (CIAETA) was found to improve antibacterial and antifungal ability; thus, it may be a potential application in the biomedical sector.

11.4 Challenges and Future Outlook

DSF has earned a place as a natural source in the composites sector because of its structure and qualities, which are almost identical to those of other natural fibres. The fibres from durian skin are currently used in a wide variety of applications, mostly in composite lines. The development of durian fibre–based products has benefited from intensive research and production efforts, which have been aided by rising customer demand for natural products, as well as a promising method of management and utilization of durian fruit wastes.

However, there are a few challenges that need to be focused on to widen the application of durian-based composites. The two most challenging aspects of using durian fibres are the availability and the extraction process of the fibre material itself. Considering that durian is a seasonal fruit, the amount of waste would be predicted to be higher during the harvesting season, normally around May to August every year. But, Malaysia is also an exporter of durian, which makes durian waste always available throughout the year. In terms of the fibre extraction process, it must be processed to obtain durian skin short particles. Normally, the durian skin is cut into small pieces using a shredder or chipper machine to reduce the size. The methods involved are mostly for small-scale production. The shredded particles will then be used for the next process. Another limitation of durian fibres is their compatibility, due to different polarities with the matrix. Many studies have been conducted to improve the interfacial bonding between the hydrophilic durian fibres and the hydrophobic matrix. Most researchers treat the fibres using NaOH to improve the fibres' properties in order to promote the interface bonding between the fibres and the matrix and thus enhance the mechanical characteristics of the composite. In addition, more research is needed on many aspects, such as toxicity, antibacterial properties, and thermal and strength characteristics, to improve durian fibre–based products for many applications, especially food packaging applications, before they may be used commercially.

11.5 Conclusions

Durian waste biomass is a by-product that is frequently left in the field and poses environmental problems. One of the most promising approaches to utilize and manage durian fruit waste is the use of DSFs in composite production. The utilization of durian fibres as a reinforcement material in thermoplastic and thermoset polymer composites was summarized, including the influential parameters in optimizing the composite properties. Numerous reinforcing characteristics, including fibre composition, hybridization, treatment, the addition of stabilizers, and the integration of nanomaterials, have been studied in order to determine how they affect the properties of the composite. For many potential applications, most research has been effective in increasing thermoplastic and thermoset properties. This review article was intended to compile information and expertise to promote more study of durian fibre–based composites.

Acknowledgements

The author is grateful to the Universiti Putra Malaysia for the access to the digital library and is deeply indebted to Noorashikin and Aleif Hakimi for their invaluable help.

References

Ali, Mohd, Mohd Maimunah, Norhashila Hashim, Samsuzana Abd Aziz, and Ola Lasekan. 2020. "Exploring the Chemical Composition, Emerging Applications, Potential Uses, and Health Benefits of Durian: A Review." *Food Control* 113: 107189. https://doi.org/10.1016/j.foodcont.2020.107189.

Anuar, Hazleen, Nur Aimi Mohd Nasir, and Yousuf El-Shekeil. 2015. "Effects of Coupling Agent on the Properties of Durian Skin Fiber Filled Polypropylene Composite." *International Journal of Materials and Metallurgical Engineering* 9(12): 1378–1382. https://doi.org/10.5281/zenodo.1110357.

Anuar, Hazleen, Mohd Syafiq Razali, Hafizul Adzim Saidin, Ammelia Fazlina Badrul Hisham, Siti Nur E'zzati Mohd Apandi, and Fathilah Ali. 2016. "Tensile Properties of Durian Skin Fiber Reinforced Plasticized Polylactic Acid Biocomposites." *International Journal of Engineering Materials and Manufacture* 1(1): 16–20. https://doi.org/10.26776/ijemm.01.01.2016.04.

Anuar, Hazleen, Mohd Apandi Siti Nur E'zzati, Abu Bakar Nur Fatin Izzati, Syed Mustafa Sharifah Nurul Inani, Fathilah Ali, and F. Ali. 2018. "Physical and Functional Properties of Durian Skin Fiber Biocomposite Films Filled with Natural Antimicrobial Agents." *BioResources* 13(4): 7255–7269. https://doi.org/10.15376/biores.13.47255-7269.

Aup-Ngoen, Kamonwan, and Mai Noipitak. 2020. "Effect of Carbon-Rich Biochar on Mechanical Properties of PLA-Biochar Composites." *Sustainable Chemistry and Pharmacy* 15: (100204): 1–4. https://doi.org/10.1016/j.scp.2019.100204.

Charoenvai, S. 2013. "New Insulating Material: Binderless Particleboard from Durian Peel." *Advances in Civil Engineering and Building Materials*: 119–123.

Charoenvai, Sarocha. 2014. "Durian Peels Fiber and Recycled HDPE Composites Obtained by Extrusion." *Energy Procedia* 56: 539–546. https://doi.org/10.1016/j.egypro.2014.07.190.

Chun, Koay Seong, Varnesh Subramaniam, Chan Ming Yeng, Pang Ming Meng, Chantara Thevy Ratnam, Tshai Kim Yeow, and Cheah Kean How. 2019. "Wood Plastic Composites Made from Post-Used Polystyrene Foam and Agricultural Waste." *Journal of Thermoplastic Composite Materials* 32(11): 1455–1466. https://doi.org/10.1177/0892705718799983.

Gan, Jia Ler, and Choon Yoong Cheok. 2021. "Enhanced Removal Efficiency of Methylene Blue and Water Hardness Using NaOH-Modified Durian and Passion Fruit Peel Adsorbents." *Progress in Energy and Environment* 16: 36–44.

Idris, S. 2011. "Introduction." *Durio of Malaysia*: 1–3. Strategic Resources Research Centre, Malaysian Agricultural Research and Development Institute (MARDI), Kuala Lumpur.

Kaiser, Mohammad Rejaul 2012. "Processing and Characterization of Durian Skin Fibers Composites: Mechanical Properties." *International Conference on Design and Innovation (ICDC)*, 7–8 November 2012: 142–146.

Khedari, Joseph, Sarocha Charoenvai, and Jongjit Hirunlabh. 2003. "New Insulating Particleboards from Durian Peel and Coconut Coir." *Building and Environment* 38(3): 435–441. https://doi.org/10.1016/S0360-1323(02)00030-6.

Khedari, Joseph, Noppanun Nankongnab, Jongjit Hirunlabh, and Sombat Teekasap. 2004. "New Low-Cost Insulation Particleboards from Mixture of Durian Peel and Coconut Coir." *Building and Environment* 39(1): 59–65. https://doi.org/10.1016/j.buildenv.2003.08.001.

Koay, Seong Chun, Varnesh Subramanian, Ming Yeng Chan, Ming Meng Pang, Kim Yeow Tsai, and Kean How Cheah. 2018. "Preparation and Characterization of Wood Plastic Composite Made Up of Durian Husk Fiber and Recycled Polystyrene Foam." *MATEC Web of Conferences* 152: 02019. https://doi.org/10.1051/matecconf/201815202019.

Liu, Hui, Jingyong Liu, Hongyi Huang, Fatih Evrendilek, Shaoting Wen, and Weixin Li. 2021. "Optimizing Bioenergy and By-Product Outputs from Durian Shell Pyrolysis." *Renewable Energy* 164: 407–418. https://doi.org/10.1016/j.renene. 2020.09.044.

Lubis, Rosliana, Sri Wahyuna Saragih, Basuki Wirjosentono, and Eddyanto Eddyanto. 2018. "Characterization of Durian Rinds Fiber (*Durio zubinthinus, murr*) from North Sumatera." *AIP Conference Proceedings* 2049(1): 020069. https://doi.org/10 .1063/1.5082474.

Lubis, Rosliana, Basuki Wirjosentono, and Athanasia Amanda Septevani. 2020. "Preparation, Characterization and Antimicrobial Activity of Grafted Cellulose Fiber from Durian Rind Waste." *Colloids and Surfaces A: Physicochemical and Engineering Aspects* 604: 125311. https://doi.org/10.1016/j.colsurfa.2020.125311.

Mamat, M. R. 2020. "Sesuka Hati Buang Kulit Durian." *Harian Metro. Malay Mail* [Online]. [Accessed 17th April 2022]. https://www.hmetro.com.my/mutakhir/ 2020/07/598030/sesuka-hati-buang-kulit-durian.

Manshor, R. M., Hazleen Anuar, Wan Busu Wan Nazri, and M. I. Fitrie. 2012. "Preparation and Characterization of Physical Properties of Durian Skin Fibers Biocomposite." *Advanced Materials Research* 576: 212–215. https://doi.org/10.4028 /www.scientific.net/AMR.576.212.

Manshor, M. R., H. Anuar, M. N. Nur Aimi, M. I. Ahmad Fitrie, W. B. Wan Nazri, S. M. Sapuan, Y. A. El-Shekeil, and M. U. Wahit. 2014. "Mechanical, Thermal and Morphological Properties of Durian Skin Fiber Reinforced PLA Biocomposites." *Materials and Design* 59: 279–286. https://doi.org/10.1016/j.matdes.2014.02.062.

Marzuki, Nor Ashikin, and Nur Hanani Daud. 2018. "Effect of Durian Husk and Palm Fiber on Performance of "Dukoffilt" to Treat Domestic Waste Water." *Modern Environmental Science and Engineering* 4(2): 169.

Masrol, Shaiful Rizal, Mohd. Halim Irwan Ibrahim, and Sharmiza Adnan. 2015. "Chemi-Mechanical Pulping of Durian Rinds." *Procedia Manufacturing* 2: 171– 180. https://doi.org/10.1016/j.promfg.2015.07.030.

Masrol, Shaiful Rizal, Mohd. Halim Irwan Ibrahim, and Sharmiza Adnan, Muhammad Syauqi Asyraf Ahmad Tajudin, Radhi Abdul Raub, Siti Nurul Aqma Abdul Razak, and Siti Nur Faeza Md Zain. 2017. "Effects of Total Chlorine Free (TCF) Bleaching on the Characteristics of Chemi Mechanical (CMP) Pulp and Paper from Malaysian Durian (*Durio zibethinus Murr.*) Rind." *Jurnal Teknologi* 79(4): 55–64.

Masrol, Shaiful Rizal, Mohd. Halim Irwan Ibrahim, and Sharmiza Adnan, Muhammad Syauqi Asyraf Ahmad Tajudin, Radhi Abdul Raub, Siti Nurul Aqma Abdul Razak, and Siti Nur Faeza Md Zain. 2018. "Effects of Beating on the Characteristics of Malaysian Durian (*Durio Zibethinus Murr.*) Rind Chemi-Mechanical (CMP) Pulp and Paper." *Jurnal Teknologi* 80(2): 9–17.

Masturi, Masturi, Suhardi Effendy, Afrianus Gelu, Hammam Hammam, and Fianti Fianti. 2018. "Analysis of the Mechanical Properties Brake Canvas with Basic Ingredients of the Durian Fruit Skin and Teak Leaves." *Jurnal Bahan Alam Terbarukan* 7(2): 149–155. https://doi.org/10.15294/jbat.v7i2.15019.

Mohd Nordin, Nurfarahin, Hazleen Anuar, Yose Fachmi Buys, Fathilah Ali, Sabu Thomas, and Nur Aimi Mohd Nasir. 2021. "Effect of Freeze-Dried Durian Skin Nanofiber on the Physical Properties of Poly (Lactic Acid) Biocomposites." *Polymer Composites* 42(2): 842–848. https://doi.org/10.1002/pc.25869.

Mossello, A. A., J. Harun, S. R. F. Shamsi, H. Resalati, P. M. Tahir, Ibrahim Rushdan, and A. Z. Mohmamed. 2010. "A Review of Literatures Related to Kenaf as an Alternative for Pulpwoods." *Agricultural Journal* 5(3): 131–138.

Muhammad, N. H. Z., S. Y. Low, S. N. S. M. Shukri, A. H. A. Samah, H. Z. H. Basri, M. H. M. Shuhaimi, H. N. Hamzah, M. A. Zahidin, M. S. A. Ariffin, And M. N. Zalipah. 2021. "Flower Visiting Bats and Durian Trees: Species Richness and Population Size." *Journal of Sustainability Science and Management* 16(5): 80–90.

Ngabura, Mohammed, Siti Aslina Hussain, Wan Azlina W. A. K. Ghani, Mohammed S. Jami, and Yen P. Tan. 2019. "Optimization and Activation of Renewable Durian Husk for Biosorption of Lead (II) From an Aqueous Medium." *Journal of Chemical Technology and Biotechnology* 94(5): 1384–1396. https://doi.org/10.1002/jctb.5882.

Nur Aimi, M. N., F. Ahmad, M. R. Kaiser, and H. Anuar 2012. "Effect of Malaeic Anhydride on Mechanical Properties of Polylactic Acid (PLA) Composites Reinforced with Durian Skin Fiber (DSF)." *1st International Conference on Design and Innovation (ICDI 2012)*: 318–324.

Nur Aimi, M. N., H. Anuar, M. R. Manshor, W. B. Wan Nazri, and S. M. Sapuan. 2014. "Optimizing the Parameters in Durian Skin Fiber Reinforced Polypropylene Composites by Response Surface Methodology." *Industrial Crops and Products* 54: 291–295. https://doi.org/10.1016/j.indcrop.2014.01.016.

Nur Aimi, M. N., H. Anuar, M. Maizirwan, S. M. Sapuan, M. U. Wahit, and S. Zakaria. 2015. "Preparation of Durian Skin Nanofiber (DSNF) and Its Effect on the Properties of Polylactic Acid (PLA) Biocomposites." *Sains Malaysiana* 44(11): 1551–1559.

Nur E'zzati, M. A., H. Anuar, R., M. S. A. Salimah, Y. F. Buys, and R. Othman 2017. "Effect of Supercritical Carbon Dioxide on Tensile Properties of Durian Skin Fiber Biocomposite." *International Journal of Current Science, Engineering & Technology*: 197–202. https://doi.org/10.30967/ijcrset.1.S1.2018.197-202

Nur E'zzati, M. A., H. Anuar, and A. R. Siti Munirah Salimah. 2018a. "Effect of Coupling Agent on Durian Skin Fiber Nanocomposite Reinforced Polypropylene." *IOP Conference Series: Materials Science and Engineering* 290(1): 012032. https://doi.org/10.1088/1757-899X/290/1/012032.

Nur E'zzati, M. A., H. Anuar, and A. R. Siti Munirah Salimah. 2018b. "Environmental Degradation of Durian Skin Nanofiber Biocomposite." *IIUM Engineering Journal* 19(1): 223–236. https://doi.org/10.31436/iiumej.v19i1.903.

Pang, Ming Meng, Yesudian Aaron, Seong Chun Koay, Jiun Hor Low, Hui Leng Choo, and Kim Yeow Tshai. 2019. "Soil Burial, Hygrothermal and Morphology of Durian Skin Fiber Filled Polylactic Acid Biocomposites." *Advances in Environmental Biology* 13(3): 21–26.

Payus, C. M., M. A. Refdin, N. Z. Zahari, A. B. Rimba, M. Geetha, C. Saroj, A. Gasparatos, K. Fukushi, and P. Alvin Oliver. 2021. "Durian Husk Wastes as Low-Cost Adsorbent for Physical Pollutants Removal: Groundwater Supply." *Materials Today: Proceedings* 42(1): 80–87. https://doi.org/10.1016/j.matpr.2020.10.006.

Penjumras, P., B. A. R. Russly, A. T. Rosnita, and A. Khalina. 2014. "Extraction and Characterization of Cellulose from Durian Rind." *Agriculture and Agricultural Science Procedia* 2: 237–243. https://doi.org/10.1016/j.aaspro.2014.11.034.

Putra, Azma, Muhammad Nur Othman, Thaynan Oliveira, M'hamed Souli, Dg Hafizah Kassim, and Safarudin Herawan. 2022. "Waste Durian Husk Fibers as Natural Sound Absorber: Performance and Acoustic Characterization." *Buildings* 12(8): 1112. https://doi.org/10.3390/buildings12081112.

Rashid, Abd, Siti Munirah Salimah, Hazleen Anuar, Siti Nur E'zzati Mohd Apandi, and Yose Fachmi Buys. 2018. "Effect of UV Radiation on Degradation of Durian Skin Fiber Composite." *International Journal of Engineering Materials and Manufacture* 3(2): 105–112. https://doi.org/10.26776/ijemm.03.02.2018.05.

San Ha, Ngoc, Guoxing Lu, DongWei Shu, and T. X. Yu. 2020. "Mechanical Properties and Energy Absorption Characteristics of Tropical Fruit Durian (*Durio zibethinus*)." *Journal of the Mechanical Behavior of Biomedical Materials* 104: 103603. https://doi.org/10.1016/j.jmbbm.2019.10.360.

Salman, Suhad Dawood, and Zulkiflle Bin Leman. 2018. "Effect of Natural Durian Skin on Mechanical and Morphological Properties of Kevlar Composites in Structural Applications." *Journal of Engineering and Sustainable Development (JEASD)* 22(2): 1–9.

Selvarajoo, A., C. W. Lee, D. Oochit, and K. H. O. Almashjary. 2021. "Bio-pellets From Empty Fruit Bunch and Durian Rinds with Cornstarch Adhesive for Potential Renewable Energy." *Materials Science for Energy Technologies* 4: 242–248. https://doi.org/10.1016/j.mset.2021.06.008.

Sembiring, Timbangen, Kerista Sebayang, Pardinan Sinuhaji, and P. Sinuhaji. 2018. "Characterization of Biocomposite Materials Based on the Durian Fiber (*Durio Zibethinus Murr*) Reinforced Using Polyester Resin." *Journal of Physics: Conference Series* 1116(3): 032031. https://doi.org/10.1088/1742-6596/1116/3/032031.

Statista. "Malaysia: durian production." Statista, https://www.statista.com/statistics/1000876/malaysia-durian-production/. Accessed 19 February 2024

Tan, P. F., S. K. Ng, T. B. Tan, G. H. Chong, and C. P. Tan. 2019. "Shelf-Life Determination of Durian (*Durio Zibethinus*) Paste and Pulp Upon High-Pressure Processing." *Food Research* 3(3): 221–230.

Tan, Yee-Huan, Chee-Chian Kerk, Chew-Tin Lee, and Choon-Yoong Cheok. 2020. "The Potential of Tropical Fruit Peels as Ion Exchangers for Water Hardness Removal." *IOP Conference Series: Earth and Environmental Science* 463(1): 012093. https://doi.org/10.1088/1755-1315/463/1/012093.

The Sun Daily. 2020, September 13. *Malaysia's Durian Production to Hit 443,000 Metric Tonnes by 2030 - Mafi*. Sun Media Corporation Sdn Bhd [Online]. [Accessed 17 May 2022]. https://www.thesundaily.my/local/m-sia-s-durian-production-to-hit-443000-metric-tonnes-by-2030-mafi-BY4001553.

Wan Nazri, W. B., Z. Ezdiani, M. Romainor, K. Syarajatul Erma, J. Jurina, and I. Z. A. Noor Fadzlina. 2014. "Effect of Fiber Loading on Mechanical Properties of Durian Skin Fiber Composite." *Journal of Tropical Agricultural Food Science* 42(2): 169–174.

Wang, Ya, Lin Zhu, Fangyan Zhu, Liangjun You, Xiangqian Shen, and Songjun Li. 2017. "Removal of Organic Solvents/Oils Using Carbon Aerogels Derived from Waste Durian Shell." *Journal of the Taiwan Institute of Chemical Engineers* 78: 351–358. https://doi.org/10.1016/j.jtice.2017.06.037.

Wang, Ligong, Baocheng Wei, Fanfan Cai, Chang Chen, and Guangqing Liu. 2022. "Recycling Durian Shell and Jackfruit Peel via Anaerobic Digestion." *Bioresource Technology* 343: 126032. https://doi.org/10.1016/j.biortech.2021.126032.

Watanapa, A., and W. Wiyaratn. 2013. "Fabrication and Physical Testing Compressed Durian Fiberboard." *Journal of Engineering and Technology* 5: 73–75.

Xu, Wenjie, Quanlin Zhao, Rufan Wang, Zhenming Jiang, Zhenzhong Zhang, Xuewen Gao, and Ye Zhengfang. 2017. "Optimization of Organic Pollutants Removal from Soil Eluent by Activated Carbon Derived from Peanut Shells Using Response Surface Methodology." *Vacuum* 141: 307–315. https://doi.org/10.1016/j.vacuum.2017.04.031.

Xue, Mingzhe, Hong Xu, Yan Tan, Chen Chen, Bing Li, and Cunman Zhang. 2021. "A Novel Hierarchical Porous Carbon Derived from Durian Shell as Enhanced Sulfur Carrier for High Performance Li-S Batteries." *Journal of Electroanalytical Chemistry* 893: 115306. https://doi.org/10.1016/j.jelechem.2021.115306.

Zddin, Z., and M. S. Risby. 2010. "Durian Husk as Potential Source for Particleboard Industry." *AIP Conference Proceedings* 1217(1): 546–553. https://doi.org/10.1063/1.3377886.

Zhan, Yuan-fei, Xiao-tao Hou, Li-li Fan, Zheng-cai Du, Soo Ee Ch'ng, Siok Meng Ng, Khamphanh Thepkaysone, Er-wei Hao, and Jia-gang Deng. 2021. "Chemical Constituents and Pharmacological Effects of Durian Shells in ASEAN Countries: A Review." *Chinese Herbal Medicines* 13(4): 461–471. https://doi.org/10.1016/j.chmed.2021.10.001.

Zhao, Guili, Xiaomei Lyu, Jaslyn Lee, Xi Cui, and Wei-Ning Chen. 2019. "Biodegradable and Transparent Cellulose Film Prepared Eco-friendly from Durian Rind for Packaging Application." *Food Packaging and Shelf Life* 21: 100345. https://doi.org/10.1016/j.fpsl.2019.100345.

12

Biomaterials Based on Plant Fibres: Sustainable for Green Composites

M. N. M. Azlin, K. Z. Hazrati, and K. Z. Hafila

12.1 Introduction

Green materials can be made from these new components, since they are based on natural fibres and resins, fulfilling the overwhelming need to address environmental and economic concerns in manufacturing new materials (Hazrati et al., 2020; Hermawan et al., 2019). Due to increasing demand, the necessity for eco-friendly products, recycling, and reusing is widely acknowledged in the scientific and business worlds. From this vantage point, green composites made from biofibres and biopolymers are appealing because they can provide the necessary characteristics and functionalities at an affordable price (Norrrahim et al., 2021). New materials based totally on renewable resources are being developed due to efforts to reduce the emission of gases that contribute to the greenhouse effect, such as carbon dioxide (CO_2) (Tarique et al., 2021). The overuse of petroleum-based polymers causes a significant reduction in landfill capacity. In addition, due to government plastic waste management legislation and rising consumer interest in eco-friendly products, businesses are increasingly investing in the research and development of sustainable materials at reasonable prices to mitigate the effects of global warming (Azlin et al., 2020; Harussani et al., 2021; Hazrati et al., 2022; Tarique et al., 2023). Consequently, interest has increased in developing entirely biodegradable materials with acceptable functionalities due to growing public awareness of the environmental impacts of non-decomposable solid wastes. Biodegradable materials have recently been the focus of widespread interest (Azlin et al., 2022; Kamaruddin et al., 2023).

Natural fibres have emerged as a significant reinforcing material category in recent decades. According to the key global markets for material consumption, the projected drive growth of biodegradable materials is anticipated to increase at a pace of almost 13% on average annually (Azlin et al., 2020; Ilyas et al., 2021; Taharuddin et al., 2022). Fully biodegradable compounds are expensive and have few useful features, restricting their wide range of applications (Ilyas et al., 2018; Syafiq et al., 2021). Therefore, various projects focused on creating environmentally friendly materials have recently emerged to address these issues and slow the depletion of natural resources. It is widely accepted that these materials will play a crucial role in all industries over the next several centuries (Hazrati et al., 2021; Jumaidin et al., 2016; Thyavihalli Girijappa et al., 2019). According to the definition, natural fibre–reinforced composites typically comprise a matrix of biodegradable polymer and natural compounds. Due to their potential to replace conventional fibre–reinforced composites, notably glass fibre–reinforced composites, natural fibre–reinforced composites have recently attracted much attention. By 2021, it is predicted that up to 28% of the market for reinforcement materials will be made up of fibres made from bio-based sources (Harussani et al., 2020; Lila, 2021).

This review focuses on the potential of green materials made from natural fibres. They have attracted great attention and spurred a growing industry in materials technology due to their features, such as high mechanical strength and outstanding biocompatibility. Although there is a substantial quantity of research on natural fibre–based biomaterials, there is a dearth of comprehensive review studies that explicitly focus on the features of natural fibre–based biomaterials. This chapter attempts to highlight the research done in the field of natural fibre–based biocomposites. Its goal is to present an overview of the advances that have been made as well as the future course of action. It will provide an overview of natural fibres and biocomposites and point researchers towards specific application areas.

12.2 Natural Fibres

In the most recent few years, there has been a rapid increase in the usage of natural fibres for the production of a new kind of composites that are beneficial to the environment. Recent developments in natural fibre development, genetic engineering, and the science of composites offer considerable prospects for the production of superior materials derived from renewable resources and providing enhanced support for global sustainability (Mohanty et al., 2002). Natural fibre composites are preferred by industry because of their higher density and lower impact on the environment in comparison to traditional composites.

12.2.1 Natural Fibre Structure

Natural fibres have a well-organized structure inside their cell walls, which can be divided into three basic structural components. The angle and arrangement of the microfibrils within the cell wall can determine the properties of the fibres. The structure of natural fibres consisting of several cells is shown in Figure 12.1. The structure of the fibres can be divided into sections, such as primary and secondary walls. The primary cell wall consists of loose and irregularly densely packed cellulose microfibrils. The secondary wall consists of three layers, the outer layer (S1), the middle layer (S2), and the inner layer (S3). Among these layers, the middle layer (S2) is the thickest and plays an important role in contributing to the mechanical properties of the fibres (Thomas et al., 2011).

The primary cell wall of a plant spreads throughout the plant and contributes to the growth of the plant. Moreover, all three layers of the secondary cell wall of a plant contain a long microfibril chain. The amount of cellulose increases consistently from S1 to S2, but the content of hemicelluloses stays the same in each layer (Vimal et al., 2015). On the other hand, the content of lignin displays a reciprocal pattern to that of cellulose. The molecules that make up hemicellulose have the shape of a net, and they bind with cellulosic fibrils. Cellulose and hemicelluloses form a network, and the adhesive qualities of lignin and pectin are provided by these components. The tensile and shear strengths of cellulosic fibres are directly attributable to their adhesive qualities. Meanwhile, the physical and mechanical strength of the fibres is determined by the secondary layer (S2) (Beaugrand et al., 2014).

Usually, the strength qualities of a material are improved by having a higher cellulose content and a lower microfibrillar angle. The specific modulus and elongation at break are better in natural fibres than in synthetic fibres, which is regarded as a key factor in polymer engineering composites.

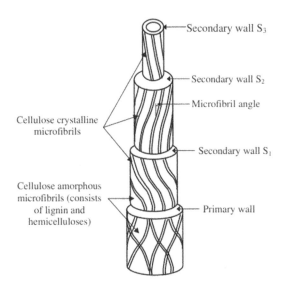

FIGURE 12.1
Structure of natural fibre. (From Kabir, M.M. et al., *Composites Part B: Engineering*, 7, 43, 2883–2892, 2012.)

In general, synthetic fibres show greater mechanical and physical properties than natural fibres, but natural fibres have better specific modulus and elongation at break than synthetic fibres.

12.2.2 Types of Natural Fibres

Natural fibres are organized into a variety of categories, and these categories can be further subdivided according to the plants from which the natural fibres originated, as depicted in the following.

12.2.2.1 Wood Fibres

Many different kinds of trees are used in the process of extracting wood fibres. Hardwood and softwood are the two primary categories that are used to describe the different types of wood fibres. The length of the fibres in softwood is typically longer than in hardwood, which is the primary distinction between the two types of wood. Besides, these fibres go through a manufacturing process in which they acquire unique properties, such as bond augmentation and alkali treatment. Additionally, the controlled manufacturing processes of these fibres ensure that the high variability in mechanical qualities and dimensional stability that are associated with the unprocessed plant-based fibres is significantly minimized (Zhou et al., 2015).

12.2.2.2 Stalk Fibres

These fibres are typically collected from plant stalks and can be obtained from plants like corn, maize, egg-plant, sugarcane, sunflower, wood, and various crop straws, including barley, wheat, rice, and others (Loganathan et al., 2020).

12.2.2.3 Leaf Fibres

Leaf fibres are coarse and brittle fibres that are taken from the leaf of the plant by hand scraping after the beating/retting process or by mechanical extraction. These fibres are mostly utilized for the production of woven products such as ropes, carpets, textiles, and mats due to their comparatively high strength. The most common types of leaf fibres include *Cymbopogon citratus* fibre, pandanus fibre, sisal, and pineapple leaf fibre (Arib et al., 2006).

12.2.2.4 Bast/Stem Fibres

Fibres obtained from the outermost part of plant stems are known as stem or bast fibres. Fibres such as ramie, banana, soya bean, jute, flax, abaca, kenaf, and hemp fall into the category of bast fibres. All these fibres have undergone the chemical and biological decomposition process of the plant stalks

called rotting. The natural properties of these fibres, which have a long filament and high mechanical strength, have led to their use as material for
the manufacture of textile products such as ropes, socks, curtains, bags, and
yarn (Bhatnagar et al., 2015).

12.2.2.5 Seed/Fruit Fibres

Fibres obtained from the outer husk of the respective fruit are called seed or
fruit fibres. Natural fibres from coconut husk, termed coir fibres, are obtained
from the outer husk of the coconut and are an example of fruit fibres. The
fibre properties, such as high tensile strength and low density, have led to
this type of fibre being used as one of the best materials for making geotex-
tiles, bags, mats, brushes, and ropes (Lazim et al., 2014).

12.2.2.6 Reed/Grass Fibres

Grass plants such as switchgrass, rye grass, elephant grass, and bamboo
plants can also produce natural fibres, which are called grass fibres. In addition, fibrous crop leftovers, including esparto, communis, bagasse, sabei,
phragmites, and canary grass, can primarily be used as fibre reinforcements
in cement-based composites.

12.3 Natural Fibre Extraction

Natural fibres comprise cellulose, hemicellulose, lignin, pectin, waxes, and
other water-soluble extractives (Kamaruddin, Jumaidin, Selamat et al., 2021).
Nonetheless, the utilization of biocomposite materials encounters certain
challenges. The utilization of natural fibres in biocomposites is restricted by
significant limitations, such as their susceptibility to moisture, inconsistent
fibre properties, and the inherent incompatibility between the hydrophilic
fibres and the hydrophobic matrix. The degree of this constraint is contin-
gent upon the fibre type, matrix characteristics, and the interaction between
the fibre and the matrix (Kamaruddin, Jumaidin, Rushdan et al., 2021).

Various treatment strategies have been devised to surmount these constraints. A potential approach involves the surface modification of fibres
through the removal of non-essential constituents that are unlikely to impact
composite properties, or alternatively, altering the structural composition of
the fibres to enhance adhesion. Additional methods involve the utilization of
compatibilizers or coupling agents in order to enhance stress transfer from
the matrix to the fibre. Chemical coupling agents are typically characterized
by the presence of two functional groups. The primary role of the first function is to interact with the hydroxyl groups present in cellulose, while the

second function involves reacting with the functional groups of the matrix. Numerous investigations have been carried out pertaining to the alteration of natural fibres for the purpose of composite fabrication (Xue Li et al., 2007). Various treatment techniques have been employed in these studies. Physical techniques are capable of altering the surface characteristics of fibres, thereby improving their adhesion to the matrix, without affecting their structural composition. Chemical techniques have been extensively developed and lead to modifications in the structure of fibres, facilitating interfacial bonding between the fibre and the matrix. The utilization of bacterial activity on the surface of fibres is a common technique in biological methodologies, which leads to a notable enhancement in the interfacial adhesion with polymeric matrices (Chen et al., 2018).

12.3.1 Fibre Surface Modification

The manufacturing of biocomposite materials poses challenges due to inherent incompatibilities between the lignocellulosic fibres and the matrix. Employing techniques that can modify the physical, chemical, or morphological attributes of fibres is essential in achieving effective adhesion between hydrophilic natural fibres and highly non-polar hydrophobic polymers (Kamaruddin, Jumaidin, Ilyas, Selamat et al., 2022a). As previously indicated, a significant constraint associated with the utilization of natural fibres pertains to their susceptibility to moisture, which results in inadequate bonding with the polymer and consequent weakness. Furthermore, the existence of waxes and pectin within the cellular wall impedes the interlocking process with the matrix through the masking of the reactive functional groups of the fibres. The conjoined impact of the aforementioned inaccessibility is a substandard adhesion across the phase boundary, leading to a feeble dispersion of force and inadequate strength properties (Wang et al., 2007). As a result, natural fibres have undergone treatment aimed at reducing their hydrophilic sites and enhancing their adhesion to matrix materials. Various treatment strategies have been devised, encompassing physical, chemical, and biological approaches.

12.3.2 Chemical Treatment

The majority of issues encountered in natural fibre composites stem from the hydrophilic properties of the fibres and the hydrophobic properties of the matrix. Consequently, there exists an innate incompatibility between the fibres and the matrix. The application of chemical treatment to reinforcing fibres has the potential to decrease their hydrophilic nature, leading to enhanced compatibility with the matrix. Numerous research endeavours have been undertaken to enhance the properties of fibres via diverse chemical treatments and their consequential impact on the properties of composites. This section provides an overview of various treatment methodologies and their impact on the characteristics of fibres and composites.

12.3.2.1 Alkaline Treatment

The utilization of NaOH for the treatment of natural fibres is a commonly employed technique for the production of thermoset and thermoplastic composites. The treatment induces a reorientation of the densely packed crystalline cellulose structure, resulting in the formation of an amorphous region through the swelling of the fibre cell wall (Kamaruddin, Jumaidin, Ilyas, Selamat et al., 2022b). This facilitates increased chemical penetration. The breakdown of alkali-sensitive hydrogen bonds present among the fibres leads to the formation of new reactive hydrogen bonds between the molecular chains of cellulose. Consequently, the hydrophilic hydroxyl groups undergo partial removal, thereby enhancing the material's resistance to moisture (Kamaruddin, Jumaidin, Rushdan et al., 2021). Additionally, it selectively removes specific fractions of hemicelluloses, lignin, pectin, and surface coatings of wax and oil (Mwaikambo et al., 2007). Consequently, the surface of the fibre is rendered more pristine. To clarify, the uniformity of the fibre surface is enhanced through the elimination of micro-voids, resulting in an improvement in the stress transfer capacity among the ultimate cells. Moreover, it diminishes the diameter of the fibre, leading to an increase in aspect ratio, consequently enhancing the interfacial adhesion between the fibre and the matrix (Wang et al., 2007) .

12.3.2.2 Silane Treatment

Silane is a type of compound that is represented by the chemical formula SiH_4. Silanes function as coupling agents that facilitate the adhesion of glass fibres to a polymer matrix, thereby imparting stability to the resulting composite material. The application of silane coupling agents has the potential to decrease the quantity of hydroxyl groups present in the fibre–matrix interface of cellulose. In the presence of moisture, the hydrolysable alkoxy group undergoes a reaction leading to the formation of silanols. The silanol moiety reacts with the hydroxyl functional group on the fibre, resulting in the formation of robust covalent bonds with the cell wall, which are chemisorbed on the fibre surface. The application of silane leads to the formation of hydrocarbon chains that effectively limit the swelling of the fibre. This is achieved by the formation of a cross-linked network facilitated by the covalent bond between the matrix and the fibre (Xue Li et al., 2007). The study conducted by Seki (Seki, 2009) aimed to investigate the effects of alkali treatment (with 5% NaOH for 2 hours) and silane treatment (with 1% oligomeric siloxane with 96% alcohol solution for 1 hour) on the flexural properties of jute-epoxy and jute-polyester composites. The application of an alkali and a silane treatment to jute-epoxy composites resulted in strength and modulus properties that were about 12% and 7% higher, respectively, compared with the alkali treatment alone. The application of comparable treatment techniques resulted in an approximate improvement of 20% and 8% for jute and polyester composites, respectively. Sever et al. (Sever et al., 2010) conducted an experiment in which they subjected jute fabric polyester composites to silane treatment at different concentrations (0.1%, 0.3%, and 0.5% γ-methacryloxypropyltrimeth oxysilane). The aim of the study was to investigate and compare the tensile, flexural, and interlaminar shear strength properties of composites that were not subjected to any treatment. The experimental results show that the composites treated with 0.3% silane exhibited a remarkable increase in their tensile, flexural, and interlaminar shear strength by about 40%, 30%, and 55%, respectively.

12.3.2.3 Benzoylation Treatment

Benzoyl chloride is used in benzoylation to reduce the hydrophilicity of the fibres and improve the adhesion of the fibre matrix, ultimately increasing the strength of the composite. It also improves the thermal stability of the fibre. Benzoylation treatment uses benzoyl chloride to reduce the hydrophilic nature of the fibre and improve the adhesion of the fibre matrix, which ultimately leads to an increase in the strength of the composite (Manikandan Nair et al., 2001). Moreover, the benzoylation treatment also improved the thermal stability of the fibre. According to Joseph et al. (Joseph et al., 1996), the application of benzoyl chloride treatment to alkali pre-treated sisal fibre resulted in increased thermal stability when compared with composites made from untreated fibres. Wang et al. (2007) conducted analogous treatment on composites of low-density polyethylene reinforced with flax fibre and documented enhancements of 6% and 33% in tensile strength and moisture resistance properties, respectively.

12.4 Natural Fibre–Reinforced Composites

Natural fibres, synthetic resin, and biodegradable polymers can be combined to create natural fibre–reinforced composites. The thermoplastic or thermoset matrix types are available for synthetic and biodegradable materials (Neto et al., 2022). The scope of application for natural fibres is expanding continuously, especially in the construction, automotive, and aviation sectors, which has significantly increased their utilization for manufacturing polymer-based composites (Abu Bakar et al., 2020). They are beginning to be utilized extensively in the building industry, the transport of biological systems, the plastics industry, construction, and the automotive sector. Natural fibres are used as reinforcement because they can lessen respiratory discomfort and tool wear and serve as substitutes for synthetic fibre composites amid the escalating global energy crisis and ecological dangers (Ashok et al., 2019; Venkategowda et al., 2021).

At least 100 years have passed since natural fibre–reinforced composites were first employed, but the last 30–40 years have seen the most commercial activity, particularly in thermoplastic composite technology. The construction industry uses natural fibre–reinforced composites for automobile interior panels and architectural mouldings, decking, and railing components (Azlina Ramlee et al., 2021; Xuemei Li et al., 2022; Rozyanty et al., 2021). Natural fibre–reinforced composites have found use in additive manufacturing recently. Poly-olefins (such as polypropylene [PP] and polyethylene [PE]), polyvinyl chloride (PVC), polyamides (such as nylon), and more recently, bio-based polylactic acid (PLA) are the most often used thermoplastics for natural fibre–reinforced composites. To process natural fibres without thermal deterioration, thermoplastic polymer matrices in natural fibre–reinforced composites seem promising due to their low cost, wide availability, and low processing temperatures, usually not more than 225 C (Kamaruddin, Jumaidin, Ilyas, & Selamat, 2022; Mayilswamy & Kandasubramanian, 2022; Sosiati et al., 2018).

Due to the growing demand for green, environmentally friendly products, new plant fibre–based materials have been developed to replace non-degradable materials generated from petroleum resources currently employed for various applications. Materials made from natural resources, such as reed grass, straw, hay, linen, hemp, or lichens, have been utilized for thermal insulation (Azlina Ramlee et al., 2021). The vehicle industry started using natural fibre–reinforced composites in several exterior and interior applications because of the significant weight reduction and inexpensive cost of raw materials (Zulnazri & Sulhatun, 2018). However, there are numerous issues with the manufacture of natural fibre–reinforced composites, including problems with processability, lack of dimensional stability, and compatibility with synthetic polymers. Many researchers have examined these difficulties in depth, since natural fibre–reinforced composites play a significant role in environmental situations and various applications (Neto et al., 2022).

A composite is created by combining matrix, fibres, and other components using various techniques. The manufacturing process affects the final properties of composites in addition to the fibre and matrix components. The following are the main objectives to be met to create composites with well-balanced mechanical properties: (i) a uniform distribution of fibres throughout the matrix; (ii) a balanced interaction between the matrix and the fibres to permit fibre pull-out; (iii) low matrix porosity; and (iv) an optimized fibre percentage, enough to reinforce the material while preserving matrix continuity. The extruder uses heat and shear force to break down the polymer and distribute the fibres. Due to natural fibre composites' sensitivity to heat, the process temperature must always be maintained at or below 200 C. Nadia et al. detailed the steps involved in making thermoplastic biocomposites, which include incorporating natural fibre in resin, impregnating inorganic fibre into the composite, and finally, dispersing the fibres to obtain the final product (Bakhori et al., 2022). Composites are typically made via injection or compression moulding following compounding. Compression moulding is favoured for natural fibre composites because its impact resistance is 50% higher than that of injection moulding (Kamaruddin, Jumaidin, Ilyas, Selamat et al., 2022a).

12.5 Industrial Application of Green Composites

The main challenge in using wood fibres (cellulose from trees) and natural plant fibres as reinforcement for polymer composites is the lack of compatibility between the hydrophilic nature of the natural fibres and the hydrophilic thermoplastic matrices during the incorporation process. This incompatibility ultimately leads to unfavourable properties of the resulting composites. Therefore, it is important to mitigate this problem by making various modifications to the fibre–polymer interface to improve the bond between the fibre and the matrix, thereby enhancing the overall performance of the composite. The construction industry, particularly in the areas of decking and cladding, and the automotive sector with respect to interior components are considered to be the main application areas for these biocomposites. An alternative approach is the use of interior door panels made of biodegradable polymers from sustainable sources such as PLA and polyhydroxyalkanoate (PHA) as opposed to conventional petroleum-based polymers such as PP, PE, or acrylonitrile-butadiene-styrene (ABS). Moreover, among the above-mentioned polymers, PHA is considered the most suitable biodegradable polymer from natural sources due to its superior properties and potential to meet the desired criteria. However, the use of biodegradable polymers as a material for the production of green composites faces the cost problem: the price of biodegradable polymers is still high, making their use economically unfeasible. Since the cost problem has affected the use of certain biodegradable polymers, blending of several biodegradable polymers is recommended. The aim is to reduce the cost of biodegradable polymers or materials to form green composites with natural fibres by introducing other, cheaper biodegradable polymers such as PLA (Loureiro & Esteves, 2018).

12.5.1 Automotive Industry

Nowadays, the use of conventional materials for the production of composite materials, such as plastics, in the automotive industry has shifted to environmentally friendly materials. This trend is due to the fact that people are aware of the importance of protecting the environment by using green products such as green composites in the automotive industry. The automotive industry has realized that the use of green composites can contribute to emission reduction, weight reduction, and fuel efficiency. These positive effects can now be seen when the various structures and components of vehicles are made from green composites, which is beneficial not only for the automotive industry but also for the environment. In fact, efforts to use green composites in the automotive industry started much earlier. There is evidence that Henry Ford introduced car parts made from bioplastics derived from soybeans in 1940 (Joshi et al., 2012). In 1957, East Germany manufactured a Trabant automobile (roof, boot lid, bonnet, fenders, and doors) from cotton fibres reinforced with thermosetting phenolic resin (Akampumuza et al., 2017). Mercedes-Benz began manufacturing door panels for the E-Class in 1996 from epoxy-polymer composite reinforced with jute fibres (Mohanty et al., 2005). Moreover, most car components today are made from natural fibre–reinforced bioplastic, which is used for interior and exterior applications because of its excellent impact resistance, nontoxicity, light weight, and insensitivity to sharp-edged fractures in the event of an impact/crash (Gholampour & Ozbakkaloglu, 2020).

Green composites, often reinforced with natural fibres such as flax (Oksman et al., 2003), hemp (Pappu et al., 2019), or jute (Kandola et al., 2018), have a high strength-to-weight ratio. Replacing heavier materials such as steel or aluminium with lightweight green composites can result in a significant reduction in vehicle weight (Stresses et al., n.d.; Loureiro & Esteves, 2018). Reducing the weight of vehicles leads to a reduction in the amount of energy required for propulsion. It has been observed that the use of green composites in the automotive industry helps to meet stricter fuel efficiency standards by improving the overall energy efficiency of the vehicle. Consumers today prefer cars with good fuel efficiency, which can reduce fuel consumption. Fuel consumption is related to the weight of the vehicle. The use of green composite materials can reduce the weight of the vehicle while contributing to lower fuel consumption.

In addition, vehicle interior components such as the instrument panel, seat structure, and door panels can also be made from green composites (Koronis et al., 2013). Green composites are known for their sustainability and light weight. At the same time, green composites could also improve durability, fuel costs, comfort, and aesthetic value for vehicle occupants. Green composites can also make the vehicle lighter, reducing the energy needed for acceleration. Fuel-efficient cars can also help reduce dependence on fossil fuels and meet stringent emissions regulations. Green composites are also used in hybrid or electric cars, as lighter properties are needed to meet the demand for fuel-efficient cars. Green composites have simultaneously promoted the development of these types of cars, which have become people's first choice due to their lower emissions and fuel consumption.

Car parts made of bioplastic or environmentally friendly material have already been successfully developed and built by Toyota. The spare wheel cover of the 2003 Toyota RAUM model was made of environmentally friendly composite materials using a PLA matrix reinforced with kenaf fibres. The PLA material was previously extracted from sweet potatoes and sugar cane (Vila et al., 2006). Moreover, the integration of natural fibres in the production of automotive exterior components is a well-established idea. Dealing with an exterior component is a more complicated task than for its interior counterparts, which are protected from the outdoor conditions (Koronis et al., 2013).

12.5.2 Aerospace Industry

The use of green composite materials is also suitable for the aerospace industry, especially for the manufacture of aircraft components. The light weight and high strength of green composites have led to these materials being used for a range of aircraft components, such as baggage compartments, flooring, interior panelling, and even structural components such as large fuselage sections and wings. Other benefits, such as reduced emissions and weight reduction, which subsequently contribute to improving fuel efficiency, also reflect the positive impact of using green composites in the aircraft industry. Boegler et al. (Boegler, O., Kling, U., Empl, D., & Isikveren, 2015) have successfully fabricated an aircraft wing box utilizing ramie fibre reinforced with PLA composites. The utilization of ramie fibre composites was found to yield a reduction in weight ranging from 12% to 14%. Initial studies by John et al. (Anandjiwala et al., 2008) have shown that sandwich-like honeycomb structures based on woven flax and phenolic resin have significant potential as aerospace materials in the foreseeable future. Although they are only partially environmentally friendly, they have a weight-saving property that is essential for the intended applications. Moreover, Haris et al. (Haris et al., 2011) reviewed the mechanical properties of several fibres. such as banana, pineapple leaf, kenaf, oil palm, and bamboo fibre. for aircraft radome application. The preliminary results show that kenaf fibre is the most suitable natural fibre for radome applications due to its low dielectric constant and stiffness, comparable to glass fibres.

12.5.3 Construction and Building Industry

Green composites are increasingly being used in the construction and building industries. They are used in a variety of structural elements, such as beams, columns, panels, and facade systems. The comparable properties of green composites with polymer composites in terms of durability, high strength, and weather resistance are the main reason why green composites have been selected for use in this industry. In addition, green composites also contribute to sustainable construction by reducing carbon dioxide emissions and energy consumption.

Green composites for structural applications, especially in construction, are associated with some limitations and challenges (Lau et al., 2018). Some of the challenges are related to thermal stability (Steuernagel & Ring, 2016), inconsistent fibre properties (Khalil et al., 2015), and high moisture absorption (Facca et al., 2006), which has led to performance uncertainties in composites. Although green composites struggle with the problem of inconsistency, several studies have shown that these composites have a bright future and can be used for primary and secondary structural components in the construction industry. Structural components such as roof panels, walls, and window and door frames have been made from green composites (Dev et al., 2021).

12.5.4 Marine Industry

Green composites are also used in the marine industry for applications such as interior components, bulkheads, platforms, and boat hulls. These applications are suitable due to the properties of green composites themselves, as they are durable, lightweight, and corrosion resistant. In addition, green composites can also contribute to the sustainability and efficiency of maritime transport.

It is said that the normal lifespan of ships and offshore structures is around two decades. This is because ships and structures on the open sea are exposed to various harsh environmental conditions. It is therefore very important to build a ship or offshore structure that can withstand these conditions. The construction materials chosen must fulfil the requirements, and the long-term durability of these structures must be properly assessed based on their applications. This evaluation is necessary to determine the appropriate degradation mechanisms (Crupi et al., 2023). The disadvantage of using natural fibres as a material for boat building is the ability to resist degradation in extreme environments such as the marine environment (Davies & Choqueuse, 2008). Dąbrowska (2022) recommended that parts of the ship such as hulls, propeller blades, tidal turbine blades, or components of offshore platforms and wind farms should be made of green composites. Although biocomposites have already been shown to be potential replacements for fibreglass/epoxy and other conventional composites for use in shipbuilding, hybrid materials are being developed because their application is still limited due to significant moisture absorption. Although promising, they should be considered as a transitional solution until biopolymers with fully vegetable oil–based resins and natural fillers are used in predominant industries, especially in the maritime sector (Pemberton & Graham-Jones, 2016).

12.5.5 Packaging Industry

Green composites are used in the packaging sector, where they are used to produce environmentally friendly containers and products. They can be used to replace polymers made from petroleum, reducing waste and protecting the environment. Green composites are becoming increasingly popular in the packaging industry as a way to solve environmental problems and meet sustainability goals. These cutting-edge materials, made from renewable resources, offer a number of benefits, including reduced dependence on fossil fuels, improved recycling, and a lower carbon footprint. Companies can support a greener and more circular economy while maintaining the functionality and performance required for packaging applications by replacing traditional packaging materials with green composites.

The main functions of packaging materials are to protect food or other packaged items from external influences and damage, and to provide customers with the appropriate ingredients and nutritional values of the product they are buying (Editor et al., n.d.). Thermoplastics are widely used for packaging because they can be easily moulded and formed into a variety of products, including plastic films, bottles, and jugs. This makes them an ideal choice for food packaging (Moustafa et al., 2019). Smart packaging, in particular, can benefit from the use of eco-friendly or green reinforcement materials at the micro and nanoscale to improve its permeability. Due to their biodegradability, biopolymers and biological reinforcing materials have emerged as promising alternatives that could be developed into sustainable food packaging (Tang et al., 2012).

As for food packaging applications, they are made from biopolymers. According to the European Bioplastics Association, biopolymers are polymers derived from renewable resources that possess the properties of biocompatibility, compostability, and biodegradability (Santagata et al., 2017). Their feasible film-forming ability and biodegradability have made them widely considered as the most promising sustainable substitutes for petroleum-based synthetic polymers in food packaging (Zhong et al., 2020).

Biopolymers can be used in many different ways to package things. They enable the production of a wide range of packaging items, such as food containers, cups, lids, and release coatings. These biopolymers can also be used to make composite films and multi-layer coatings that help to preserve food for longer. In addition, other beneficial ingredients, such as probiotics, additives, nutraceuticals, antioxidants, and antimicrobials, can also be used with biopolymers (Rangaraj et al., 2021).

12.6 Future Trends of Green Composites

Various development efforts and ongoing research are aimed at improving the effectiveness and properties of green composites, particularly in the area of advanced materials. The development phase included matrix materials, production methods, and also innovative fibre reinforcements. Researchers have conducted numerous experiments to expand the application possibilities of green composites by improving the durability and mechanical properties of the composites in use (Avecilla-ram & Edgar, 2021). The use of green composites in numerous industries has increased worldwide due to their sustainability properties and fulfilling the requirements for certified products. Sectors such as aerospace, automotive, construction, and packaging are expected to utilize more and more green composites in their operations. This trend will help to reduce carbon emissions, meet sustainability goals, and fulfil regulatory requirements.

The incorporation of nanotechnology into green composites is a promising development. Today, the use of nanofillers such as nanocellulose, nanotubes, and nanoparticles in the production of green composites can be seen to have a positive impact on the environment. The improvements relate to electrical conductivity, increased tensile strength, and also barrier properties. However, the incorporation of nanotechnology into green composites can help improve their performance and functionality (Bharti & Kumar, 2021).

In addition, bio-based resins and matrix materials have gained attention, especially in the development of green composites. These materials are derived from sources such as vegetable oil, starch, and lignin, which are renewable. By using bio-based materials, the negative impact on the environment can be reduced, dependence on non-renewable sources can be reduced, and better biodegradability can be achieved (Pokharel et al., 2022).

Product recycling is a key component of the circular economy, which has become an important concept in the development of sustainable composites. There are various recycling techniques to recycle green composites, and the goals are to minimize waste for reuse and improve the effectiveness of recycling methods. Moreover, it can be said that the success of green composite recycling depends on the incorporation of composite recycling technologies and techniques (Jean-luc et al., 2015; Krauklis et al., 2021).

Several researchers have attempted to improve the properties of green composites by incorporating features such as smart cellulose (Liu et al., 2019), thermal insulation (Abdallah & Abu-jdayil, 2022), electrical conductivity (Papadopoulou et al., 2019), thermal conductivity (Takagi et al., 2007), self-healing capability (Kim & Netravali, 2018), and antibacterial properties (Tran et al., 2016). With improved properties of the green composites, their use in various applications and industries can be further expanded.

Efforts are currently underway to improve the cost-effectiveness and scalability of green composites. Improvements in manufacturing processes, material sourcing, and production techniques are aimed at reducing production costs and increasing the commercial feasibility of green composites, thereby expanding their availability for a wider range of industries and applications.

These trends indicate that there is an increasing focus on improving the efficiency, environmental compatibility, and commercial feasibility of green composites. With advances in research and development, green composites are expected to make a significant contribution to solving environmental problems and promoting sustainable innovation in various sectors.

12.7 Conclusion

Green composites using natural and renewable fibres as reinforcement for polymer composites face challenges due to incompatibility between hydrophilic natural fibres and hydrophobic thermoplastic matrices. However, various modifications to the fibre–polymer interface can improve bonding and overall composite performance. The automotive industry is using green composites to reduce weight, improve fuel efficiency, and meet emissions regulations. The aerospace, construction, marine, and packaging industries also use these composites for their lightweight, durable, and sustainable properties. Ongoing research is focused on developing innovative fibre reinforcements and bio-based resins, and incorporating nanotechnology to improve the properties and functionality of green composites. Efforts are also underway to develop recycling techniques and explore multifunctional applications. Improving cost efficiency and scalability is also a key area of development to make green composites accessible to all industries. These trends underscore the growing importance of green composites in achieving sustainability goals and addressing environmental challenges.

References

Al Abdallah, H., & Abu-jdayil, B. (2022). The Effect of alkaline treatment on poly (lactic acid)/ date palm wood green composites for thermal insulation. *Polymers, 14*(6), 1143.

Abu Bakar, M. S., Salit, M. S., Mohamad Yusoff, M. Z., Zainudin, E. S., & Ya, H. H. (2020). The crashworthiness performance of stacking sequence on filament wound hybrid composite energy absorption tube subjected to quasi-static compression load. *Journal of Materials Research and Technology, 9*(1), 654–666. https://doi.org/10.1016/j.jmrt.2019.11.006

Akampumuza, O., Wambua, P. M., Ahmed, A., Li, W., & Qin, X. (2017). Review of the applications of biocomposites in the automotive industry. *Polymer Composites, 38*(11), 2553–2569. https://doi.org/10.1002/pc.23847

Anandjiwala, R. D., John, M. J., Wambua, P., Chapple, S., Klems, T., Doecker, M., Goulain, M., & Erasmus, L. (2008). Bio-based structural composite materials for aerospace applications. *2 Nd SAIAS Symposium*, September, 14–16.

Arib, R. M. N., Sapuan, S. M., Ahmad, M. M. H. M., Paridah, M. T., & Khairul Zaman, H. M. D. (2006). Mechanical properties of pineapple leaf fibre reinforced polypropylene composites. *Materials and Design, 27*(5), 391–396. https://doi.org/10.1016/j.matdes.2004.11.009

Ashok, R. B., Srinivasa, C. V., & Basavaraju, B. (2019). Dynamic mechanical properties of natural fiber composites—A review. In *Advanced Composites and Hybrid Materials* (Vol. 2, Issue 4, pp. 586–607). https://doi.org/10.1007/s42114-019-00121-8

Avecilla-ram, A. M., & Edgar, V. (2021). *Green composites and their contribution toward sustainability : A review*. https://doi.org/10.1177/09673911211009372

Azlin, M. N. M., Sapuan, S. M., Ilyas, R. A., Zainudin, E. S., & Zuhri, M. Y. M. (2020). Natural polylactic acid-based fiber composites: A review. In *Advanced Processing, Properties, and Applications of Starch and Other Bio-based Polymers*. Elsevier Inc.. https://doi.org/10.1016/B978-0-12-819661-8.00003-2

Azlin, M., Nor, M., Sapuan, S. M., Zuhri, M., Yusoff, M., & Zainudin, E. S. (2022). Mechanical , thermal and morphological properties of woven kenaf fiber reinforced polylactic acid (PLA) composites. *23*(10), 2875–2884. https://doi.org/10 .1007/s12221-022-4370-2

Azlina Ramlee, N., Jawaid, M., Abdul Karim Yamani, S., Syams Zainudin, E., & Alamery, S. (2021). Effect of surface treatment on mechanical, physical and morphological properties of oil palm/bagasse fiber reinforced phenolic hybrid composites for wall thermal insulation application. *Construction and Building Materials, 276*, 122239. https://doi.org/10.1016/j.conbuildmat.2020.122239

Bakhori, S. N. M., Hassan, M. Z., Bakhori, N. M., Rashedi, A., Mohammad, R., Md Daud, M. Y., Aziz, S. A., Ramlie, F., Kumar, A., & Naveen, J. (2022). Mechanical properties of PALF/Kevlar-reinforced unsaturated polyester hybrid composite laminates. *Polymers, 14*(12). https://doi.org/10.3390/polym14122468

Beaugrand, J., Nottez, M., Konnerth, J., & Bourmaud, A. (2014). Multi-scale analysis of the structure and mechanical performance of woody hemp core and the dependence on the sampling location. *Industrial Crops and Products, 60*, 193–204. https://doi.org/10.1016/j.indcrop.2014.06.019

Bharti, B., & Kumar, R. (2021). *Advanced Applications and Current Status of Green Nanotechnology in the Environmental Industry* (Issue August). https://doi.org/10 .1016/B978-0-12-823137-1.00012-9

Bhatnagar, R., Gupta, G., & Yadav, S. (2015). A review on composition and properties of bagasse fibers. *International Journal of Scientific and Engineering Research, 6*(5), 143–148.

Boegler, O., Kling, U., Empl, D., & Isikveren, A. T. (2015). *Potential of Sustainable Materials in Wing Structural Design*. Deutsche Gesellschaft für Luft-und Raumfahrt-Lilienthal-Oberth eV.

Chen, H., Zhang, W., Wang, X., Wang, H., Wu, Y., Zhong, T., & Fei, B. (2018). Effect of alkali treatment on wettability and thermal stability of individual bamboo fibers. *Journal of Wood Science, 64*(4), 398–405. https://doi.org/10.1007/s10086-018-1713-0

Crupi, V., Epasto, G., Napolitano, F., Palomba, G., Papa, I., & Russo, P. (2023). Green composites for maritime engineering : A review. *Journal of Marine Science and Engineering, 11*(3), 599.

Dąbrowska, A. (2022). Plant-oil-based fibre composites for boat hulls. *Materials, 15*(1699), 1–29. https://doi.org/10.3390/ma15051699

Davies, P., & Choqueuse, D. (2008). *Ageing of composites in marine vessels*. https://doi .org/10.3390/ma15051699

Dev, S., Kopparthy, S., & Netravali, A. N. (2021). Composites part C : Open access review : Green composites for structural applications. *Composites Part C, 6*, 100169. https://doi.org/10.1016/j.jcomc.2021.100169

Editor, S., Giles, G. A., & GP Management (n.d.). *Food packaging technology*.

Facca, A. G., Kortschot, M. T., & Yan, N. (2006). Predicting the elastic modulus of natural fibre reinforced thermoplastics, *37*(10), 1660–1671. https://doi.org/10.1016/j .compositesa.2005.10.006

Gholampour, A., & Ozbakkaloglu, T. (2020). A review of natural fiber composites: Properties, modification and processing techniques, characterization, applications. *Journal of Materials Science 55*(Issue 3). https://doi.org/10.1007/s10853-019 -03990-y

Haris, M. Y., Laila, D., Zainudin, E. S., Mustapha, F., Zahari, R., & Halim, Z. (2011). Preliminary review of biocomposites materials for aircraft radome application. *Key Engineering Materials*, 471–472, 563–567. https://doi.org/10.4028/www.sci-entific.net/KEM.471-472.563

Harussani, M. M., Sapuan, S. M., Abdan, K., Rashid, U., Syafiq, R. M. O., Nazrin, A., Sherwani, S. F. K., Tarique, J., Hazrol, M. D., Hazrati, K. Z., Azlin, M. N. M., & Abotbina, W. (2021). Recent development of carbonaceous materials char as nanofillers in fuel biocomposite briquettes : A review. *International Conference on Sugar Palm and Allied Fibre Polymer Composites 2021*(December), 61–64.

Harussani, M. M., Sapuan, S. M., Khalina, A., Ilyas, R. A., & Hazrol, M. D. (2020). Review on green technology pyrolysis for plastic wastes. *7th Postgraduate Seminar on Natural Fibre Reinforced Polymer Composites 2020*(December), 50–53.

Hazrati, K. Z., Sapuan, S. M., Jumaidin, R., Hafila, K. Z., Tarique, J., Azlin, M. N. M., & Syafiq, R. M. O. (2022). Mechanical properties of Dioscorea hispida fibre and other natural fibre starch-based biocomposites film: A review. *Composite Sciences and Technology International Conference 2022*, 200–202. https://doi.org/10 .3390/polym14030514.K

Hazrati, K. Z., Sapuan, S. M., Zuhri, M. Y. M., & Jumaidin, R. (2020). Ubi Gadong (Dioscorea hispida) as potential biocomposites material; A comprehensive review. In R. Jumaidin, S. M. Sapuan, & H. Ismail (Eds.), *Biofiller-Reinforced Biodegradable Polymer Composites* (1st editio, pp. 169–184). CRC Press, Taylor and Francis.

Hazrati, K. Z., Sapuan, S. M., Zuhri, M. Y. M., & Jumaidin, R. (2021). Extraction and characterization of potential biodegradable materials based on Dioscorea his-pida tubers. *Polymers*, 13(4), 1–19. https://doi.org/10.3390/polym13040584

Hermawan, D., Hazwan, C. M., Owolabi, F. A. T., Gopakumar, D. A., Hasan, M., Rizal, S., Sri Aprilla, N. A., Mohamed, A. R., & Khalil, H. P. S. A. (2019). Oil palm microfiber-reinforced handsheet-molded thermoplastic green composites for sustainable packaging applications. *Progress in Rubber, Plastics and Recycling Technology*, 35(4), 173–187. https://doi.org/10.1177/1477760619861984

Ilyas, R. A., Sapuan, S. M., Harussani, M. M., Hakimi, M. Y. A. Y., Haziq, M. Z. M., Atikah, M. S. N., Asyraf, M. R. M., Ishak, M. R., Razman, M. R., Nurazzi, N. M., Norrrahim, M. N. F., Abral, H., & Asrofi, M. (2021). Polylactic acid (PLA) biocomposite: Processing, additive manufacturing and advanced applications. *Polymers*, 13(8), 1326. https://doi.org/10.3390/polym13081326

Ilyas, R. A., Sapuan, S. M., Ishak, M. R., Zainudin, E. S., Atikah, M. S. N., & Huzaifah, M. R. M. (2018). Water barrier properties of biodegradable films reinforced with nanocellulose for food packaging application : A review. *6th Postgraduate Seminar on Natural Fiber Reinforced Polymer Composites 2018*(December), 55–59.

Joseph, K., Thomas, S., & Pavithran, C. (1996). Effect of chemical treatment on the ten-sile properties of short sisal fibre-reinforced polyethylene composites. *Polymer*, 37(23), 5139–5149. https://doi.org/10.1016/0032-3861(96)00144-9

Joshi, A. S., Barhanpurkar, S., Paharia, A., & Maloo, T. (2012). Bio-composite materials as alternatives to glass fibre reinforced composites for automotive applications. *Man-Made Textiles in India*, 40(11), 386–390.

Jumaidin, R., Sapuan, S. M., Jawaid, M., Ishak, M. R., & Sahari, J. (2016). Characteristics of thermoplastic sugar palm Starch/Agar blend: Thermal, tensile, and physical properties. *International Journal of Biological Macromolecules, 89,* 575–581. https://doi.org/10.1016/j.ijbiomac.2016.05.028

Kabir, M. M., Wang, H., Lau, K. T., & Cardona, F. (2012). Chemical treatments on plant-based natural fibre reinforced polymer composites: An overview. *Composites Part B: Engineering, 7*(43), 2883–2892. https://doi.org/10.1016/j.compositesb.2012.04.053

Kamaruddin, Z. H., Jumaidin, R., Ilyas, R. A., & Selamat, M. Z. (2022). Cymbopogan citratus fiber-reinforced, Thermoplastic Cassava Starch – Palm Wax composites. *Polymers, 14*(July), 1–26. https://doi.org/10.3390/polym14142769

Kamaruddin, Z. H., Jumaidin, R., Ilyas, R. A., Selamat, M. Z., Alamjuri, R. H., & Yusof, F. A. M. (2022a). Biocomposite of cassava starch-Cymbopogan citratus fibre: Mechanical, thermal and biodegradation properties. *Polymers, 14*(3), 1–19. https://doi.org/10.3390/polym14030514

Kamaruddin, Z. H., Jumaidin, R., Ilyas, R. A., Selamat, M. Z., Alamjuri, R. H., & Yusof, F. A. M. (2022b). Influence of alkali treatment on the mechanical, thermal, water absorption, and biodegradation properties of Cymbopogan citratus fiber-reinforced, Thermoplastic Cassava starch–Palm Wax composites. *Polymers, 14*(2769), 1–26. https://doi.org/10.3390/polym14142769

Kamaruddin, Z. H., Jumaidin, R., Kamaruddin, Z. H., Muhammad Asyraf, M. R., Razman, M. R., & Khan, T. (2023). Effect of Cymbopogan citratus fibre on physical and impact properties of Thermoplastic Cassava Starch/Palm Wax composites. *Polymers, 15*(10), 2364. https://doi.org/10.3390/polym15102364.

Kamaruddin, Z. H., Jumaidin, R., Rushdan, A. I., Selamat, M. Z., & Alamjuri, R. H. (2021). Characterization of natural cellulosic fiber isolated from Malaysian Cymbopogan citratus leaves. *BioResources, 16*(4), 7729–7750. https://doi.org/10.15376/biores.16.4.7729-7750

Kamaruddin, Z. H., Jumaidin, R., Selamat, M. Z., & Ilyas, R. A. (2021). Characteristics and properties of lemongrass (Cymbopogan citratus): A comprehensive review. *Journal of Natural Fibers,* 1–18. https://doi.org/10.1080/15440478.2021.1958439

Kandola, B. K., Mistik, S. I., Pornwannachai, W., & Anand, S. C. (2018). Natural fibre-reinforced thermoplastic composites from woven-nonwoven textile preforms: Mechanical and fire performance study. *Composites Part B: Engineering, 153*(August), 456–464. https://doi.org/10.1016/j.compositesb.2018.09.013

Khalil, H. P. S. A., Hossain, S., Rosamah, E., Azli, N. A., Saddon, N., Davoudpoura, Y., Islam, N., & Dungani, R. (2015). The role of soil properties and it ' s interaction towards quality plant fi ber : A review. *Renewable and Sustainable Energy Reviews, 43,* 1006–1015. https://doi.org/10.1016/j.rser.2014.11.099

Kim, J. R., & Netravali, A. N. (2018). *Self-Healing Green Polymers and Composites.* John Wiley & Sons, Inc., Hoboken, NJ, USA, pp 135–185.

Koronis, G., Silva, A., & Fontul, M. (2013). Green composites: A review of adequate materials for automotive applications. *Composites Part B: Engineering, 44*(1), 120–127. https://doi.org/10.1016/j.compositesb.2012.07.004

Krauklis, A. E., Karl, C. W., & Gagani, A. I. (2021). Composite material recycling technology — State-of-the-art and sustainable development for the 2020s. *Journal of Composites Science, 5*(1), 28.

Lau, K., Hung, P., Zhu, M., & Hui, D. (2018). Properties of natural fi bre composites for structural engineering applications. *Composites Part B, 136*(September 2017), 222–233. https://doi.org/10.1016/j.compositesb.2017.10.038

Lazim, Y., Salit, S. M., Zainudin, E. S., Mustapha, M., & Jawaid, M. (2014). Effect of alkali treatment on the physical, mechanical, and morphological properties of waste betel nut (Areca catechu) husk fibre. *BioResources, 9*(4), 7721–7736. https://doi.org/10.15376/biores.9.4.7721-7736

Li, X., Tabil, L. G., & Panigrahi, S. (2007). Chemical treatments of natural fiber for use in natural fiber-reinforced composites: A review. *Journal of Polymers and the Environment, 15*(1), 25–33. https://doi.org/10.1007/s10924-006-0042-3

Li, X., Qin, D., Hu, Y., Ahmad, W., Ahmad, A., Aslam, F., & Joyklad, P. (2022). A systematic review of waste materials in cement-based composites for construction applications. *Journal of Building Engineering, 45*(October 2021), 103447. https://doi.org/10.1016/j.jobe.2021.103447

Lila, M. K. (2021). Extraction and characterization of Munja fibers and its potential in the biocomposites. *Journal of Natural Fibers*, 1–19. https://doi.org/10.1080/15440478.2020.1821287

Liu, Y., Liu, L., Li, Z., Zhao, Y., & Yao, J. (2019). Green and facile fabrication of smart cellulose composites assembled by graphene nanoplates for dual sensing. *Cellulose*, 0123456789. https://doi.org/10.1007/s10570-019-02735-z

Loganathan, T. M., Sultan, M. T. H., Ahsan, Q., Jawaid, M., Naveen, J., Md Shah, A. U., & Hua, L. S. (2020). Characterization of alkali treated new cellulosic fibre from Cyrtostachys renda. *Journal of Materials Research and Technology, 9*(3), 3537–3546. https://doi.org/10.1016/j.jmrt.2020.01.091

Loureiro, N. C., & Esteves, J. L. (2018). Green composites in automotive interior parts: A solution using cellulosic fibers. *Green Composites for Automotive Applications.* Elsevier Ltd. https://doi.org/10.1016/B978-0-08-102177-4.00004-5

Manikandan Nair, K. C., Thomas, S., & Groeninckx, G. (2001). Thermal and dynamic mechanical analysis of polystyrene composites reinforced with short sisal fibres. *Composites Science and Technology, 61*(16), 2519–2529. https://doi.org/10.1016/S0266-3538(01)00170-1

Mayilswamy, N., & Kandasubramanian, B. (2022). Green composites prepared from soy protein, polylactic acid (PLA), starch, cellulose, chitin: A review. *Emergent Materials*, 0123456789. https://doi.org/10.1007/s42247-022-00354-2

Mohanty, A. K., Misra, M., & Drzal, L. T. (2002). Sustainable bio-composites from renewable resources in green materials world. *Journal of Polymers and the Environment, 10*(April), 19–26. https://doi.org/10.1023/A:1021013921916

Mohanty, A. K., Misra, M., & Drzal, L. T. (Eds.). (2005). Natural fibers, biopolymers, and biocomposites. (1st ed.). CRC Press.

Moustafa, H., Youssef, A. M., Darwish, N. A., & Abou-kandil, A. I. (2019). Eco-friendly polymer composites for green packaging : Future vision and challenges. *Composites Part B, 172*(March), 16–25. https://doi.org/10.1016/j.compositesb.2019.05.048

Mwaikambo, L. Y., Tucker, N., & Clark, A. J. (2007). Mechanical properties of hemp-fibre-reinforced euphorbia composites. *Macromolecular Materials and Engineering, 292*(9), 993–1000. https://doi.org/10.1002/mame.200700092

Neto, J., Queiroz, H., Aguiar, R., Lima, R., Cavalcanti, D., & Banea, M. D. (2022). A review of recent advances in hybrid natural fiber reinforced polymer composites. *Journal of Renewable Materials, 10*(3), 561–589. https://doi.org/10.32604/jrm.2022.017434

Norrrahim, M. N. F., Huzaifah, M. R. M., Farid, M. A. A., Shazleen, S. S., Misenan, M. S. M., Yasim-Anuar, T. A. T., Naveen, J., Nurazzi, N. M., Rani, M. S. A., Hakimi, M. I., Ilyas, R. A., & Jenol, A. (2021). Greener pretreatment approaches for the valorisation of natural fibre biomass into bioproducts. *Polymers, 13*(17), 2971.

Oliveux, G., Dandy, L. O., Jean-luc, B., Mantaux, O., & Leeke, G. A. (2015). Future directions in the recycling of composite materials. *ICCM International Conferences on Composite Materials, 2015-July*(July).

Oksman, K., Skrifvars, M., & Selin, J. F. (2003). Natural fibres as reinforcement in polylactic acid (PLA) composites. *Composites Science and Technology, 63*(9), 1317–1324. https://doi.org/10.1016/S0266-3538(03)00103-9

Papadopoulou, E. L., Basnett, P., Paul, U. C., Marras, S., Ceseracciu, L., Roy, I., & Athanassiou, A. (2019). *Green composites of poly (3-hydroxybutyrate) Containing graphene nanoplatelets with desirable electrical conductivity and oxygen barrier properties.* https://doi.org/10.1021/acsomega.9b02528

Pappu, A., Pickering, K. L., & Thakur, V. K. (2019). Manufacturing and characterization of sustainable hybrid composites using sisal and hemp fibres as reinforcement of poly (lactic acid) via injection moulding. *Industrial Crops and Products, 137*(November 2018), 260–269. https://doi.org/10.1016/j.indcrop.2019.05.040

Pemberton, R., & Graham-Jones, J. (2016). Application of composite materials to yacht rigging. In *Marine Applications of Advanced Fibre-Reinforced Composites*. Elsevier Ltd. https://doi.org/10.1016/B978-1-78242-250-1.00012-0

Pokharel, A., Falua, K. J., & Babaei-ghazvini, A. (2022). *Biobased polymer composites : A review. Journal of Composites Science, 6*(9), 255.

Rangaraj, V. M., Rambabu, K., Banat, F., & Mittal, V. (2021). Food bioscience natural antioxidants-based edible active food packaging : An overview of current advancements. *Food Bioscience, 43*(May), 101251. https://doi.org/10.1016/j.fbio.2021.101251

Rozyanty, A. R., Zhafer, S. F., Shayfull, Z., Nainggolan, I., Musa, L., & Zheing, L. T. (2021). Effect of water and mechanical retting process on mechanical and physical properties of kenaf bast fiber reinforced unsaturated polyester composites. *Composite Structures, 257*(November 2020), 113384. https://doi.org/10.1016/j.compstruct.2020.113384

Santagata, G., Valerio, F., Cimmino, A., Dal, G., Masi, M., Di, M., Malinconico, M., Lavermicocca, P., & Evidente, A. (2017). Chemico-physical and antifungal properties of poly (butylene succinate)/ cavoxin blend : Study of a novel bioactive polymeric based system. *European Polymer Journal, 94*(April), 230–247. https://doi.org/10.1016/j.eurpolymj.2017.07.004

Seki, Y. (2009). Innovative multifunctional siloxane treatment of jute fiber surface and its effect on the mechanical properties of jute/thermoset composites. *Materials Science and Engineering A, 508*(1–2), 247–252. https://doi.org/10.1016/j.msea.2009.01.043

Sever, K., Sarikanat, M., Seki, Y., Erkan, G., & Erdoğan, Ü. H. (2010). The mechanical properties of γ-methacryloxypropyltrimethoxy silane-treated jute/polyester composites. *Journal of Composite Materials, 44*(15), 1913–1924. https://doi.org/10.1177/0021998309360939

Sosiati, H., Shofie, Y. A., & Nugroho, A. W. (2018). Tensile properties of kenaf/E-glass reinforced hybrid polypropylene (PP) composites with different fiber loading. *Evergreen, 5*(2), 1–5. https://doi.org/10.5109/1936210

Steuernagel, L., & Ring, J. (2016). Natural fibre/PA6 composites with flame retardance properties: Extrusion and characterisation. *Composites Part B.* https://doi.org/10.1016/j.compositesb.2016.10.012

Stresses, R. Materials, C., Joints, A. C. Composites, P., & Industry, A. (n.d.). *Lightweight Composite Structures in Transport.*

Syafiq, R., Sapuan, S. M., Zuhri, M., Othman, S. H., Ilyas, R. A., Nazrin, A., Sherwani, S. F. K., Tarique, J., Harussani, M. M., Hazrol, M. D., & Hazrati, K. Z. (2021). Water transport properties on different plasticizer type and concentration of bionanocomposite films incorporated with essential oil for food packaging application: A review. *International Conference on Sugar Palm and Allied Fibre Polymer Composites 2021(December)*, 91–93.

Taharuddin, N. H., Jumaidin, R., Ilyas, R. A., Kamaruddin, Z. H., Mansor, M. R., Md Yusof, F. A., Knight, V. F., & Norrrahim, M. N. F. (2022). Effect of Agar on the mechanical, thermal, and moisture absorption properties of thermoplastic sago starch composites. *Materials*, 15(24). https://doi.org/10.3390/ma15248954

Takagi, H., Kako, S., & Kusano, K. (2007). Thermal conductivity of PLA-bamboo fiber composites. *July 2013*, 37–41. https://doi.org/10.1163/156855107782325186

Tang, X. Z., Kumar, P., Alavi, S., & Sandeep, K. P. (2012). Recent advances in Biopolymers and biopolymer-based nanocomposites for food packaging materials recent advances in Biopolymers and biopolymer-based nanocomposites. *Critical Reviews in Food Science and Nutrition*, 5(52), 426–442. https://doi.org/10.1080/10408398.2010.500508

Tarique, J., Sapuan, S. M., Khalina, A., Sherwani, S. F. K., & Yusuf, J. (2021). Recent developments in sustainable arrowroot (Maranta arundinacea Linn) starch biopolymers , fibres , biopolymer composites and their potential industrial applications : A review. *Journal of Materials Research and Technology*, 13, 1191–1219. https://doi.org/10.1016/j.jmrt.2021.05.047

Tarique, J., Sapuan, S. M., Zainudin, E. S., Khalina, A., Ilyas, R. A., Hazrati, K. Z., & Aliyu, I. (2023). A comparative review of the effects of different fibre concentrations on arrowroot fibre and other fibre-reinforced composite films. *Materials Today: Proceedings*, 74, 411–414. https://doi.org/10.1016/j.matpr.2022.11.136

Thomas, S., Paul, S. A., Pothan, L. A., & Deepa, B. (2011). Natural fibres: Structure, properties and applications. *Cellulose Fibers: Bio-and Nano-Polymer Composites: Green Chemistry and Technology*, 3–42.

Thyavihalli Girijappa, Y. G., Mavinkere Rangappa, S., Parameswaranpillai, J., & Siengchin, S. (2019). Natural fibers as sustainable and renewable resource for development of eco-friendly composites: A comprehensive review. *Frontiers in Materials*, 6(September), 1–14. https://doi.org/10.3389/fmats.2019.00226

Tran, C. D., Prosenc, F., Franko, M., & Benzi, G. (2016). Synthesis, structure and antimicrobial property of green composites from cellulose, wool, hair and chicken feather. *Carbohydrate Polymers*. https://doi.org/10.1016/j.carbpol.2016.06.021

Venkategowda, T., Manjunatha, L. H., & Anilkumar, P. R. (2021). Dynamic mechanical behavior of natural fibers reinforced polymer matrix composites – A review. *Materials Today: Proceedings*. https://doi.org/10.1016/j.matpr.2021.09.465

Vila, C., Santos, V., Cunha, A. M., Campos, A. R., & Cristova, C. (2006). Sustainable materials in automotive applications, 35(6), 233–241. https://doi.org/10.1179/174328906X146487

Vimal, R., Subramanian, K. H. H., Aswin, C., Logeswaran, V., & Ramesh, M. (2015). Comparisonal study of succinylation and Phthalicylation of jute fibres: Study of mechanical properties of modified fibre reinforced epoxy composites. *Materials Today: Proceedings*, 2(4–5), 2918–2927. https://doi.org/10.1016/j.matpr.2015.07.254

Wang, B., Panigrahi, S., Tabil, L., & Crerar, W. (2007). Pre-treatment of flax fibers for use in rotationally molded biocomposites. *Journal of Reinforced Plastics and Composites*, 26(5), 447–463. https://doi.org/10.1177/0731684406072526

Zhong, Y., Godwin, P., Jin, Y., & Xiao, H. (2020). Advanced Industrial and Engineering Polymer Research Biodegradable polymers and green-based antimicrobial packaging materials : A mini-review. *Advanced Industrial and Engineering Polymer Research, 3*(1), 27–35. https://doi.org/10.1016/j.aiepr.2019.11.002

Zhou, Y., Fan, M., Chen, L., & Zhuang, J. (2015). Lignocellulosic fibre mediated rubber composites: An overview. *Composites Part B: Engineering, 76*, 180–191. https://doi.org/10.1016/j.compositesb.2015.02.028

Zulnazri, & Sulhatun (2018). Effect of palm oil bunches microfiller on ldpe-recycled composite tensile strength through melt blending process. *Emerald Reach Proceedings Series, 1*, 503–509. https://doi.org/10.1108/978-1-78756-793-1-00039

13

Carbonization of Biomass and Waste into Biosourced Carbon (BC) Nanofillers for Advanced Composite Applications

M. M. Harussani and S. M. Sapuan

13.1 Introduction

Massive global production of plastic waste and underutilization of agricultural waste have sparked interest from researchers and industrial practitioners. Annual worldwide plastic demand has risen from 1.5 million metric tons in 1950, the beginning of the plastic industry, to 3 billion metric tons in 2018. The globe produces 381 million tons of plastic waste per year, which is anticipated to quadruple by 2034 (Ritchie & Roser, 2018). As stated by Tripathi et al. (2019), agricultural and forestry activities generate a lot of waste as a result of harvestable output. The global annual production of biomass waste exceeds 140 Gt, posing considerable management challenges, since disposed biomass might have detrimental environmental consequences. Thus, conversion of waste into carbonaceous materials for a plethora of engineering applications is the best possible alternative to inhibit the underutilization of the waste, as illustrated in Figure 13.1.

Various carbonaceous materials, including graphene, graphite, carbon nanotubes (CNTs), activated carbon (AC) and char, are unique and versatile, highly competent to form different architectures at the nanoscale regime. These materials are obtained by the thermal decomposition (pyrolysis) of plastic and biomass feedstocks (Harussani et al., 2022; Lopez et al., 2018). There has been significant attention and enthusiasm surrounding the utilization of the inherent characteristics of carbon chars for a wide range of applications, including, but not limited to, composites, automotive, aeronautics, electronics, sensors, and various engineering industries. Composites made from carbonaceous materials have a restricted range of uses and are currently largely in the experimental or development stages. The basic concept behind these nanostructure materials is that they may replace substantial loads of mineral fillers and fibres in a composite (up to 50%) and still achieve equivalent performance in tiny amounts, leading to

DOI: 10.1201/9781003408215-13

FIGURE 13.1
Conversion of waste into char products.

lighter consumer goods. However, the cost–performance balance in composite materials is the fundamental obstacle to the adoption of nanofillers.

In recent years, researchers have primarily focused on investigating the use of carbon nanofillers to enhance the properties of composites in engineering applications. However, to the best of the authors' knowledge, the utilization of char for fuel briquettes has received relatively less attention and exploration. However, several researchers have discovered this briquette manufacture to utilize various sources of biomass waste, i.e. mesocarp fibre (Safana et al., 2018), mangrove wood (Safana et al., 2018), faecal (Ward et al., 2014), sawdust (Prasityousil & Muenjina, 2013), and agricultural bagasse (Teixeira et al., 2011; Zanella et al., 2016), with different types of natural binders, i.e. cassava starch (Teixeira et al., 2009), corn starch (Zanella et al., 2016), sugar palm starch (Harussani et al., 2021), etc., which are mainly useful in fuel sectors. Also, the available literature has focused on the fuel applications of biochar briquettes, with less research on plastic-derived char within this spectrum. Therefore, the current research aims to examine the unique features exhibited by composites reinforced with char nanofillers, which represent a highly promising alternative for advancing fuel energy development.

13.2 Biosourced Carbon and Its Various Manufacturing Technologies

In recent years, carbon-based nanomaterials have captured the interest of a significant portion of the scientific community due to their exceptional strength and versatility in composite material applications (Aisyah et al., 2019; Harussani et al., 2021; Nam et al., 2017; Ubertini et al., 2014). There are two categories of carbonaceous materials: (i) conventional carbon materials like char/biochar, carbon blacks (CBs), and AC, and (ii) nano-structured carbons, which encompass graphene, graphite, fullerenes, and CNTs. Char, charcoal, and activated carbons are three carbon compounds that share many similarities in terms of structure and processing mechanisms (Harussani & Sapuan, 2022). Char quality and properties are determined by pyrolysis parameters, including heating duration, maximum temperature, pressure, and oxygen content, which can vary with different feedstocks (Ge et al., 2021; Harussani et al., 2022). According to the findings of Ward et al. (2014), the properties and qualities of char primarily rely on the composition of the waste material subjected to pyrolysis. Consequently, this composition has an impact on the surface area and porosity of the char. Furthermore, as mentioned by Du et al. (2021), the presence of various metal ions within the char particles enables them to serve as catalysts for condensation, cracking of volatiles, and polymerization reactions.

13.2.1 Biocarbon

Researchers have long been looking for methods to reduce our reliance on fossil fuel, and one of the methods is reducing the use of polypropylene (PP). In order to achieve this, the incorporation of biobased fillers has come under great consideration, which – along with a host of other reasons – has led to ever-increasing interest in biosourced carbons (BCs) (Watt et al., 2020). BC is a biobased material produced when biomass undergoes thermal conversion in the pyrolysis process. To put it briefly, pyrolysis involves the heating of a low-energy organic material (biomass in this case), in an oxygen-free environment, to produce high-energy products, including BCs (Jubinville et al., 2020).

BCs have been shown to have desirable qualities all the way from pre-production to their applications. For example, with biomass as the primary source, BCs can be produced in excess due to the relatively low cost and availability of the required raw materials (Jubinville et al., 2020). This gives them a significant advantage over petroleum-based carbons and other filler materials, in that they can use a seemingly unlimited source of waste to sustain their production; along with this comes the bonus of clearing the environment of post-industrial and post-consumer waste products, i.e. sawdust, corn cobs, and spent tea (Chang, Rodriguez-Uribe et al., 2021). Moreover, due to the nature of their varying properties during production (which will be discussed shortly), BCs have a wide variety of applications, making them a very versatile material that can be used as a reinforcing agent or a multifunctional filler for polymer composite uses.

13.2.2 Properties of Biosourced Carbon

BC is a carbon-rich and highly porous product formed from the thermal decomposition of biomass (Chang, Rodriguez-Uribe et al., 2021). It has proven to be valuable in many fields due to its properties, with one such area being agriculture; because of its effective cation exchange capacity, porous structure, and hydrophobicity, BC provides a significant boost to crop cultivation (Das & Sarmah, 2015; Mukherjee et al., 2016). Given that BCs may be derived from waste crop products (i.e. rice husk and corn stalk) and are also used in crop cultivation, they are capable of fostering a sustainable resource utilization cycle within the agricultural and agronomic sectors, thereby making the material more lucrative for those looking for a green and sustainable cycle within these sectors.

Another area in which BCs have proven their effectiveness through research is their viability as a practical alternative to mineral fillers as a reinforcing agent in polymer composites (de Gortari et al., 2020; Snowdon et al., 2019). For example, composites made from BCs have achieved the same standards (in relation to their tensile and flexural stress, and moduli) as talc-based composites, and it has even been reported that a relatively greater impact strength can be accomplished by reducing the particle sizes of the BC (<1 μm), as shown in Figure 13.2. The majority of studies on BC primarily concentrate on enhancing the mechanical, thermal, and electrical properties of composites. Research has demonstrated that BCs exhibit superior performance compared with their precursors, such as rice husk and wood.

FIGURE 13.2
Field emission scanning electron micrographs of char particles at 100× magnification.

BCs have been shown to significantly enhance the tensile strength, bending strength, water absorption capabilities, and thermal stability of biobased fillers in composite materials (Chang, Rodriguez-Uribe et al., 2021).

Moreover, BCs possess a relatively lower density, typically ranging from 1.3 to 1.4 g/cm^3, which can vary based on the pyrolysis temperature and the source material used. This relatively low density gives them a crucial advantage over other fillers such as talc, which has a density of 2.6–2.7 g/cm^3, and carbon black, with a density of 1.8–2.1 g/cm^3 (Chang, Rodriguez-Uribe et al., 2021; Watt et al., 2020). In the context of this application, lower density is an advantage, as more filler content may be added without drastically increasing the weight of the synthetic polymer; this was exemplified in a paper published by Chang et al. (2021), where it was stated that the weight of the final product after compounding 50 wt.% of BC onto polypropylene composites was lighter than both carbon black and talc composites at the same filler loading. The benefits of lightweight BC composites can be seen cascading into real-world advantages; for instance, within the automobile industry, vehicles that use composite parts produced with BC have improved fuel economy, which would in turn allow direct cost savings for the consumer.

The conditions of the pyrolysis process play a significant role in determining the physical structure and properties of the BC produced. For example, pyrolysis carried out at 500 °C produces BC containing 3D networks of functional groups in its structure (Quosai et al., 2018). However, pyrolysis at 900 °C produces volatiles, which eliminates the functionality of the BC surface and increases the surface area of the material due to increased porosity (Chen et al., 2017). Furthermore, studies have indicated that biochar pyrolysed at a higher temperature of 900 °C demonstrated significantly improved performance in terms of mechanical properties, thermal properties, and coefficient of linear thermal expansion when compared with biochar pyrolysed at lower temperatures (650 °C) (Abdelwahab et al., 2019). This improvement was attributed to the lack of functional groups on the surface of the material, resulting in a stronger affinity with the polymer matrix. Therefore, given the variable surface chemistry of the BCs, their compatibility with polymers may be substantially improved when the right conditions are met.

According to a study, the differences in surface chemistry of BC can be attributed to the deposition of condensed gases on its surface. The study further suggests that as the carbonization temperature increases, the surface functionality decreases, while the presence of pyrogenic nano-pores (porosity) significantly increases (Gray et al., 2014). A study showed that BCs produced at a temperature of 900 °C exhibited higher compatibility with an ethylene-octene copolymer modified polypropylene matrix compared with BCs produced at 500 °C. This was attributed to the reduced surface functionality of BCs at higher temperatures. Moreover, when it comes to getting the desired BC properties, the temperature of the pyrolysis process must be adjusted accordingly, but there is usually a trade-off; at high temperatures, the BC's affinity to the non-polar polymeric chains of the modified polypropylene matrix increases, while the overall mechanical strength of the synthetic polymer decreases. However, at lower temperatures, the BCs produced were relatively less compatible but returned a higher yield, which made them more practical for industrial use (Behazin et al., 2017).

As previously stated, BCs produced at high pyrolysis temperatures tend to lose their functionality. This loss in functionality may be attributed to the presence of fewer surface functional groups as the biomass is deoxygenated and dehydrated (Behazin et al., 2017). This signifies that the available dominant functional groups on the BC surface, C=O and C–H, also decrease due to this dehydration. The functional groups that are commonly present on BC surfaces include carboxyl, carbonyl, pyrone, and ether, to name just a few (Tomczyk et al., 2020).

In addition to temperature, the properties of the biomass used as feedstock and the rate of biomass heating also influence the characteristics of the resulting BCs, particularly in terms of BC yield. Various biomass sources contain different levels of moisture, lignin, and cellulose composition, which play a crucial role in determining the yield of the final BCs. Furthermore, the rate of heating also has a significant effect on the yield of the BCs produced; Cha et al. (2016) pointed out that slow pyrolysis (1–10 °C/min) was more beneficial for BC production due to the high yield obtained, while fast pyrolysis, which occurs well above the previously mentioned heating rate, is better suited for bio-oil production.

13.2.3 BC and Commercialized Carbon Black (CB)

Given that BCs have proven to be effectively implemented into some (if not most) of carbon black (CB)'s area of applications, i.e. polymer composites fabrication, it has been seen as a potential replacement. This could mainly be attributed to its waste utilization (use of biomass as "source material"), which supports its sustainability. The use of carbon black, on the other hand, may not be considered sustainable due to its use of fossil fuels as a source. The term CB refers to a group of industrial products that consist of near-spherical particles that are obtained from partial combustion or thermal decomposition of hydrocarbons (Donnet, 2018).

The substance is available as a soft, fine powder, characterized by a surface area that can vary from 10 to 2000 m^2/g. It consists of over 90% carbon. There are three types of carbon black, each named after its specific production process: thermal black, furnace black, and channel black (Chang, Rodriguez-Uribe et al., 2021). CB finds its primary applications in rubber reinforcement, as a black pigment, and in the production of electrically conductive materials and composites (Baan, 2007). CB is also a highly popular carbon-based filler that has been widely utilized in plastic and rubber composites. Its extensive use in the tyre industry has contributed to its market success, with a price of USD 1.2/kg as of 2020 (Bartoli et al., 2020). The usage breakdown of carbon black reveals that 70% is employed in the tyre industry, 20% in rubber products, and 10% in non-rubber products. Table 13.1 depicts the difference between BC and commercialized CB.

A lot of studies have been conducted to investigate the extent to which BC could replace carbon black – with one study pointing out that one factor would be BC's natural black colour (Paleri et al., 2021). When it comes to thermal stability, BC is just as stable as carbon black, which enables its processing and compounding without temperature restrictions. This trait gives both of the aforementioned fillers a major advantage over other bio-fillers and natural fibres (Rodriguez-Uribe et al., 2021). Furthermore, BCs have a more porous structure (relative to that of carbon black) and so are usually less dense. Chang et al. (2021) further emphasizes this by indicating that BCs on average are 30% lighter than carbon blacks; a difference that further carries on to the weight difference between the composites made using these materials. In the context of the aforementioned publication, it was stated that the weight of a PP–BCs composite (compounded with 50 wt.% of BCs) was lower than that of a PP–carbon black composite with a similar filler loading.

Moreover, another study found that epoxy infused with BCs sourced from coffee waste was more electrically conductive when compared with CB-infused epoxy. This was credited to the improved dispersibility of BCs within the epoxy (Giorcelli & Bartoli, 2019; Noori et al., 2020). However, this contradicts a study by Snowdon et al. (2019), which demonstrated that BCs sourced from carbonized lignin displayed inferior electrical conductivity (9.5 S/m less than CB) but a thermal conductivity 36% greater than that of CB. However, this conclusion was not without its complexities. For instance, the electrical conductivity of fillers is influenced by the presence of impurities on the surface of these carbon-based materials. A higher concentration of impurities leads to decreased conductivity. Therefore, the presence

TABLE 13.1

Comparison between BC and CB

	BC	CB
Feedstock	Biomass	Petroleum, coal tar, and asphalt
Pre-processing method	Pyrolysis at moderate to high temperatures	Incomplete combustion in an oxygen-deficient environment
Carbon content	Ranges from 40% to 90% (varying based on the pyrolysis conditions and final temperature)	>95%
Structure	Amorphous carbon with microcrystalline structures, accompanied by abundant surface functional groups and porosity	Microcrystals or amorphous carbon particles

Source: Chang, B.P. et al., *Renewable and Sustainable Energy Reviews*, 152, 111666, 2021 and Donnet, J.-B., *Carbon Black: Science and Technology*, Oxfordshire: Routledge, 2018.

of more residual oxygen on the surface of BCs results in reduced electrical conductivity (Pantea et al., 2003). Moreover, it was also mentioned that the difference in heat conductivity was due to the size difference of the particles; larger particles, which in this study belonged to the BCs, had better heat transfer capabilities due to the reduced proximity to each other and the increased number of pores and cavities, which further contributed to the radiant heat (Snowdon et al., 2014). As a result, the thermal resistance was diminished, leading to enhanced thermal conductance. However, it is essential to acknowledge that the properties of the BCs were influenced by the conditions of the feedstock, which varied. Consequently, this disparity in feedstock conditions contributed to inconsistent results when comparing their electrical conductivity capabilities.

Another study reported that BCs possess the ability to increase the modulus of natural rubber, which is one of the many factors contributing to carbon black's popularity in the tyre industry. It was found that BC/natural rubber composites had better curing capabilities than CB-reinforced composites (Lay et al., 2020). It was further stated that BC/natural rubber composites performed well; this was attributed to the synergistic effects produced by the crosslinking of Si–O–C bonds. However, it should be noted that this study only took into account BCs sourced from rice husk, as they contain carbon and silica elements, which are required for rubber formation (Xue et al., 2019).

13.2.4 BC Manufacturing Methods

13.2.4.1 Pyrolysis

The pyrolysis process involves the direct thermal decomposition of organic matter, such as biomass, while carefully controlling the oxygen concentration to prevent excessive combustion (Bridgwater, 1996). Biomass used in pyrolysis consists of various structural components, including hemicellulose, cellulose, lignin, and extractives. Each of these components undergoes pyrolysis at different temperatures, following distinct mechanisms and pathways. Additionally, the rate and extent of their degradation depend on several factors, including reactor type, particle size, heating rate, and temperature (Vamvuka, 2011). During pyrolysis, a range of simultaneous reactions occur, including dehydration, cracking, isomerization, coking, aromatization, and condensation. These reactions result in the production of water, carbon dioxide, charcoal, tars, organic compounds, polymers, and other gases (Balat, 2008).

The biomass is converted into three main end products: bio-oils (liquid), pyro-gas (gaseous), and BC (solid) (Braghiroli et al., 2020). The heating rate, the temperature in the reaction vessel, and the residence time, the time taken for the material to enter and exit the reaction vessel, dictate the yield or rather, the relative proportions of each of the products (Williams & Nugranad, 2000).

Based on the relationship between the heating rate and the residence time of the solid in the heated reactor zone, the pyrolysis processes can be categorized into different classes (Li et al., 2013). These classifications include slow pyrolysis, intermediate pyrolysis, fast pyrolysis, and flash pyrolysis.

13.2.4.1.1 Slow Pyrolysis

Slow pyrolysis involves heating biomass to approximately 500 °C, and the residence times within the reaction chamber span several hours. Slow pyrolysis is distinguished by its lower pyrolysis temperature, gradual heating rate, and extended residence times. At temperatures below 500 °C, the conditions of slow pyrolysis promote the maximization of char yield through secondary coking and re-polymerization reactions (Williams & Besler, 1996), see Table 13.2. In short, slow pyrolysis promotes secondary reactions between the volatile components, which in turn gives a high charcoal yield but lower liquid (bio-oil) yields (Bridgwater, 1996). It is typically performed at 1–10 °C/min and is the preferred method in the industrial production of BC due to the aforementioned high yield produced.

Studies have indicated that in the slow pyrolysis process, raising the temperature (with a constant heating rate) leads to a reduction in char yield while increasing the yields of oil and gas. Additionally, increasing the heating rate (at a constant temperature) has been observed to enhance char yield at lower temperatures (around 300 °C), but it has the opposite effect at higher temperatures (>420 °C). It has been inferred that the decrease in char yield (accompanied by an increase in oil and gas yield) resulting from higher temperature and heating rate is attributed to the volatilization and thermal degradation of high-molecular-weight hydrocarbons present in the char.

To alleviate any concerns regarding sample size, it is important to point out that while some of the information was omitted, the general trend being discussed is consistent with the original; the omitted information was

TABLE 13.2

Product Yield of the Slow Pyrolysis of Wood

Temperature (°C)	Heating Rate (K/min)	Residence Time (s)	Char (wt.%)	Oil (wt.%)	Gas (wt.%)
300	5	7200	53.8	10.6	14.6
300	20	7200	55.6	10.1	14.0
300	80	7200	60.8	6.4	11.2
420	5	7200	29.7	12.4	26.4
420	20	7200	27.2	12.2	23.0
420	80	7200	25.2	11.9	26.0
600	5	7200	24.4	12.4	26.4
600	20	7200	22.6	12.8	27.0
600	80	7200	18.7	14.6	29.1

Source: Williams, P.T. and Besler, S., *Renewable Energy*, 7, 3, 233–250.

considered irrelevant to the context at hand. Furthermore, the remaining wt.% consists of the aqueous phase, which was not displayed in Table 13.2.

13.2.4.1.2 *Intermediate Pyrolysis*

This type of process is performed at temperatures within the 400–500 °C range, with relatively slow heating rates (1–1000 °C/s) and average residence times (typically 5–10 min) (Yang et al., 2013) in the absence of oxygen or with limited oxygen supply. It is an intermediate step between low-temperature pyrolysis and high-temperature gasification or combustion.

Intermediate pyrolysis offers several advantages over other thermal conversion processes. It operates at lower temperatures compared with high-temperature gasification or combustion, which helps preserve the energy content of the feedstock and reduces the formation of undesired by-products. The process also allows the production of a range of valuable products, including biochar, bio-oil, and gases, which can be used as renewable energy sources or as feedstocks for various industries. From this process, biochar is produced and can be used as agricultural fertilizer or as a component to facilitate energy production through co-combustion in thermal power plants.

13.2.4.1.3 *Fast Pyrolysis*

In order to increase the yield of bio-oil, pyrolysis is carried out extremely fast so that the number of secondary reactions that may take place is minimized. This type of process requires that some precautions are to be taken into account – these include: (i) having very high heating and heat transfer rates; (ii) controlled pyrolysis temperature within the 425–600 °C range (other studies reported 450–550 °C as the ideal temperature; Heo et al., 2010); (iii) short vapour residence times (<2 s); and (iii) rapid cooling and water quenching of pyrolysis vapours and aerosols. Multiple studies of fast pyrolysis have all exhibited the same overall trend of having higher oil yields relative to the char and gas yields, as shown in Table 13.3; this is regardless of the feedstock source, as the first two data points, corresponding to corn cob and straw, respectively, still have the oil as the highest-yield product, albeit less than expected.

TABLE 13.3

Weight Fractions of Different Fast Pyrolysis Processes from Differing Studies

Temperature (°C)	Heating Rate (°C/min)	Residence Time (s)	Char (wt.%)	Oil (wt.%)	Gas (wt.%)	Ref.
500	–	<2	~23	~41	30	(Heo et al., 2010)
	–	<2	~35	~35	39	
	–	<2	15	65–75	10	(Jamradloedluk & Lertsatitthanakorn, 2014)
450–550	10–1000	1.5	10–15	50–70	10–15	(Chen & Pilla, 2022)

It should be noted that feeds with a high initial moisture content are ill suited for fast pyrolysis due to the volumes of water vapour generated being detrimental to the reactor temperature. However, with the right conditions, fast pyrolysis is said to consistently produce 60–75 wt.% bio-oil, 15–25 wt.% solid char, and 10–20 wt.% of non-condensable gases, all of which are dictated by the specifics of the feedstock content (Mohan et al., 2006).

13.2.4.1.4 Flash Pyrolysis

Flash pyrolysis is typically characterized by its high temperatures (90–300 °C) and extremely fast heating rates (>1000 °C/s). This process usually results in high yields of bio-oil with a conversion efficiency of up to 70% and low water content (Harussani & Sapuan, 2022; Ighalo et al., 2022). Furthermore, the residence times are extremely short, with most reactions lasting <0.5 s. In order for these conditions to be viable, the feedstock particle sizes have to be minuscule, usually in the range of 105–250 µm (Gerçel, 2002).

13.2.4.1.5 Torrefaction

Torrefaction can be described as a pre-treatment method where the decomposition reaction occurring at the established temperature level results in the biomass being completely dried and its tenacious and fibrous structure degenerated. This enables the grindability of the subjected biomass to improve significantly (Abdulyekeen et al., 2021). This process is very similar to the pyrolysis process, with the exception of the lower operating temperatures – typically between 200 and 300 °C (Conag et al., 2017). The process is carried out at near atmospheric pressures in the absence of oxygen. It is also characterized by its relatively low heating rates (<50 °C/min). As with pyrolysis, the energy and mass yields are highly dependent on the torrefaction temperature, reaction time, and biomass constituents. Torrefaction is a fairly slow process with a residence time of 5–15 min; this diminishes the influence that particle size has relative to the faster pyrolysis processes.

While the products of torrefied biomass usually have similar properties (due to most biomass being built from the same lignocellulose polymer), the mass and energy yield – as previously stated – will vary wildly for different biomass types due to differing polymeric compositions and reactivity. This observation stems from a study comparing the torrefaction of beech, willow, straw, and larch (Abdulyekeen et al., 2021). The primary objective of torrefaction is to enhance the quality of solid biomass as a substitute for coal (Chen et al., 2012). It is also employed as an effective pre-treatment method for producing bio-oil through pyrolysis. Meng et al. (2012) further elaborate that bio-oils derived from the pyrolysis of torrefied biomass exhibit reduced water content and increased carbon content compared with those obtained from raw biomass pyrolysis. Solid biomass is not an ideal fuel source due to its high moisture content, low bulk and energy density, poor grindability, and inferior storability. In light of these limitations, torrefaction has garnered significant interest, as it can greatly enhance both the biomass material itself and its biocarbon derivatives, such as biochar. Similarly to pyrolysis, torrefaction also encompasses its own classification system.

13.2.4.1.6 Dry Torrefaction

Dry torrefaction can be classified into two distinct conditions: oxidative and non-oxidative torrefaction, which are determined by the presence or absence of oxygen, respectively. In non-oxidative torrefaction, carrier gases like nitrogen and carbon dioxide are used, with nitrogen being commonly employed for sweeping biomass materials during thermal pre-treatment (Chen & Kuo, 2010; Thanapal et al., 2014). Conversely, oxidative torrefaction utilizes air, flue gas, or other gases with varying oxygen concentrations as carrier gases for biomass pre-treatment.

Due to the exothermic reactions that occur during thermal degradation and the presence of oxygen, oxidative torrefaction exhibits a higher reaction rate and consequently requires a shorter torrefaction time compared with non-oxidative torrefaction (see Table 13.4). However, oxidative torrefaction leads to a lower solid yield compared with non-oxidative torrefaction. Additionally, it has been observed that the heating value of torrefied biomass decreases as the oxygen concentrations increase, particularly at temperatures below 300 °C.

13.2.4.1.7 Wet Torrefaction

In contrast to dry torrefaction, which involves upgrading biomass in a dry environment, wet torrefaction entails the pre-treatment of biomass in a wet medium, such as water or dilute acid (Lynam et al., 2011). Wet torrefaction typically occurs within the temperature range of 180–260 °C, with a reaction time spanning 5–240 minutes. Studies have indicated that biomass does not undergo significant reactions in hot compressed water (or in other hydrothermal media) when the temperature remains below 180 °C (Chen et al., 2012). Furthermore, due to the alterations in water properties, such as dielectric constant, ion products, density, viscosity, and diffusivity, at high

TABLE 13.4

Advantages and Disadvantages of Oxidative and Non-Oxidative Torrefaction

Non-Oxidative Torrefaction	Oxidative Torrefaction
Advantages • High mass and energy yield • Easier temperature maintenance	Advantages • Lower operating cost • Lower heat supply • Faster reaction
Disadvantages • Higher heat requirement • Lower reaction rate • Requires extraction of nitrogen (from air)	Disadvantages • Lower mass and energy yield (at higher temperatures) • Relatively difficult temperature maintenance

temperatures, wet torrefaction is conducted under conditions close to the subcritical state. This is because extremely high temperatures would increase the likelihood of biomass degradation.

Furthermore, the incorporation of acids, such as sulfuric acid or acetic acid, and additives like lithium chloride has demonstrated an overall enhancement in the performance of wet torrefaction. It has been observed that as biomass undergoes degradation in hot compressed water, volatile acids are released into the solution, thereby further improving the torrefaction process (Yan et al., 2010). This phenomenon is exemplified by Shi et al. (2020), who highlight that as the treatment temperature reaches 180 °C, aldehydes and furfural derivatives are generated, subsequently serving as the primary source of humins. This leads to the progressive formation of solid products.

Wet torrefaction comes with a couple of advantages over dry torrefaction, such as similar improvements in the caloric value of fuels at much lower temperatures, with the implication that wet torrefaction can obtain higher energy densities at an identical mass yield. Moreover, while it can also be considered as more energy-saving due to the redundancy of pre-heating, it is more accurate to regard this aspect as more of a trade-off due to the requirements in the harvesting product processes of wet torrefaction (Bach et al., 2017). Additionally, wet torrefaction features reduced ash content in its products (due to dissolution of minerals in ash into the aqueous phase), which prevents issues such as deposition, corrosion, fouling, agglomeration, and slagging, which are usually encountered in hydrochar conversion processes.

13.2.4.1.8 Steam Torrefaction

This process torrefies biomass using high-pressure and high-temperature steam (Balat, 2008). The process operates within a temperature range of 200–260 °C and has a residence time of 5–10 minutes. By introducing steam into a sealed chamber and subsequently releasing it, the lignocellulosic structure of the feedstock expands, resulting in the separation of individual fibres (Mabee et al., 2006). This process entails a slight loss of feedstock, particularly low-molecular-weight volatiles, leading to an increase in the caloric value and carbon content of the product. Additionally, this phenomenon reduces the bulk density, mean particle size, and equilibrium moisture content of the resulting products (Lam et al., 2013). Studies have indicated that steam torrefaction outperforms dry torrefaction, as it is capable of achieving higher caloric value, carbon content, and hydrophobicity levels at lower operating temperatures and shorter treatment durations. Table 13.5 summarizes the differences between those three kinds of torrefaction techniques.

13.2.4.1.9 Hydrothermal Carbonization

Hydrothermal carbonization (HTC) can be described as a pre-treatment for the formation of a carbon-rich, homogenized, and energy-dense fuel from lignocellulosic biomass, all within a water medium (Lynam et al., 2011; Reza

TABLE 13.5

Comparison between Dry, Wet, and Steam Torrefaction

	Dry Torrefaction	Wet Torrefaction	Steam Torrefaction
Operating conditions	200–300 °C, 10–240 min, 1atm	180–260 °C , 5–240 min, 1–200 atm	200–400 °C , 5–120 min, 1–40 atm
Advantages	Easier operation No post-drying Continuous production	Lower reaction temperature No pre-drying required By-products in liquids Lower ash content in hydrochar	Suitable for wet biomass use No pre-drying required Higher-pelletability solid products
Disadvantages	Higher ash content in biochar Pre-drying required	High-pressure operation Corrosion of reactor by inorganic salts Challenge in continuous production Post-drying	Higher energy consumption High-pressure operation

et al., 2013). The reaction chamber operates at relatively low temperatures in the range of 180–400 °C. The process is carried out at saturation pressures of 2–25 MPa (depending on the temperature range) in order to keep the water in its liquid state. HTC has seen a lot of interest, as it is relatively more efficient when it comes to processing biomass with a high moisture content. In fact, the moisture contained within the biomass is actually beneficial for this process; this is attributed to the pressure increase (and consequent temperature rise) that comes from the formation of vapour after heating (Cheng & Li, 2018). The process has even demonstrated some advantages over conventional pyrolysis, including no drying process required, less volatile gas emission, and lower energy consumption.

From the carbonization of biomass via this process, the solid product (hydrochar) – a biocarbon derivative – is obtained at 41–90 wt.%, while the gas product (consisting mainly of CO_2) makes up only about 10% of the original biomass. The aqueous extractives make up the rest of the mass and consist of sugars, acetic acid, and other organic acids (Hoekman et al., 2011; Reza et al., 2013). One of the major issues that arise with processes like these is the effects that inorganic materials have on the efficacy of the process. Biomass obtained from grasses or other agricultural residues contains more inorganic content than woody biomass (Gholizadeh et al., 2019). These inorganic elements are usually found in the ash content and typically have high melting points; the presence of these elements is detrimental to biomass firing, and it is essential that they are removed; otherwise, they would increase the complexities of co-firing and decrease the efficiency of the boiler. Elements such as sodium, calcium, potassium, and other metals are the root cause of slagging and fouling, which are the two major issues stemming from these inorganic elements. Table 13.6 summarized the several techniques of graphene fabrication.

TABLE 13.6

Several Techniques of Graphene Fabrication

Process	Mechanism	Benefits	Drawbacks
Mechanical exfoliation	Tape (mechanical) exfoliation is used to separate the graphene layers before adding them to a substrate.	- High-quality crystals are obtained.	- Poor scalability.
Chemical vapour deposition	Metallic material is exposed to different hydrocarbons (precursor) at high temperatures; the decomposed carbons are deposited on the metal substrate's surface.	- High structural quality, monolayer graphene is produced. - Wide range of precursors may be used (besides hydrocarbons). - Good scalability.	- Inertness of graphene may cause defects and wrinkles (chance of low-quality produce). - High energy consumption. - Complex process.
Liquid phase exfoliation	Initially graphite is dispersed in a solvent, followed by exfoliation using ultrasound and subsequent centrifugation to control size of graphene flakes, and lastly, purification to remove the unnecessary impurities.	- Excellent scalability. - Relatively simple and fast process. - Yield is easily adjustable. - Moderate quality.	- Costly due to the need for a highly reactive solvent. - Graphene produced may have low electrical conductivity.
Electrochemical exfoliation	Graphite electrode in an electrolyte (with a current) used to produce graphene. Cathodic graphene – high-quality graphene with few layers. Anodic graphene – low yield, low quality (depending on application).	- Ease of operation. - Relatively fewer resources required. - Good upscaling potential. - Eco-friendly process.	- Increased cost primarily due to expensive ionic liquid required. - Crumpled morphology may limit application.
Chemical reduction of graphene oxide	Graphene oxide is exfoliated into a single-layer sheet, then subsequently reduced into mono-layered graphene.	- Good upscaling potential. - Relatively low-cost process.	- Low-quality produces due to excessive defects, - Trade-off between quality and eco-friendly approaches.
Bottom-up synthesis	Graphene molecules are built up from small and atomically precise building blocks containing the necessary coupling sites (for assembly of structural units).	- Excellent quality.	- Limited upscaling.

Slagging can best be described as the accumulation of molten or partially fused deposits on the furnace walls, while fouling is the formation of deposits on convection surfaces, i.e. reheaters and superheaters (Saddawi et al., 2012). These issues arise when side reactions occur to form the by-products that are usually found in the ash content. For example, when alkali metals react with silica (within the reaction chamber), alkali silicates are formed, which soften at temperatures below 700 °C, and thus, slagging occurs. Another example is the reactions between sulphur and alkali metals, where the by-product – alkali sulphate – accumulates in heat transfer surfaces, thereby reducing efficiencies in the combustion process. Despite these challenges, a study conducted by Becker et al. (2019) demonstrated the efficiency of the HTC process in cleaning up the impurities; up to 90% of the various inorganic elements (i.e. Ca, Si, P, etc.) were eliminated from the biomass, where a large percentage of them were converted into their respective oxides (i.e. CaO, SiO_2, P_2O_5, etc.).

13.2.4.1.10 Fischer–Tropsch

The Fischer–Tropsch (FT) process is a method whereby solid or gaseous carbon-based energy sources are converted into fuels and chemicals. The process is operated at 150–300 °C (which limits methane by-product formation) and at pressures ranging from one to several thousand atmospheres. Furthermore, it was reported that increased pressures would result in higher conversion rates and favour the formation of the much-desired long-chain alkanes. In Lappas and Heracleous (2016), most of the FT facilities were only used for coal-to-liquid or gas-to-liquid conversions.

While biomass has not been utilized as extensively as coal or methane, its potential as a substitute for fossil fuels has been widely discussed in numerous publications. Ali et al. (2021) even assert that despite large-scale plants predominantly relying on methane reforming or coal gasification systems, biomass-based FT plants should not encounter significant technical hurdles. The combination of biomass gasification and the FT process presents a promising and encouraging option for the production of environmentally friendly liquid fuels. The general process involves biomass gasification, where the resulting bio-syngas is utilized in the FT synthesis stage to produce long-chain hydrocarbons, which can further be converted into fractions such as green diesel.

The fuels produced from this process (with a biomass source) may potentially replace fossil fuels due to their climate change mitigation potential and clean combustion properties (Ail & Dasappa, 2016). Moreover, these fuels are directly compatible with current distribution infrastructures and engine technologies and are easily adjustable to suit specific applications (which require a certain fuel composition). Another advantage of FT would be its versatility, as it can produce hydrocarbons of varying lengths from carbon-containing feedstock, i.e. coal, natural gas, and biomass. Furthermore, FT-produced synthetic fuels are virtually free of sulphur, nitrogen, and aromatics, therefore giving them an upper hand over crude-refined fuels where the environmental impact is concerned (Lappas & Heracleous, 2016).

However, despite its intriguing benefits, the FT process has to overcome the hurdles of economics. The high energy consumption and large capital costs required to set up FT plants greatly hamper the development of this process. Consequently, this means that with crude oil being its major competitor, the economic viability of the process is largely tied to the state of the crude oil market. This trend was indicated by Lappas and Heracleous (2016), where in the 1990s, the interest in FT was largely sparked by the world having significantly higher reserves of natural gas and coal than crude oil, which has led to an increase in the price of crude oil. Since high-quality clean biofuels can be generated using the FT process using a variety of biomass resources and are compatible with current infrastructure and vehicle technology, global warming and the widespread attempts to reduce CO_2 emissions have reignited interest in FT technology.

13.2.4.2 Plasma Technology

Plasma technology (PT) is a convenient but challenging method of gasification that can also be implemented into the pyrolysis process and enhance it. It offers the unique advantage of providing high temperatures and heating rates (relative to other thermal methods) and even allows the decomposition of biomass at temperatures above 2000 °C in the absence of oxygen. The extremely high temperatures break down the organic compounds in the biomass into their constituent compounds and elements (i.e. hydrogen, carbon dioxide, methane, and water vapour) at an extremely fast rate (Tendler et al., 2005). Existing plasma generator designs can be classified into two groups: (i) direct current plasma torches and (ii) alternating current plasma torches. Between the two, it is the direct current plasma technology that has seen more widespread use (mostly in waste destruction); this is due to the additional difficulties brought about by variations of electrical parameters used by plasma generation with relation to time (Grigaitienė et al., 2011).

This technology has proven to be very practical and even has the potential to solve some of the problems found in conventional pyrolysis (Tendler et al., 2005). For example, gasification carried out by plasma allows better management of unwanted product quantities (i.e. excess carbon dioxide, tar, and/or hydrocarbons) and more control over the composition of synthetic gas (Van Oost et al., 2006). This indicates that when it comes to the production of biogas (at relatively high yields), the use of plasma technologies is a lot more convenient than the current technologies used in conventional pyrolysis. Other advantages offered by this technology include much higher temperatures, resulting in shorter residence times and reduction of undesired pollutants, as previously mentioned, and it does not require additional fuel to run. Furthermore, conventional pyrolysis is known to produce a concerning amount of tar vapour within the gaseous phase, which leads to equipment corrosion; utilizing plasma technologies (plasma pyrolysis) allows the circumvention of this issue, as the gases produced have a low tar content while retaining their high heating value.

While plasma technologies seem lucrative and have proven to be practical, they have presented some challenges that need tackling before they can be employed for widespread industrial use. These challenges include the need for constant maintenance and limited life span of the direct current plasma generators, high capital cost, the requirement for a large refractory–armoured furnace for transportation of the hot carrier gases, and the need for large quantities of gases (i.e. argon) in order to stabilize the plasma. Moreover, the thermal energy supplied by the plasma is greater than necessary, and a lot of it is lost to the environment; this, coupled with the enormous energy demand, indicates large and unnecessary losses of energy when carrying out this process. Additionally, the immense energy demand limits the research that can be conducted to study this technology, thus hampering its development.

13.3 Development and Performance of Carbon-Based Nanofillers in Composites

13.3.1 Graphene

Graphene is an atom-thick sheet of graphite made up of carbon–carbon bonds with a length of 0.142 nm. The material has a strength of 130 GPa, a Young's modulus of 1 TPa, thermal conductivity of around 50,000 W/mK, and an elective conductivity of 6000 S/cm (Choi et al., 2010; Norizan et al., 2020). Graphite has attracted a lot of interest in the nanocomposite field due to its unique structure and properties. Based on literature studies, it is reported that the different technique of graphene fabrication would also lead to different properties of graphene generated. A brief summary relating to the various technique of graphene fabrication depicted in Figure 13.3. Composite materials produced with these graphene fillers are well known for their significantly improved tensile strength, elastic modulus, and electrical and thermal conductivity (Du & Cheng, 2012). Moreover, the source material for these fillers can be found in great abundance, comes at a low cost, and can be applied in a variety of areas (multifunctional) (Sadasivuni et al., 2020). These fillers have also proven to have better gas impermeability, greater surface area, and distinctive functional properties, all of which may be attributed to their structure. Graphene structures are characterized by 2p orbitals that form π state bands that delocalize over the carbon sheets. It is this remarkable structure that gives graphene its desirable properties,

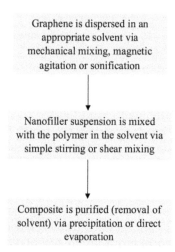

FIGURE 13.3

This method typically involves three steps.

which include high stiffness, excellent thermal conductivity, impermeability, and low resistance to charge carriers, all the while remaining optically transparent.

Graphene potential as a filler has brought about comparisons with other carbon-based fillers such as CNTs. While CNT-polymer composites are far superior in some areas, graphene-polymer composites still remain a lucrative option, primarily due to their abundance of graphene and low cost (Du & Cheng, 2012; Galpayage Dona et al., 2012). Moreover, graphene fillers have a higher surface-to-volume ratio due to their structure and the separation between the inner nanotube surface and polymer molecules; this makes graphene more convenient when it comes to enhancing the properties of polymer matrices, i.e. mechanical, electrical, and microwave absorption properties. The results of comparisons between CNT and graphene polymer composites, however, have not been very consistent. It was inferred that factors such as the dispersion rate of graphene in the polymer matrix and interfacial interactions, the network structure, and the wrinkling of the graphene were the cause of the inconsistent results.

From a study carried out by Du and Cheng (2012), it was stated that the main advantage graphene fillers have over other compositions is their ability to greatly improve the mechanical properties (among others) of rubber matrices, a type of elastomer. Additionally, graphene fillers may also be modified such that their interfacial interactions with elastomer chains are improved. One study exemplified this (Sadasivuni et al., 2014), where acid-modified expandable graphite reinforced styrene butadiene rubber (SBR) nanocomposite demonstrated superior performance over unmodified graphite and carbon black. Another study (Romasanta et al., 2011) reported that functionalized graphene (graphene modified by the addition of a functional group) had the ability to improve the dielectric properties of elastomer nanocomposites roughly tenfold while still maintaining its low dielectric losses and excellent tensile strength.

Several main processing methods of graphene-reinforced composites utilized in previous literature are as follows.

13.3.1.1 Solution Mixing

Considered to be the most commonly used technique due to its being readily applicable to small sample sizes and effective at dispersing nanoparticles (filler) in low-viscosity conditions, solution mixing is regarded as an effective means of preparing composites with uniformly dispersed graphene and its derivatives. Figure 13.3 illustrates the main steps involved in the method.

13.3.1.2 Melt Blending Technique

While the melt blending technique is considered to be relatively more economical and scalable than solvent mixing and even in-situ polymerization, the thermal stability of most chemically modified graphene has so far limited the practicality (and study) of this method. The process is carried out under high shear pressures and temperatures to blend the filler (graphene) and matrix (polymer) and is regarded as being more environmentally friendly than solvent mixing.

Melt blending, however, does come with its disadvantages. The low bulk density of the thermally exfoliated graphene makes extruder feeding difficult, resulting in poor filler dispersion. Moreover, the high shear pressure may cause breaking of the graphene sheets, which further adds to the drawbacks of this process; these setbacks call for the challenging task of optimizing the melting process in order to obtain high-quality polymer composites.

13.3.1.3 In-Situ Polymerization

When the polymer matrices are insoluble or thermally unstable, the in-situ technique is utilized, as melt and solution processing would be impractical in such situations. This method is very valuable in producing a homogeneous dispersion of graphene and in cultivating strong interfacial interactions between the graphene and the polymer matrix (see Figure 13.4).

13.3.2 Carbon Black

Carbon black (CB), widely recognized as the most commonly used powdered carbon filler, encompasses a range of carbon materials utilized for rubber reinforcement, black pigmenting, and the production of electrically conductive composites (Baan, 2007). Different types of CB can be generated, such

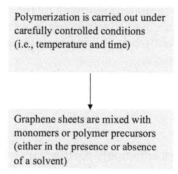

FIGURE 13.4
Facile procedures for in-situ polymerization.

as thermal black, furnace black, and channel black, each named after the specific process employed in its production.

A majority of CB production plants are located close to automotive and rubber industries; this is due to the properties of CB making it difficult to transport, and hence, logistics are costly and complex. The Americas, Europe, and Asia take the largest shares of CB production, contributing 27%, 41%, and 24%, respectively. The rubber industry itself is said to take up to 90% of all carbon black produced; the rest is distributed between inking and plastic industries, with a smaller proportion going to black paint and paper production.

13.3.2.1 Production of Carbon Black

The majority of today's carbon black is produced by incomplete combustion (thermal-oxidative decomposition) of hydrocarbons; it is estimated that up to 98% of all CB produced worldwide undergoes this process. The other process involves thermal decomposition, where CB is produced in the absence of oxygen (much like pyrolysis). Of the two processes used, thermal-oxidative decomposition is considered to be the more convenient method due to its greater yield (see Figure 13.5).

Furnace black refers to the CB collected from the closed-systems furnace (in which incomplete combustion occurs), whereas CB collected from the open-system thermal-oxidative process is produced by the gas black and channel black processes. Lastly, CB from the thermal decomposition of hydrocarbons in a closed system is classified as thermal black and acetylene black, with each process having differing feedstocks.

As indicated, hydrocarbons are the raw materials from which most – if not all – CBs are produced. The hydrocarbons usually come in the form of natural gases, with the exception of the gas black and acetylene black processes, which use coal-tar distillates and acetylene feedstocks. From the morphology of the CB particles, it was deduced that the particles are formed in the

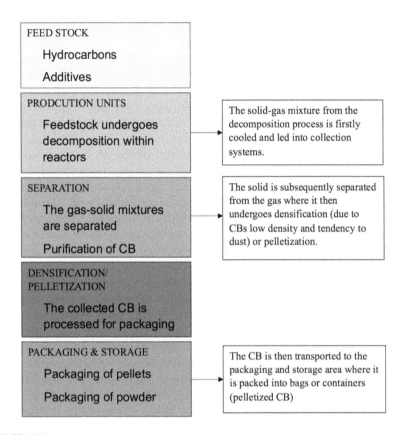

FIGURE 13.5
CB production process. The production of CB is typically split into six areas of focus: (i) feed-stock storage, (ii) CB production units, (iii) purification of CB, (iv) processing of CB, (v) storage of CB, and (vi) recycling of gaseous by-products.

gaseous phase. Therefore, it is crucial that the feedstock be in gaseous form (or at least be able to attain this phase easily), and as a result, all of the gas and liquid feedstocks used today can be vaporized under the right conditions.

13.3.3 Activated Carbon

Activated carbons are characterized by their highly porous structure and large internal surface area. They are widely employed as adsorbents to effectively remove both organic and inorganic pollutants from water and air environments (Ioannidou & Zabaniotou, 2007). The char residue obtained from thermal decomposition processes like pyrolysis serves as a precursor for the production of AC through activation. In fact, any material that possesses a high carbon content and low inorganic content can be used as a raw material for AC. Agricultural waste, in particular, is extensively utilized in this field due to its abundant availability and low cost, making it highly desirable for various manufacturing industries (Njoku & Hameed, 2011). Agricultural by-products have demonstrated the ability to produce AC with excellent adsorption capacity, significant mechanical strength, and low carbon content.

Activated carbons have seen extensive use in wastewater treatment, primarily due to their high absorbing abilities, low cost, and availability. Adsorption, in contrast to absorption, refers to the process of chemicals physically adhering to the surface of a solid material (Lewoyehu, 2021). The adsorption characteristics of a solid material derive from its extensively developed porosity, expansive surface area, variable surface chemistry, and highly reactive surface. This property makes it suitable for applications such as serving as a catalyst or catalyst support in the elimination of pollutants from both liquid and gaseous phases, as well as in the purification and recovery of chemicals Its broad usage in the purification of media is also due to its versatility, as its physical and chemical properties can be altered (during the production process) to produce specific characteristics.

A property that is recurringly mentioned (in this context) is porosity, which plays a crucial role in AC adsorption qualities. The International Union of Pure and Applied Chemistry (IUPAC) divided pore sizes into three categories: macropores (>50 nm), micropores (<2 nm) and mesopores, which exist between micropores. The sizes and quality of the pores determine the adsorption capabilities of the AC produced. Mesopores are very effective at absorbing gases and small molecules; it has even been claimed that pore sizes should be less than 1.0 nm to be suitable for carbon dioxide adsorption (Azmi et al., 2022). The mechanism of gas adsorption was briefly explained by Kaveeshwar et al. (2018), who disclosed that the localization (of pores) responsible for the overlapping of van der Waal forces results in a strong electrostatic attraction between AC and the gas (in their context, carbon dioxide), which facilitates physisorption. This phenomenon allows the adsorption of multiple layers of gases onto the surface, thus increasing adsorption capacities.

In the production of AC, there are two main steps: (i) thermal decomposition of a carbon-rich raw material in the absence of oxygen (otherwise known as carbonization) at temperatures below 800 °C and (ii) the activation of solid char products using chemical agents. The processes of activating char can be divided into three categories, namely, physical, chemical, and steam activation, as shown in Table 13.7.

TABLE 13.7

Different AC Production Methods

Process	Steps	Requirements
Physical	2	Oxidizing agent, i.e. carbon dioxide
Chemical	1	Chemical agents (dehydrating agents and oxidants), i.e. $ZnCl_2$, KOH, H_3PO_4, K_2CO_3
Steam	1	Steam/Pure steam

13.3.3.1 Physical Activation

This two-step process involves the carbonization of char, which is subsequently followed by the activation stage. The first stage is basically pyrolysis, which functions to remove heteroatoms and volatiles while leaving behind chars with a high carbon content (which increases with an increasing temperature). The second stage is termed gasification (Lewoyehu, 2021), which involves the removal of reactive carbon atoms to improve the quality of the material (char) porosity. This process is carried out in the presence of oxidizing gases – typically carbon dioxide, due to its relative cleanliness and ease of handling – and allows more control over the process due to the slow reaction rate at temperatures in the 800 °C range. While the carbonization temperatures usually lie between 400 and 850 °C, the activation stage has a differing range of 600–900 °C (Ioannidou & Zabaniotou, 2007). However, gasification temperatures are highly dependent on the gasification agent used; agents such as water vapour and carbon dioxide operate in the 700–900 °C range, while pure oxygen and/or air (as agents) operate at much lower temperature ranges (300–450 °C), as they are highly reactive. As a side note, a study carried out by Molina-Sabio et al. (1996) compared carbon dioxide and steam as gasification agents. It was found that using carbon dioxide as an agent (as opposed to steam) resulted in an AC with higher micropore volume but lower meso- and macropore volumes. Additionally, it should be noted that extremely reactive agents (i.e. oxygen) limit the quality of porosity formation, as the process is relatively difficult to control; hence carbon dioxide's popularity in this area.

13.3.3.2 Chemical Activation

While this route might be considered a two-step process (much like physical activation), the steps are usually carried out concurrently. The process involves the precursors being blended with chemical agents, i.e. dehydrating agents and oxidants, such that both the carbonization and activation steps occur simultaneously (Ioannidou & Zabaniotou, 2007). However, a more comprehensive view would show that there are three different stages within this process. First, the precursor is impregnated with the activating agent (chemical agent). This stage involves various techniques for impregnating the precursor material, such as immersing it in an aqueous solution of the activating agent or physically mixing the ingredients (precursor and agents), which is generally considered a simpler and more straightforward approach. The second phase consists of subjecting the material to thermal treatment in an inert atmosphere, with the temperature varying depending on the specific activating agent employed. During this stage, de-polymerization, dehydration, and condensation reactions occur, leading to a higher carbon yield compared with physical activation. This is attributed to the restriction of tar and volatile formation imposed by this particular activation process. The last

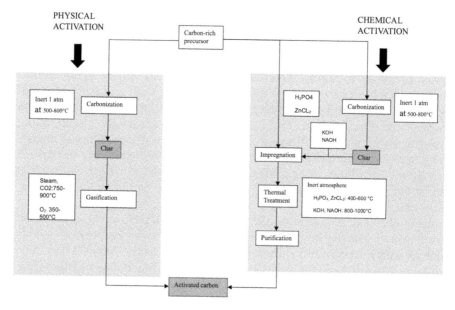

FIGURE 13.6
AC production.

stage involves the purification of the AC, where the contaminants from the chemical agents and reaction by-products are removed. Additionally, when treating biomass with strong base agents such as NaOH and KOH, an initial carbonization step is required such that the organic matter of the precursors does not dissolve.

Chemical activation exhibits a number of advantages over physical activation, some of which include lower activation temperature (<500 °C) relative to physical activation temperatures of 500–900 °C, higher yield, improved porosity, and a shorter activation temperature (Caturla et al., 1991; Delhaes, 2002). Moreover, due to the restrictions on volatile and tar formation, the AC produced has a higher yield and greater surface area than with physical activation.

13.3.4 Carbon Nanotubes

Carbon nanotubes are simply graphene sheets rolled into cylinders with a diameter in the nanometre scale. Two types of CNT exist: single-walled carbon nanotubes (SWCNTs) and multiwalled carbon nanotubes (MWCNTs). As opposed to the single layer for the SWCNTs, MWCNTs consist of several graphene shells layered on top of each other with an adjacent shell separation of 0.34 nm. The diameter of the SWNT was estimated to be 1.4 nm (Dai, 2001) while other sources give the range as 0.4 to 3 nm (Iijima, 1991); moreover, according to calculations, nanotubes smaller than 0.4 nm in diameter will experience too much strain from their high degree of curvature to be thermodynamically stable. MWCNTs, on the other hand, have been found to have a Kijima range of 2–10 nm (Ebbesen & Ajayan, 1992). Within the structures of CNTs, the carbon atoms in the CNT walls are covalently bonded, while the bonding between CNTs (i.e. MWCNTs) is through van der Waal forces. It was also found that the spacing between the shells in MWCNTs decreases as the nanotube's outer diameter increases. This counterintuitive discovery is a result of greater repulsive forces associated with the high curvature of nanotube walls at smaller diameters.

Lattice vector values (n, m) are used to study the structure (and corresponding attributes) of the "rolled-up" CNTs; in the finished nanotube structure, the tail and head of the vector created by adding the integer lattice vectors m and n together are stacked on top of one another. These values are what determine the diameter and chirality of the CNT structure (Ebbesen & Ajayan, 1992; Ganesh, 2013). The lattice vectors are utilized in the symmetrical classification of different CNT structures; these include armchair type, chiral type, and zigzag type as the main classification groups. Armchair types are termed "achiral nanotubes", where achiral is defined as a tube with a structure that is a mirror image of the original. Similarly, zigzag types are also termed "achiral nanotubes", with the exception of m = 0, resulting in the (n, 0) lattice vectors. Additionally, armchair types are said to have a chiral angle of 30°, whereas zigzag types have 0°; essentially, turning an armchair-type CNT structure 30° would result in it turning into a zigzag-type structure. Finally, there is the chiral type, and as the name suggests, these are termed "chiral nanotubes", meaning that the nanotubes have a spiral symmetry that does not allow identically structured mirror images as in the case of achiral nanotubes. These types have the general n and m values with the lattice vector (n, m) (Dresselhaus et al., 1995); see Figure 13.7.

The intrigue of CNTs stems from their exceptional physical and chemical properties (Wu et al., 2022). CNTs are extremely strong and resilient, with a Young's modulus of 1.2 TPa, a tensile strength of 800 GPa, and a density 1/6

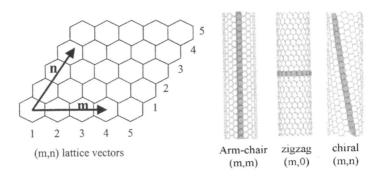

FIGURE 13.7
Different forms of CNTs. (From Sinnott, S. et al., *Chemical Physics Letters*, 315, 1–2, 25–30, 1999.)

TABLE 13.8

Methods of Fabrication of CNTs

Method	Benefits	Drawbacks
Arc discharge	➢ Moderate cost (relative) ➢ Few defects	➢ Discontinuous process ➢ Complex post-processing
Laser ablation	➢ Few shells (MWCT) ➢ Easy post-processing	➢ Poor scalability ➢ Low yield ➢ Complex equipment
Chemical vapour deposition	➢ Sustainable raw material ➢ High yield ➢ High purity	➢ Expensive raw material ➢ High costs
Catalytic pyrolysis	➢ Low cost ➢ Sustainable raw material ➢ Highly scalable	➢ Complex post-processing ➢ High impurities ➢ Discontinuous process

Source: Wu, L. et al., *Fuel Processing Technology*, 229, 107171, 2022.

that of steel – all of which allow the material to withstand enormous strains before mechanical failure. Moreover, CNTs also have exceptional electrically conductive capabilities. The current carrying capacity of CNTs (more specifically SWNTs) was found to be 109 A/cm^2 – 1000 times higher than that of copper (Dresselhaus et al., 1995). Additionally, research carried out to study the electrical resistance within SWNTs found that for metallic SWNTs, the resistance rose from 12 to 20 KΩ for a given temperature rise of 2 to 300 K, respectively. For the semiconductor SWNTs, the resistance was found to be in the 200–500 KΩ range; it was also discovered that CNTs with smaller diameters had a greater resistance (>1000 KΩ). According to Table 13.8, the main production processes of CNTs are as follows.

13.3.4.1 Arc Discharge Evaporation

The arc discharge method was initially used by Iijima (1991) in the synthesis of fullerenes, and it was during these studies that CNT was discovered, as shown in Figure 13.8. Carbon-based needles were found and later discovered to consist of MWCNTs with varying diameters in the 4–10 nm range. These structures grew on the cathode within an argon-filled vessel. The hexagonal carbon atoms on each tube were organized in a helical pattern around the axis of the needle. Within a single needle, the helical pitch varied from needle to needle and from tube to tube. Typically, the needles' tips were covered by conical, polygonal, or curved caps.

While the initial method primarily produced MWCNTs, it was later found that a variation of the arc discharge method was also practical in the synthesis of SWCNTs. This was discovered in two studies (carried out around the same time) where two electrodes were placed in a chamber containing methane (10 Torr) and argon (40 Torr); a direct current of 200 A was used

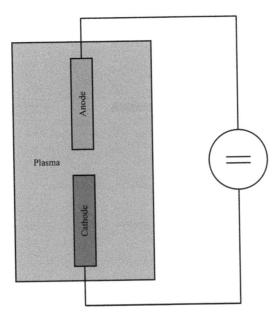

FIGURE 13.8
Simple representation of the arc discharge set-up. (From Popov, V.N., *Materials Science and Engineering: R: Reports*, 43, 3, 61–102, 2004.)

with a potential difference of 20 V applied across the graphitic electrodes. It was during this study that it was discovered that argon, iron, and methane were crucial in the synthesis of SWCNTs. Moreover, it was reported that the SWCNTs formed were entangled in a bundle; upon further inspection, it was found that the SWCNTs had diameters ranging from 0.7 to 1.65 nm (Iijima & Ichihashi, 1993).

The arc discharge method also demonstrated potential scalability in a study carried out by Ebbesen and Ajayan (1992). The experiment was carried out in a chamber containing helium (500 Torr), and a potential difference of 18 V (DC) was applied across the electrodes. It was reported that a yield of 75% (of the original graphitic material) was obtained, with the nanotubes having a diameter in the 2–20 nm range. After the production of MWCNTs, the presence of carbon nanoparticles mixed within the CNTs necessitates the need for purification. One method involves placing the cathode deposit onto a quartz specimen holder in the presence of air, then subsequently irradiating the deposit at 500 °C (30 min) using an infrared heating system.

13.3.4.2 Laser Ablation

This method involves heating and vaporizing a graphitic rod (with metal catalysts) at 1200 °C (Thess et al., 1996), as demonstrated in Figure 13.9. This method demonstrated a remarkable yield of over 70% of the original

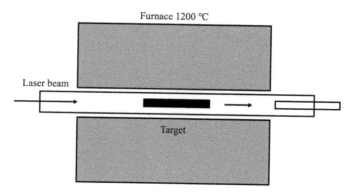

FIGURE 13.9
Simplified representation of the laser ablation set-up.

rod, with the synthesized nanotubes exhibiting highly uniform diameters. Analysis using X-ray diffraction and transmission electron microscopy (TEM) revealed that these nanotubes formed bundles, or ropes, measuring 5–20 nm in diameter and ranging from tens to hundreds of micrometres in length. These ropes formed a two-dimensional (2D) triangular lattice through van der Waals bonding, with a lattice constant of 1.7 nm (Popov, 2004). The growth mechanism for these nanotubes, known as the "scooter" process, involves the chemisorption of a single Ni or Co atom onto the open edge of a nanotube. The metal atom must possess both high catalytic efficiency for nanotube growth and sufficient electronegativity to prevent the formation of fullerenes. The metal atom "scoots" around the open end of the tube, absorbing small carbon molecules and transforming them into graphite-like sheets. The tube continues to grow until an excessive accumulation of catalyst atoms occurs at the end of the nanotube. At this point, larger particles either detach or become heavily coated in carbon, which hampers catalysis and results in the nanotube terminating with a catalyst particle or a fullerene-like tip.

The laser ablation method has many similarities to the arc discharge methods, such as high production yield of CNTs, reliance on the evaporation of solid targets at high temperatures, the entangling of produced CNTs, which complicates the post-processing of the material, and the use of a metal catalyst in the production of SWNTs. However, arc discharge has the advantage of being much cheaper than the laser ablation method.

As previously mentioned, the presence of contaminants necessitates the need for post-processing (purification). Liu et al. (1998) proposed an alternative approach for purifying SWCNTs. In this method, the impure SWCNTs are subjected to reflux in 2.6 M nitric acid and subsequently transferred to water with a pH of 10. The mixture is then filtered using a crossflow filtration system. The purified SWCNTs are further processed by passing them through a polytetrafluoroethylene filter and undergo a cutting step.

According to the researchers, the most effective cutting technique involved subjecting the purified SWCNTs to prolonged sonication in a mixture of concentrated sulfuric and nitric acids at a temperature of 40 °C.

13.3.4.3 Catalytic Growth

This method involves the thermal or chemical vapour decomposition of hydrocarbons in the presence of a catalyst. José-Yacamán et al. (1993), Ivanov et al. (1994), and Amelinckx et al. (1994) were the first to grow MWCNTs using this method, which was inspired by the method used to produce carbon fibres and filaments. According to Figure 13.10, the chemical vapour deposition (CVD) method commonly involves the synthesis of carbon nanotubes by cooling the system and allowing their formation on the catalyst. This process occurs alongside the catalyst-assisted decomposition of hydrocarbons, typically ethylene or acetylene, within a tube reactor operating at temperatures ranging from 550 to 750 °C. Fe, Ni, and Co nanoparticles were identified as the most effective catalysts, which aligns with their usage in previous methods of carbon nanotube production, such as arc discharge and laser ablation. The consistency of catalyst preference across these different approaches suggests a shared mechanism for nanotube growth. Additionally, it has been hypothesized that the type of interaction between the catalyst particles and the substrate determines whether the nanotubes grow from the tips or the base of the catalyst nanoparticles lodged within the pores.

Li et al. (1996) demonstrated the scalability of their method by successfully synthesizing aligned carbon nanotubes using iron as a catalyst and the CVD process. In their experiment, a substrate containing iron nanoparticles embedded in mesoporous silica was placed in the reaction chamber. A mixture of 9% acetylene and nitrogen was introduced into the chamber at a flow rate of 110 cm^3/min. The carbon atoms generated by the breakdown of

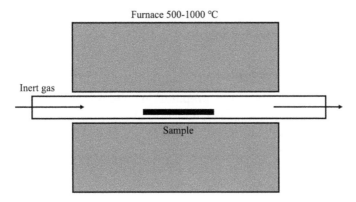

FIGURE 13.10
Simple representation of a catalytic growth set-up.

acetylene at 700 °C were deposited on the substrate, where carbon nanotubes were formed. According to the findings, the nanotubes formed arrays with a spacing of 100 nm, which corresponded to the distance between the pores on the substrate. The outer diameter of the nanotubes was reported to be 30 nm, with 40 shells. However, it was not definitively established whether the growth mechanism involved tip growth or base growth, as the two possibilities could not be clearly distinguished.

As previously mentioned, the catalytic growth method has two available routes (which may also be considered as two separate methods): chemical vapour decomposition or thermal decomposition. The thermal decomposition route, also referred to as catalytic pyrolysis, is a straightforward method with a controllable process. This approach utilizes transition metal catalysts such as Fe, Co, and Ni, which are similar to the catalysts employed in laser ablation and arc discharge methods, to produce CNTs. In order to enhance the catalyst activity, rare earth elements are incorporated alongside these catalysts (Shah & Tali, 2016). Within catalytic pyrolysis lie different approaches in carrying out the process: conventional, plasma, and microwave pyrolysis. Conventional and plasma are similar with respect to their mode of heat transfer, where heat convection, conduction, and radiation are employed. Essentially, the target is heated from the outside in. Microwave pyrolysis, on the other hand, evenly heats the target by utilizing the internal friction heat released by high-frequency vibrations and rotations of molecules in the microwave field.

In order to ensure that the benefits of CNTs as reinforcements within composites are maximized, it is crucial that the aggregation of these fillers is minimized – basically, dispersion must be maximized to enhance the interfacial interactions between the filler and the matrix.

Different processing methods have been explored and utilized for different types of CNT/polymer composites, i.e. composites with thermoplastic or thermosetting matrices. All the available methods are effective in tackling key issues such as exfoliation of CNT bundles and ropes, homogeneous dispersion of individual CNTs into their respective matrices, and improved alignment and interfacial bonding.

13.3.4.4 Solution Processing

Considered the most common preparation method, this route involves mixing both components – CNT and polymer precursors – followed by evaporation of the solvent to form a composite film. The steps are shown in Figure 13.11.

13.3.4.5 Other Approaches

The bulk mixing method involves the utilization of the milling process to generate local high pressures from the collisions of the nanoparticles throughout

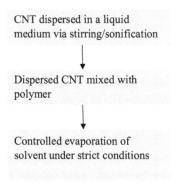

FIGURE 13.11
Methodology for solution processing.

the grinding media. In one instance, a variation termed "high energy ball milling" was used to mix CNTs into polymer matrices, which resulted in a homogeneous dispersion of the CNT into the matrix. Another approach is the melt-mixing method. When the polymer to be processed is not suited for solution techniques (due to insolubility in the common solvents), the melt-mixing method becomes very convenient. It typically involves the blending of the polymer melt with CNTs via extreme shear force. While solution mixing is usually considered in most CNT/polymer composite processes, it was reported that melt-mixing is more suitable for processes with higher CNT concentrations (>5%).

13.3.5 Carbon Quantum Dots

Carbon quantum dots (CQDs) are simply described as nanomaterials with semiconducting properties (Xu et al., 2014). They have dimensions of less than 10 nm, are spherically symmetrical, and can have either amorphous or crystalline structures. Interesting characteristics of both amorphous and crystalline structures include photoluminescence, wavelength-dependent emission, high solubility, low toxicity, simplicity of functionalization, and biocompatibility (Chen et al., 2021). The increasing range of applications in areas such as bioimaging, biosensing, photonics, optoelectronics, and other technologies has made CQDs and other luminescent carbon-based nanomaterials highly attractive as potential alternatives to organic dyes and semiconductor quantum dots (QDs) based on heavy metals like CdSe, PbS, and others.

One drawback of CQDs is their low fluorescence quantum yield and emission of blue-green wavelengths, which limits their practicality in industrial and biological applications, one instance being the low tissue penetration and autofluorescence of visible light being detrimental to their potential application within these fields (Masha & Oluwafemi, 2022). Additionally, under ultraviolet (UV) light, the majority of the reported CQDs exhibited blue-green luminescence, and the wavelengths of the excitation light had the strongest influence on the emission colours. As a consequence, CQDs' yellow or red fluorescence could only be seen when they were exposed to green or yellow laser light, which severely limited the use of CQDs in bioimaging. Moreover, UV light irradiation damages cells and tissues and is unsuitable for bioimaging because it causes significant tissue autofluorescence that interferes with CQD signals (Ding et al., 2017). The CQDs do, however, have a tuneable quantum confinement effect that can be altered by doping with hetero atoms, allowing them to emit wavelengths in the red spectrum and thus making them viable for biological applications. Production methods for CQDs from biobased sources are as follows.

13.3.5.1 Chemical Ablation

While this method comes with its intricacies due to its harsh reaction environment (from use of strong chemicals), other techniques have been devised to make the process more viable for wide usage. In a study conducted by Peng and Travas-Sejdic (2009), a simple method for producing luminescent CQDs in an aqueous solution was described. The process involved dehydrating carbohydrates using concentrated H_2SO_4, followed by dissolving the resulting carbonaceous materials into individual CQDs using HNO_3. To ensure the photoluminescence (PL) of these CQDs, passivation was performed using amine-terminated compounds, specifically 4,7,10-trioxa-1,13-tridecanediamine. This passivation step was found to be crucial for achieving the desired luminescent properties. Furthermore, by adjusting the starting material and the duration of nitric acid treatment, it was possible to modify the emission wavelength of the CQDs.

An alternative synthesis method was introduced by Ding et al. (2017) in order to produce the modified red-emitting CQDs. In this method, a clear solution was prepared by dissolving 1.2 g of citric acid and 2 mL of ethylenediamine in 50 mL of formamide. The solution was transferred into a stainless-steel autoclave with Teflon lining and placed in a muffle furnace at 180 °C for four hours. Once the heating process was complete, the solution was allowed to cool to room temperature within the autoclave. The resulting solution appeared dark red and was filtered through a Whatman filter paper and then through 0.22 μm syringe filters to remove larger particles. To precipitate the carbon-based quantum dots (CBQDs), 30 mL of acetone was added, followed by centrifugation at 9500 rpm for ten minutes. The resulting CBQDs were then dried after undergoing two rounds of cleaning with a mixture of acetone and ethanol (1:1). From this study, it was deduced that the red wavelength emission of the CQDs was due to the nitrogen sources used, namely, ethanediamine and formamide; it was reported that the intense red luminescence is caused by nitrogen-derived structures in the carbon cores and oxygen-related surface states of R-CDs. Furthermore, Ding et al. (2017)

claimed that to significantly advance the sectors in which CQDs are applied, the presented study offered a simple and quick method for synthesizing red-emissive CQDs with a high production yield, great emission efficiency, consistent optical quality, and strong biocompatibility.

13.3.5.2 Electrochemical Carbonization

Electrochemical soaking has proven to be a reliable method for the preparation of CQDs from a diverse selection of bulk carbon sources. Deng et al. (2014) suggest that low-molecular-weight alcohols can be electrochemically carbonized to produce CQDs. Two Pt sheets were utilized as the working and auxiliary electrodes, and a calomel electrode set on a freely adjustable Luggin capillary was employed as the reference electrode. Subsequently, under standard conditions, the alcohols underwent electrochemical carbonization to become CQDs; it was also reported that increasing the potential difference increased the size and degree of graphitization of these CQDs. Without the need for complex purification and passivation processes, the resulting CQDs were obtained with an amorphous core displaying outstanding excitation- and size-dependent PL characteristics.

13.3.5.3 Laser Ablation

This method typically involves irradiating a carbon-rich target with a laser after the submersion of said target in a solvent. The first methodology presented by Sun et al. (2006) involved the laser ablation of a carbon-rich target in the presence of water vapour, with argon – at 900 °C and 75 kPa – acting as the carrier gas. This was followed by refluxing the products in HNO_3 for 12 h, then subsequently passivating the surface by attaching organic species to produce CQDs with a bright luminescence emission. Another methodology presented by Hu et al. (2009) reported on the production of CQDs via laser ablation of carbon materials suspended in an organic solvent. The organic solvents that were chosen allowed the alteration of the surface states of the CQDs, which resulted in QDs with modifiable light-emitting capabilities. From this study, it was deduced that the luminescence was highly dependent on the ligands present on the CQDs' surface – or simply put, the CQDs' surface state.

13.3.5.4 Microwave Irradiation

This method is described as a simple yet inexpensive method of CQD synthesis. It typically involves irradiating an organic compound within a reaction medium with microwaves. One study presented by Liu et al. (2014) involved using sucrose as the carbon source and diethylene glycol (DEG) as the reaction medium; this resulted in green luminous CQDs being produced under microwave irradiation in less than a minute. Another study by Zhai et al. (2012) employed a different technique, where "microwave -mediated pyrolysis" of citric acid (mixed with various amine molecules) was carried out to produce CQDs with intense luminescent emissions. It was stated that the role of the amine molecules was both as a surface passivating agent and as an N-doping precursor. The produced CQDs were documented as being highly biocompatible, therefore displaying great potential applicability in the biomedical field.

13.3.5.5 Hydrothermal/Solvothermal Treatment

This is described as a low-cost, environmentally friendly, and non-toxic route in the production of CQDs from various carbon-rich sources. The methodology typically involves reacting an organic solution/precursor in a sealed hydrothermal reactor at high temperatures. This process has proved to be elementary, with one instance where Yang et al. (2012) documented a one-step synthesis of CQDs via hydrothermal carbonization of chitosan (organic precursor) at 180 °C for 12 hours. The CQDs produced were amino-functionalized and highly luminescent, which further supported the statement regarding the role of amine acids in the process of CQD synthesis (Zhai et al., 2012).

Another technique involves heating the carbon-rich precursors in an organic solvent with a high boiling point, followed by subsequent extraction and concentration processes. One study by Bhunia et al. (2013) documented the synthesizing of two different types of CQDs – hydrophobic and hydrophilic – by carbonizing carbohydrates within different conditions. The hydrophilic ones were synthesized via mixing carbohydrates (of varying amounts) with octadecylamine and octadecene before heating the solution for 10–13 min at 70–300 °C. On the other hand, the hydrophilic CQDs were produced by heating an aqueous solution of carbohydrates within a wide range of pHs or with concentrated phosphoric acid for 60 min at 90 °C.

In CQD/polymer nanocomposites, the polymer is commonly considered as the continuous phase, while the CQDs are seen as nanofillers or agents that enhance properties. Consequently, the core structure and properties of CQD/polymer nanocomposites are predominantly determined by the polymer matrix. The polymer matrix can be either thermoplastic (linear or branched) or thermoset (crosslinked polymer), and the CQDs can be incorporated into the matrix through either covalent attachment or non-covalent blending, depending on the specific application. The chemical structure of the polymer or thermoset resin influences the strength of secondary interactions between the polymer and CQDs, as well as the potential for covalently bonding the CQDs, which in turn affects their dispersion within the composite material. Physical blending is one of the current methods for constructing CQD-reinforced nanocomposites.

13.3.5.6 Physical Blending

Typically considered a facile method of CQD/polymer nanocomposite processing, physical blending involves the mixing of CQDs – in miniscule amounts – with a polymer matrix via solution blending or melt processing techniques. The CQDs and the polymer matrix are bonded via non-covalent interactions, i.e. hydrogen bonding, electrostatic interactions, or π–π interactions (Konwar et al., 2015; Malik et al., 2020). Table 13.9 summarizes the different interactions exist within CQD composites.

13.3.5.7 Chemical Grafting

Chemical enhancement of CQD–polymer nanocomposites is made possible by the abundant functional groups found on the surface of CQDs, where covalent bonds are formed via processes such as esterification, acylation, and epoxidation (De et al., 2015; Hazarika & Karak, 2016). Relative to physical blending, chemical grafting has been reported to produce nanocomposites with superior mechanical strength and resilience (over time), with this being attributed to the covalent bonds formed between the CQDs and polymer chains. Moreover, it was also reported that this processing method produced relatively more homogeneous nanocomposites. However, this method comes with the drawbacks of being a complicated process with

TABLE 13.9

Different Interactions within CQD Composites

Interactions	Mechanism	Reference
Hydrogen bonding	Involves intermolecular bonding between a partially positively charged hydrogen atom and a partially negatively charged nitrogen, oxygen, or fluorine atom (from a neighbouring molecule).	(Feng et al., 2021)
Electrostatic interactions	Involves the attraction/repulsion between charged molecules and is considered the main driving force for a high-quality, homogeneous CQD–polymer nanocomposite (formed from this processing method).	(Feng et al., 2021)
π–π interactions	Typically contain conjugated sp2 domains, can form π–π interactions with neighbouring molecules that have aromatic groups. This includes polymers such as polystyrene, as well as polymers that are functionalized with p orbital–rich groups like phenyl, pyrene, and dipyrene. These interactions occur between the aromatic groups present in the CQDs and the aromatic moieties in the polymer, facilitating their association in the nanocomposite system.	(Malik et al., 2020)

multiple steps and usually incorporating the use of toxic reagents and solvents in its process.

13.3.5.8 In-Situ Growth

Another method, which incorporates the facile aspect of physical blending with the superior properties of chemical enhancement, is the in-situ growth method. This processing method utilizes both the chemical and physical interactions in the bonding of the nanocomposite components. Typically, the process may be completed via one-pot thermal treatments such as hydrothermal, pyrolysis, and heating of a CQD precursor–polymer solution mixture at low temperatures (Fernandes et al., 2020). Within this process, the creation of CQDs and the formation of the composites occur simultaneously; the CQDs nucleate and grow on the active sites – found on the polymer – to produce a stable and homogeneous system.

13.4 Conclusion

This book chapter has provided a comprehensive review of the potential of biosourced carbon (BC) as fillers in composite materials, based on previous published studies. The analysis has highlighted the versatility and promising properties of BC when used as a reinforcing agent and multifunctional filler in advanced composites. The development of BC as a nanofiller for biocomposite applications from the perspective of a low-carbon economy is encouraged by the enormous energy consumption and environmental concerns over the manufacturing of petroleum-based carbon products. The reviewed literature demonstrates that BC offers unique advantages in composite applications. Its partly graphitic nature, cost-effectiveness, and innovative carbonaceous characteristics make it an attractive alternative to traditional carbon-based fillers. BC has shown great potential for enhancing the mechanical properties, electrical conductivity, antibacterial properties, and fire retardancy of composite materials. Furthermore, the chapter has shed light on the state-of-the-art production techniques and characterization methods employed in BC-filled biocomposites. These advancements in production and characterization have contributed to a better understanding of the performance and behaviour of BC-filled composites, enabling researchers and engineers to optimize the material properties for specific applications. Thus, it demonstrates that the integration of BC into composite materials represents a significant step towards achieving sustainable and eco-friendly solutions in various industries.

References

Abdelwahab, M. A., Rodriguez-Uribe, A., Misra, M., & K. Mohanty, A. (2019). Injection molded novel biocomposites from polypropylene and sustainable biocarbon. *Molecules, 24*(22), 4026.

Abdulyekeen, K. A., Umar, A. A., Patah, M. F. A., & Daud, W. M. A. W. (2021). Torrefaction of biomass: Production of enhanced solid biofuel from municipal solid waste and other types of biomass. *Renewable and Sustainable Energy Reviews, 150*, 111436.

Ail, S. S., & Dasappa, S. (2016). Biomass to liquid transportation fuel via Fischer Tropsch synthesis–Technology review and current scenario. *Renewable and Sustainable Energy Reviews, 58*, 267–286.

Aisyah, H., Paridah, M., Sapuan, S., Khalina, A., Berkalp, O., Lee, S., Lee, C., Nurazzi, N., Ramli, N., & Wahab, M. (2019). Thermal properties of woven kenaf/carbon fibre-reinforced epoxy hybrid composite panels. *International Journal of Polymer Science, 2019*, 1–8.

Ali, S. S., Elsamahy, T., Koutra, E., Kornaros, M., El-Sheekh, M., Abdelkarim, E. A., Zhu, D., & Sun, J. (2021). Degradation of conventional plastic wastes in the environment: A review on current status of knowledge and future perspectives of disposal. *Science of the Total Environment, 771*, 144719.

Amelinckx, S., Zhang, X., Bernaerts, D., Zhang, X., Ivanov, V., & Nagy, J. (1994). A formation mechanism for catalytically grown helix-shaped graphite nanotubes. *Science, 265*(5172), 635–639.

Azmi, N. Z. M., Buthiyappan, A., Raman, A. A. A., Patah, M. F. A., & Sufian, S. (2022). Recent advances in biomass based activated carbon for carbon dioxide capture-A review. *Journal of Industrial and Engineering Chemistry, 116*, 1–20.

Baan, R. A. (2007). Carcinogenic hazards from inhaled carbon black, titanium dioxide, and talc not containing asbestos or asbestiform fibers: Recent evaluations by an IARC Monographs Working Group. *Inhalation Toxicology, 19*(sup1), 213–228.

Bach, Q.-V., Chen, W.-H., Lin, S.-C., Sheen, H.-K., & Chang, J.-S. (2017). Wet torrefaction of microalga Chlorella vulgaris ESP-31 with microwave-assisted heating. *Energy Conversion and Management, 141*, 163–170.

Balat, M. (2008). Mechanisms of thermochemical biomass conversion processes. Part 3: Reactions of liquefaction. *Energy Sources Part A, 30*(7), 649–659.

Bartoli, M., Giorcelli, M., Jagdale, P., & Rovere, M. (2020). Towards traditional carbon fillers: Biochar-based reinforced plastic. In *Fillers*. IntechOpen.

Becker, G., Wüst, D., Köhler, H., Lautenbach, A., & Kruse, A. (2019). Novel approach of phosphate-reclamation as struvite from sewage sludge by utilising hydrothermal carbonization. *Journal of Environmental Management, 238*, 119–125.

Behazin, E., Misra, M., & Mohanty, A. K. (2017). Sustainable biocarbon from pyrolyzed perennial grasses and their effects on impact modified polypropylene biocomposites. *Composites Part B: Engineering, 118*, 116–124.

Bhunia, S. K., Saha, A., Maity, A. R., Ray, S. C., & Jana, N. R. (2013). Carbon nanoparticle-based fluorescent bioimaging probes. *Scientific Reports, 3*(1), 1473.

Braghiroli, F. L., Bouafif, H., Neculita, C. M., & Koubaa, A. (2020). Influence of pyrogasification and activation conditions on the porosity of activated biochars: A literature review. *Waste and Biomass Valorization, 11*(9), 5079–5098.

Bridgwater, A. (1996). Production of high grade fuels and chemicals from catalytic pyrolysis of biomass. *Catalysis Today*, 29(1–4), 285–295.

Caturla, F., Molina-Sabio, M., & Rodriguez-Reinoso, F. (1991). Preparation of activated carbon by chemical activation with ZnCl2. *Carbon*, 29(7), 999–1007.

Cha, J. S., Park, S. H., Jung, S.-C., Ryu, C., Jeon, J.-K., Shin, M.-C., & Park, Y.-K. (2016). Production and utilization of biochar: A review. *Journal of Industrial and Engineering Chemistry*, 40, 1–15.

Chang, B. P., Abdelwahab, M. A., Kiziltas, A., Mielewski, D. F., Mohanty, A. K., & Misra, M. (2021). Effect of a small amount of synthetic fiber on performance of biocarbon-filled nylon-based hybrid biocomposites. *Macromolecular Materials and Engineering*, 306(5), 2000680.

Chang, B. P., Rodriguez-Uribe, A., Mohanty, A. K., & Misra, M. (2021). A comprehensive review of renewable and sustainable biosourced carbon through pyrolysis in biocomposites uses: Current development and future opportunity. *Renewable and Sustainable Energy Reviews*, 152, 111666.

Chen, L., Wang, S., Meng, H., Wu, Z., & Zhao, J. (2017). Synergistic effect on thermal behavior and char morphology analysis during co-pyrolysis of paulownia wood blended with different plastics waste. *Applied Thermal Engineering*, 111, 834–846.

Chen, N., & Pilla, S. (2022). A comprehensive review on transforming lignocellulosic materials into biocarbon and its utilization for composites applications. *Composites Part C: Open Access*, 7, 100225.

Chen, W.-H., & Kuo, P.-C. (2010). A study on torrefaction of various biomass materials and its impact on lignocellulosic structure simulated by a thermogravimetry. *Energy*, 35(6), 2580–2586.

Chen, W.-H., Ye, S.-C., & Sheen, H.-K. (2012). Hydrothermal carbonization of sugarcane bagasse via wet torrefaction in association with microwave heating. *Bioresource Technology*, 118, 195–203.

Chen, X., Wu, W., Zhang, W., Wang, Z., Fu, Z., Zhou, L., Yi, Z., Li, G., & Zeng, L. (2021). Blue and green double band luminescent carbon quantum dots: Synthesis, origin of photoluminescence, and application in white light-emitting devices. *Applied Physics Letters*, 118(15), 153102.

Cheng, F., & Li, X. (2018). Preparation and application of biochar-based catalysts for biofuel production. *Catalysts*, 8(9), 346.

Choi, W., Lahiri, I., Seelaboyina, R., & Kang, Y. S. (2010). Synthesis of graphene and its applications: A review. *Critical Reviews in Solid State and Materials Sciences*, 35(1), 52–71.

Conag, A. T., Villahermosa, J. E. R., Cabatingan, L. K., & Go, A. W. (2017). Energy densification of sugarcane bagasse through torrefaction under minimized oxidative atmosphere. *Journal of Environmental Chemical Engineering*, 5(6), 5411–5419.

Dai, H. (2001). Nanotube growth and characterization. In *Carbon Nanotubes*, M. S. Dresselhaus, G. Dresselhaus, & P. Avouris, Eds. (pp. 29–53). Springer.

Das, O., & Sarmah, A. K. (2015). The love–hate relationship of pyrolysis biochar and water: A perspective. *Science of the Total Environment*, 512, 682–685.

De, B., Kumar, M., Mandal, B. B., & Karak, N. (2015). An in situ prepared photo-luminescent transparent biocompatible hyperbranched epoxy/carbon dot nanocomposite. *RSC Advances*, 5(91), 74692–74704.

de Gortari, M. G., Rodriguez-Uribe, A., Misra, M., & Mohanty, A. K. (2020). Insights on the structure-performance relationship of polyphthalamide (PPA) composites reinforced with high-temperature produced biocarbon. *RSC Advances*, *10*(45), 26917–26927.

Delhaes, P. (2002). Chemical vapor deposition and infiltration processes of carbon materials. *Carbon*, *40*(5), 641–657.

Deng, J., Lu, Q., Mi, N., Li, H., Liu, M., Xu, M., Tan, L., Xie, Q., Zhang, Y., & Yao, S. (2014). Electrochemical synthesis of carbon nanodots directly from alcohols. *Chemistry–A European Journal*, *20*(17), 4993–4999.

Ding, H., Wei, J.-S., Zhong, N., Gao, Q.-Y., & Xiong, H.-M. (2017). Highly efficient red-emitting carbon dots with gram-scale yield for bioimaging. *Langmuir*, *33*(44), 12635–12642.

Donnet, J.-B. (2018). *Carbon Black: Science and Technology*. Routledge.

Dresselhaus, M., Dresselhaus, G., & Saito, R. (1995). Physics of carbon nanotubes. *Carbon*, *33*(7), 883–891.

Du, J., & Cheng, H. (2012). The fabrication, properties, and uses of graphene/polymer composites. *Macromolecular Chemistry and Physics*, *213*(10–11), 1060–1077.

Du, S., Dong, Y., Guo, F., Tian, B., Mao, S., Qian, L., & Xin, C. (2021). Preparation of high-activity coal char-based catalysts from high metals containing coal gangue and lignite for catalytic decomposition of biomass tar. *International Journal of Hydrogen Energy*, *46*(27), 14138–14147.

Ebbesen, T. W., & Ajayan, P. M. (1992). Large-scale synthesis of carbon nanotubes. *Nature*, *358*(6383), 220–222.

Feng, Z., Adolfsson, K. H., Xu, Y., Fang, H., Hakkarainen, M., & Wu, M. (2021). Carbon dot/polymer nanocomposites: From green synthesis to energy, environmental and biomedical applications. *Sustainable Materials and Technologies*, *29*, e00304.

Fernandes, D., Heslop, K., Kelarakis, A., Krysmann, M., & Estevez, L. (2020). In situ generation of carbon dots within a polymer matrix. *Polymer*, *188*, 122159.

Galpayage Dona, D. G., Wang, M., Liu, M., Motta, N., Waclawik, E., & Yan, C. (2012). Recent advances in fabrication and characterization of graphene-polymer nanocomposites. *Graphene*, *1*(2), 30–49.

Ganesh, E. (2013). Single walled and multi walled carbon nanotube structure, synthesis and applications. *International Journal of Innovative Technology and Exploring Engineering*, *2*(4), 311–320.

Ge, S., Shi, Y., Xia, C., Huang, Z., Manzo, M., Cai, L., Ma, H., Zhang, S., Jiang, J., & Sonne, C. (2021). Progress in pyrolysis conversion of waste into value-added liquid pyro-oil, with focus on heating source and machine learning analysis. *Energy Conversion and Management*, *245*, 114638.

Gerçel, H. F. (2002). Production and characterization of pyrolysis liquids from sunflower-pressed bagasse. *Bioresource Technology*, *85*(2), 113–117.

Gholizadeh, M., Hu, X., & Liu, Q. (2019). A mini review of the specialties of the bio-oils produced from pyrolysis of 20 different biomasses. *Renewable and Sustainable Energy Reviews*, *114*, 109313.

Giorcelli, M., & Bartoli, M. (2019). Development of coffee biochar filler for the production of electrical conductive reinforced plastic. *Polymers*, *11*(12), 1916.

Gray, M., Johnson, M. G., Dragila, M. I., & Kleber, M. (2014). Water uptake in biochars: The roles of porosity and hydrophobicity. *Biomass and Bioenergy*, *61*, 196–205.

Grigaitienė, V., Snapkauskienė, V., Valatkevičius, P., Tamošiūnas, A., & Valinčius, V. (2011). Water vapor plasma technology for biomass conversion to synthetic gas. *Catalysis Today, 167*(1), 135–140.

Harussani, M., & Sapuan, S. (2022). Development of kenaf biochar in engineering and agricultural applications. *Chemistry Africa, 5*, 1–17.

Harussani, M., Sapuan, S., Rashid, U., & Khalina, A. (2021). Development and characterization of polypropylene waste from personal protective equipment (PPE)-derived char-filled sugar palm starch biocomposite briquettes. *Polymers, 13*(11), 1707.

Harussani, M., Sapuan, S., Rashid, U., Khalina, A., & Ilyas, R. (2022). Pyrolysis of polypropylene plastic waste into carbonaceous char: Priority of plastic waste management amidst COVID-19 pandemic. *Science of the Total Environment, 803*, 149911.

Hazarika, D., & Karak, N. (2016). Biodegradable tough waterborne hyperbranched polyester/carbon dot nanocomposite: Approach towards an eco-friendly material. *Green Chemistry, 18*(19), 5200–5211.

Heo, H. S., Park, H. J., Park, Y.-K., Ryu, C., Suh, D. J., Suh, Y.-W., Yim, J.-H., & Kim, S.-S. (2010). Bio-oil production from fast pyrolysis of waste furniture sawdust in a fluidized bed. *Bioresource Technology, 101*(1), S91–S96.

Hoekman, S. K., Broch, A., & Robbins, C. (2011). Hydrothermal carbonization (HTC) of lignocellulosic biomass. *Energy and Fuels, 25*(4), 1802–1810.

Hu, S.-L., Niu, K.-Y., Sun, J., Yang, J., Zhao, N.-Q., & Du, X.-W. (2009). One-step synthesis of fluorescent carbon nanoparticles by laser irradiation. *Journal of Materials Chemistry, 19*(4), 484–488.

Ighalo, J. O., Iwuchukwu, F. U., Eyankware, O. E., Iwuozor, K. O., Olotu, K., Bright, O. C., & Igwegbe, C. A. (2022). Flash pyrolysis of biomass: A review of recent advances. *Clean Technologies and Environmental Policy, 24*(8), 2349–2363.

Iijima, S. (1991). Helical microtubules of graphitic carbon. *Nature, 354*(6348), 56–58.

Iijima, S., & Ichihashi, T. (1993). Single-shell carbon nanotubes of 1-nm diameter. *Nature, 363*(6430), 603–605.

Ioannidou, O., & Zabaniotou, A. (2007). Agricultural residues as precursors for activated carbon production—A review. *Renewable and Sustainable Energy Reviews, 11*(9), 1966–2005.

Ivanov, V., Nagy, J., Lambin, P., Lucas, A., Zhang, X., Zhang, X., Bernaerts, D., Van Tendeloo, G., Amelinckx, S., & Van Landuyt, J. (1994). The study of carbon nanotubules produced by catalytic method. *Chemical Physics Letters, 223*(4), 329–335.

Jamradloedluk, J., & Lertsatitthanakorn, C. (2014). Characterization and utilization of char derived from fast pyrolysis of plastic wastes. *Procedia Engineering, 69*, 1437–1442.

José-Yacamán, M., Miki-Yoshida, M., Rendon, L., & Santiesteban, J. (1993). Catalytic growth of carbon microtubules with fullerene structure. *Applied Physics Letters, 62*(6), 657–659.

Jubinville, D., Abdelwahab, M., Mohanty, A. K., & Misra, M. (2020). Comparison in composite performance after thermooxidative aging of injection molded polyamide 6 with glass fiber, talc, and a sustainable biocarbon filler. *Journal of Applied Polymer Science, 137*(17), 48618.

Kaveeshwar, A. R., Kumar, P. S., Revellame, E. D., Gang, D. D., Zappi, M. E., & Subramaniam, R. (2018). Adsorption properties and mechanism of barium (II) and strontium (II) removal from fracking wastewater using pecan shell based activated carbon. *Journal of Cleaner Production, 193*, 1–13.

Konwar, A., Gogoi, N., Majumdar, G., & Chowdhury, D. (2015). Green chitosan–carbon dots nanocomposite hydrogel film with superior properties. *Carbohydrate Polymers, 115,* 238–245.

Lam, P. S., Lam, P. Y., Sokhansanj, S., Bi, X. T., & Lim, C. (2013). Mechanical and compositional characteristics of steam-treated Douglas fir (Pseudotsuga menziesii L.) during pelletization. *Biomass and Bioenergy, 56,* 116–126.

Lappas, A., & Heracleous, E. (2016). Production of biofuels via Fischer–Tropsch synthesis: Biomass-to-liquids. In *Handbook of Biofuels Production,* R. Luque, Carol Sue Ki Lin, K. Wilson, & J. Clark, Eds. (pp. 549–593). Elsevier.

Lay, M., Rusli, A., Abdullah, M. K., Hamid, Z. A. A., & Shuib, R. K. (2020). Converting dead leaf biomass into activated carbon as a potential replacement for carbon black filler in rubber composites. *Composites Part B: Engineering, 201,* 108366.

Lewoyehu, M. (2021). Comprehensive review on synthesis and application of activated carbon from agricultural residues for the remediation of venomous pollutants in wastewater. *Journal of Analytical and Applied Pyrolysis, 159,* 105279.

Li, L., Rowbotham, J. S., Greenwell, C. H., & Dyer, P. W. (2013). *An Introduction to Pyrolysis and Catalytic Pyrolysis: Versatile Techniques for Biomass Conversion.* Elsevier.

Li, W., Xie, S., Qian, L. X., Chang, B., Zou, B., Zhou, W., Zhao, R., & Wang, G. (1996). Large-scale synthesis of aligned carbon nanotubes. *Science, 274*(5293), 1701–1703.

Liu, J., Rinzler, A. G., Dai, H., Hafner, J. H., Bradley, R. K., Boul, P. J., Lu, A., Iverson, T., Shelimov, K., & Huffman, C. B. (1998). Fullerene pipes. *Science, 280*(5367), 1253–1256.

Liu, Y., Xiao, N., Gong, N., Wang, H., Shi, X., Gu, W., & Ye, L. (2014). One-step microwave-assisted polyol synthesis of green luminescent carbon dots as optical Nanoprobes. *Carbon, 68,* 258–264.

Lopez, G., Artetxe, M., Amutio, M., Alvarez, J., Bilbao, J., & Olazar, M. (2018). Recent advances in the gasification of waste plastics. A critical overview. *Renewable and Sustainable Energy Reviews, 82,* 576–596.

Lynam, J. G., Coronella, C. J., Yan, W., Reza, M. T., & Vasquez, V. R. (2011). Acetic acid and lithium chloride effects on hydrothermal carbonization of lignocellulosic biomass. *Bioresource Technology, 102*(10), 6192–6199.

Mabee, W. E., Gregg, D. J., Arato, C., Berlin, A., Bura, R., Gilkes, N., Mirochnik, O., Pan, X., Pye, E. K., & Saddler, J. N. (2006). Updates on Softwood-to-Ethanol Process Development. *Twenty-seventh symposium on biotechnology for fuels and chemicals, 129-132,* (pp. 55–70). Humana Press.

Malik, R., Lata, S., Soni, U., Rani, P., & Malik, R. S. (2020). Carbon quantum dots intercalated in polypyrrole (PPy) thin electrodes for accelerated energy storage. *Electrochimica Acta, 364,* 137281.

Masha, S., & Oluwafemi, O. S. (2022). Cost-effective synthesis of red-emitting carbon-based quantum dots and its photothermal profiling. *Materials Letters, 323,* 132590.

Meng, J., Park, J., Tilotta, D., & Park, S. (2012). The effect of torrefaction on the chemistry of fast-pyrolysis bio-oil. *Bioresource Technology, 111,* 439–446.

Mohan, D., Pittman Jr, C. U., & Steele, P. H. (2006). Pyrolysis of wood/biomass for bio-oil: A critical review. *Energy and Fuels, 20*(3), 848–889.

Molina-Sabio, M., Gonzalez, M., Rodriguez-Reinoso, F., & Sepúlveda-Escribano, A. (1996). Effect of steam and carbon dioxide activation in the micropore size distribution of activated carbon. *Carbon, 34*(4), 505–509.

Mukherjee, A., Majumdar, S., Servin, A. D., Pagano, L., Dhankher, O. P., & White, J. C. (2016). Carbon nanomaterials in agriculture: A critical review. *Frontiers in Plant Science, 7,* 172.

Nam, I., Park, S. M., Lee, H.-K., & Zheng, L. (2017). Mechanical properties and piezoresistive sensing capabilities of FRP composites incorporating CNT fibers. *Composite Structures, 178,* 1–8.

Njoku, V., & Hameed, B. (2011). Preparation and characterization of activated carbon from corncob by chemical activation with H3PO4 for 2, 4-dichlorophenoxyacetic acid adsorption. *Chemical Engineering Journal, 173*(2), 391–399.

Noori, A., Bartoli, M., Frache, A., Piatti, E., Giorcelli, M., & Tagliaferro, A. (2020). Development of pressure-responsive polypropylene and biochar-based materials. *Micromachines, 11*(4), 339.

Norizan, M. N., Moklis, M. H., Demon, S. Z. N., Halim, N. A., Samsuri, A., Mohamad, I. S., Knight, V. F., & Abdullah, N. (2020). Carbon nanotubes: Functionalisation and their application in chemical sensors. *RSC Advances, 10*(71), 43704–43732.

Paleri, D., Rodriguez-Uribe, A., Misra, M., & Mohanty, A. (2021). Preparation and characterization of eco-friendly hybrid biocomposites from natural rubber, biocarbon, and carbon black. *eXPRESS Polymer Letters, 15*(3), 236–249.

Pantea, D., Darmstadt, H., Kaliaguine, S., & Roy, C. (2003). Electrical conductivity of conductive carbon blacks: Influence of surface chemistry and topology. *Applied Surface Science, 217*(1–4), 181–193.

Peng, H., & Travas-Sejdic, J. (2009). Simple aqueous solution route to luminescent carbogenic dots from carbohydrates. *Chemistry of Materials, 21*(23), 5563–5565.

Popov, V. N. (2004). Carbon nanotubes: Properties and application. *Materials Science and Engineering: R: Reports, 43*(3), 61–102.

Prasityousil, J., & Muenjina, A. (2013). Properties of solid fuel briquettes produced from rejected material of municipal waste composting. *Procedia Environmental Sciences, 17,* 603–610.

Quosai, P., Anstey, A., Mohanty, A. K., & Misra, M. (2018). Characterization of biocarbon generated by high-and low-temperature pyrolysis of soy hulls and coffee chaff: For polymer composite applications. *Royal Society Open Science, 5*(8), 171970.

Reza, M. T., Lynam, J. G., Uddin, M. H., & Coronella, C. J. (2013). Hydrothermal carbonization: Fate of inorganics. *Biomass and Bioenergy, 49,* 86–94.

Ritchie, H., Samborska, V., & Roser, M. (2018). Plastic pollution. Published online at OurWorldInData.org. Retrieved from: 'https://ourworldindata.org/plastic-pollution' [Online Resource].

Rodriguez-Uribe, A., Snowdon, M. R., Abdelwahab, M. A., Codou, A., Misra, M., & Mohanty, A. K. (2021). Impact of renewable carbon on the properties of composites made by using three types of polymers having different polarity. *Journal of Applied Polymer Science, 138*(10), 49948.

Romasanta, L. J., Hernández, M., López-Manchado, M. A., & Verdejo, R. (2011). Functionalised graphene sheets as effective high dielectric constant fillers. *Nanoscale Research Letters, 6*(1), 1–6.

Sadasivuni, K. K., Ponnamma, D., Thomas, S., & Grohens, Y. (2014). Evolution from graphite to graphene elastomer composites. *Progress in Polymer Science, 39*(4), 749–780.

Sadasivuni, K. K., Saha, P., Adhikari, J., Deshmukh, K., Ahamed, M. B., & Cabibihan, J. (2020). Recent advances in mechanical properties of biopolymer composites: A review. *Polymer Composites, 41*(1), 32–59.

Saddawi, A., Jones, J., Williams, A., & Le Coeur, C. (2012). Commodity fuels from biomass through pretreatment and torrefaction: Effects of mineral content on torrefied fuel characteristics and quality. *Energy and Fuels, 26*(11), 6466–6474.

Safana, A., Abdullah, N., & Sulaiman, F. (2018). Bio-char and bio-oil mixture derived from the pyrolysis of mesocarp fibre for briquettes production. *Journal of Oil Palm Research, 30*(1), 130.e40.

Shah, K. A., & Tali, B. A. (2016). Synthesis of carbon nanotubes by catalytic chemical vapour deposition: A review on carbon sources, catalysts and substrates. *Materials Science in Semiconductor Processing, 41*, 67–82.

Shi, N., Liu, Q., Cen, H., Ju, R., He, X., & Ma, L. (2020). Formation of humins during degradation of carbohydrates and furfural derivatives in various solvents. *Biomass Conversion and Biorefinery, 10*(2), 277–287.

Sinnott, S., Andrews, R., Qian, D., Rao, A. M., Mao, Z., Dickey, E., & Derbyshire, F. (1999). Model of carbon nanotube growth through chemical vapor deposition. *Chemical Physics Letters, 315*(1–2), 25–30.

Snowdon, M. R., Mohanty, A. K., & Misra, M. (2014). A study of carbonized lignin as an alternative to carbon black. *ACS Sustainable Chemistry and Engineering, 2*(5), 1257–1263.

Snowdon, M. R., Wu, F., Mohanty, A. K., & Misra, M. (2019). Comparative study of the extrinsic properties of poly (lactic acid)-based biocomposites filled with talc versus sustainable biocarbon. *RSC Advances, 9*(12), 6752–6761.

Sun, Y.-P., Zhou, B., Lin, Y., Wang, W., Fernando, K. S., Pathak, P., Meziani, M. J., Harruff, B. A., Wang, X., & Wang, H. (2006). Quantum-sized carbon dots for bright and colorful photoluminescence. *Journal of the American Chemical Society, 128*(24), 7756–7757.

Teixeira, E. de M., Pasquini, D., Curvelo, A. A., Corradini, E., Belgacem, M. N., & Dufresne, A. (2009). Cassava bagasse cellulose nanofibrils reinforced thermoplastic cassava starch. *Carbohydrate Polymers, 78*(3), 422–431.

Teixeira, S. R., Souza, A., Peña, A. F. V., Lima, R., & Miguel, Á. G. (2011). Use of charcoal and partially pirolysed biomaterial in fly ash to produce briquettes: Sugarcane bagasse. *Alternative Fuel, 346*, 177–200.

Tendler, M., Rutberg, P., & Van Oost, G. (2005). Plasma based waste treatment and energy production. *Plasma Physics and Controlled Fusion, 47*(5A), A219.

Thanapal, S. S., Chen, W., Annamalai, K., Carlin, N., Ansley, R. J., & Ranjan, D. (2014). Carbon dioxide torrefaction of woody biomass. *Energy and Fuels, 28*(2), 1147–1157.

Thess, A., Lee, R., Nikolaev, P., Dai, H., Petit, P., Robert, J., Xu, C., Lee, Y. H., Kim, S. G., & Rinzler, A. G. (1996). Crystalline ropes of metallic carbon nanotubes. *Science, 273*(5274), 483–487.

Tomczyk, A., Sokołowska, Z., & Boguta, P. (2020). Biochar physicochemical properties: Pyrolysis temperature and feedstock kind effects. *Reviews in Environmental Science and Bio/Technology, 19*(1), 191–215.

Tripathi, N., Hills, C. D., Singh, R. S., & Atkinson, C. J. (2019). Biomass waste utilisation in low-carbon products: Harnessing a major potential resource. *NPJ Climate and Atmospheric Science, 2*(1), 35.

Ubertini, F., Materazzi, A. L., D'Alessandro, A., & Laflamme, S. (2014). Natural frequencies identification of a reinforced concrete beam using carbon nanotube cement-based sensors. *Engineering Structures, 60,* 265–275.

Vamvuka, D. (2011). Bio-oil, solid and gaseous biofuels from biomass pyrolysis processes—An overview. *International Journal of Energy Research, 35*(10), 835–862.

Van Oost, G., Hrabovsky, M., Kopecky, V., Konrad, M., Hlina, M., Kavka, T., Chumak, A., Beeckman, E., & Verstraeten, J. (2006). Pyrolysis of waste using a hybrid argon–water stabilized torch. *Vacuum, 80*(11–12), 1132–1137.

Ward, B. J., Yacob, T. W., & Montoya, L. D. (2014). Evaluation of solid fuel char briquettes from human waste. *Environmental Science and Technology, 48*(16), 9852–9858.

Watt, E., Abdelwahab, M. A., Snowdon, M. R., Mohanty, A. K., Khalil, H., & Misra, M. (2020). Hybrid biocomposites from polypropylene, sustainable biocarbon and graphene nanoplatelets. *Scientific Reports, 10*(1), 10714.

Williams, P. T., & Besler, S. (1996). The influence of temperature and heating rate on the slow pyrolysis of biomass. *Renewable Energy, 7*(3), 233–250.

Williams, P. T., & Nugranad, N. (2000). Comparison of products from the pyrolysis and catalytic pyrolysis of rice husks. *Energy, 25*(6), 493–513.

Wu, L., Liu, J., Reddy, B. R., & Zhou, J. (2022). Preparation of coal-based carbon nanotubes using catalytical pyrolysis: A brief review. *Fuel Processing Technology, 229,* 107171.

Xu, Y., Jia, X.-H., Yin, X.-B., He, X.-W., & Zhang, Y.-K. (2014). Carbon quantum dot stabilized gadolinium nanoprobe prepared via a one-pot hydrothermal approach for magnetic resonance and fluorescence dual-modality bioimaging. *Analytical Chemistry, 86*(24), 12122–12129.

Xue, B., Wang, X., Sui, J., Xu, D., Zhu, Y., & Liu, X. (2019). A facile ball milling method to produce sustainable pyrolytic rice husk bio-filler for reinforcement of rubber mechanical property. *Industrial Crops and Products, 141,* 111791.

Yan, W., Hastings, J. T., Acharjee, T. C., Coronella, C. J., & Vásquez, V. R. (2010). Mass and energy balances of wet torrefaction of lignocellulosic biomass. *Energy and Fuels, 24*(9), 4738–4742.

Yang, Y., Brammer, J., Ouadi, M., Samanya, J., Hornung, A., Xu, H., & Li, Y. (2013). Characterisation of waste derived intermediate pyrolysis oils for use as diesel engine fuels. *Fuel, 103,* 247–257.

Yang, Y., Cui, J., Zheng, M., Hu, C., Tan, S., Xiao, Y., Yang, Q., & Liu, Y. (2012). One-step synthesis of amino-functionalized fluorescent carbon nanoparticles by hydrothermal carbonization of chitosan. *Chemical Communications, 48*(3), 380–382.

Zanella, K., Gonçalves, J., & Taranto, O. (2016). Charcoal briquette production using orange bagasse and corn starch. *Chemical Engineering Transactions, 49,* 313–318.

Zhai, X., Zhang, P., Liu, C., Bai, T., Li, W., Dai, L., & Liu, W. (2012). Highly luminescent carbon nanodots by microwave-assisted pyrolysis. *Chemical Communications, 48*(64), 7955–7957.

14

Conducting Biopolymer Composite Films

Noor Fadhilah Rahmat, Mohd Shaiful Sajab, and A. Atiqah

14.1 Introduction

Conducting biopolymer films represent a remarkable intersection of materials science, biology, and electronics. These films are composed of biopolymers, which are natural polymers derived from living organisms, and possess the unique ability to conduct electricity. This fusion of biological and electronic properties opens up new avenues for various applications, including flexible electronics, biosensors, and sustainable energy devices. Traditionally, conducting polymers have been used in electronics due to their electrical conductivity and ease of processing. However, these synthetic polymers often involve non-renewable resources and can contribute to environmental issues. Conducting biopolymer films offer a more sustainable alternative, as they can be sourced from renewable biomass, like cellulose, chitosan, starch, and proteins. The introduction of conducting biopolymer films has led to several key advancements in different fields:

Flexible Electronics: These films enable the creation of flexible, lightweight, and biocompatible electronic devices. They can be used for flexible displays, wearable sensors, and even implantable medical devices.

Biosensors: Conducting biopolymer films provide a biocompatible platform for developing biosensors that can detect specific biological molecules. These sensors are utilized in several domains, encompassing medical diagnostics, environmental monitoring, and the assurance of food safety.

Energy Devices: Biopolymer films can be integrated into energy storage and conversion devices. For instance, they can be used in supercapacitors and batteries, enhancing their performance and sustainability.

DOI: 10.1201/9781003408215-14

Tissue Engineering: The biocompatibility of these films makes them suitable for tissue engineering applications. They can serve as substrates for cell growth and differentiation, aiding in the development of functional tissues and organs.

Environmental Sustainability: As these films are derived from renewable sources, they contribute to reducing the environmental impact of electronic waste and non-renewable resource consumption.

The development of conducting biopolymer films comes with its challenges, such as controlling their electrical properties, optimizing fabrication processes, and ensuring long-term stability. Researchers are continuously working to address these issues and unlock the full potential of these materials. In conclusion, conducting biopolymer films represent a groundbreaking advancement in materials science, merging the benefits of natural polymers with electrical conductivity. Their applications span across electronics, healthcare, energy, and sustainability, offering innovative solutions to various real-world challenges. As research and development continue to advance, it is expected that there will be more intriguing and significant breakthroughs and applications in the future.

14.2 Biopolymer Films

Biopolymer films refer to thin sheets or coatings composed of polymers produced from natural sources, which provide several advantageous characteristics, like biodegradability, biocompatibility, and the possibility for sustainable applications. The unique features of these films make them suitable for a diverse variety of applications across many industries. Microorganisms are primarily responsible for the decomposition of biopolymers, such as polysaccharides, proteins, and lipids, into methane (CH_4), carbon dioxide (CO_2), water (H_2O), and inorganic chemicals. A significant quantity of agricultural by-products generated annually have the potential to be utilized in food packaging applications as a raw material that is biodegradable, renewable, and cost-competitive. Biopolymer films predominantly consist of polymers derived from biological origins, encompassing proteins, polysaccharides, and nucleic acids. Polymers are frequently derived from several sources, including plants, animals, and microorganisms. Currently, a significant proportion of biobased biopolymers are produced from primary sources. These feedstock materials include edible biomass such as starch, sugar, and plant oils. Additionally, non-consumable sources like natural rubber have also been utilized as pioneering biopolymers within this field of study (Abhilash & Thomas, 2017).

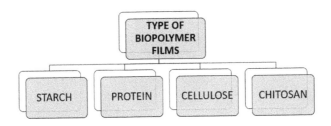

FIGURE 14.1
Types of biopolymer films.

14.2.1 Types of Biopolymer Films

Types of biopolymer films as shown in Figure 14.1 consisted of starch, protein, cellulose and chitosan.

i. *Starch-Based Films*

Starch, derived from sources like corn, potatoes, and cassava, can be processed into films that are suitable for packaging and biodegradable disposable products.

ii. *Cellulose-Based Films*

Cellulose, extracted from plant sources like wood pulp, can be used to create transparent and strong films that find applications in food packaging and biomedical devices.

iii. *Chitosan Films*

Chitosan, derived from chitin found in crustacean shells, possesses antimicrobial properties and is used in wound dressings, food packaging, and controlled-release drug delivery.

iv. *Protein-Based Films*

Proteins like collagen, soy protein, and zein (from corn) can be used to create biopolymer films with various functionalities, including barrier properties and mechanical strength.

14.2.2 Applications of Biopolymer Films

Biopolymer films find extensive use in diverse industries because of their biocompatibility, sustainability, and versatile characteristics. These films are derived from natural sources and offer an eco-friendly alternative to conventional plastics and materials. Here are some applications of biopolymer films, as shown in Figure 14.2:

Packaging: Biopolymer films can be used as packaging materials for food and non-food products. They can protect products from moisture, gases, and UV light while also being compostable or biodegradable, reducing the environmental impact of packaging waste.

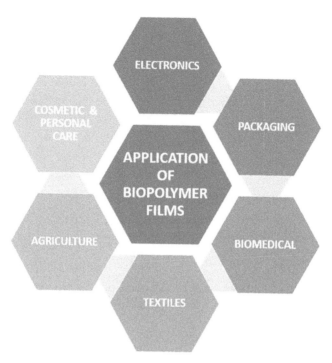

FIGURE 14.2
Applications of biopolymer films.

Agriculture: Biopolymer films can be used as mulch films to cover soil in agricultural fields. These films can control weed growth, regulate soil temperature, and reduce water evaporation. They can also be designed to degrade in the soil after the growing season.

Textiles: Biopolymer films can be applied as coatings on textiles to improve properties such as water repellency, breathability, and durability. This can be useful in outdoor apparel and sportswear.

Cosmetics: Biopolymer films can be used in cosmetic products, such as face masks and patches. They can deliver active ingredients to the skin while providing a biodegradable alternative to conventional synthetic materials.

Biomedical: Biopolymer films are used in tissue engineering and regenerative medicine. They can provide scaffolding for growing tissues and organs and can be engineered to degrade as new tissue forms.

Electronics: Biopolymer films can serve as flexible substrates in electronic devices. They can be used in displays, sensors, and wearable electronics due to their lightweight nature and potential for environmentally friendly disposal.

These applications highlight the versatility and potential of biopolymer films to address a wide range of societal and industrial needs while promoting environmental sustainability. The development of new biopolymer materials and their continuous innovation contribute to creating a more sustainable future.

14.3 Conductive Biopolymer Films and Conductive Additives

Conductive biopolymer films are a specialized type of biopolymer film that have been modified to possess electrical conductivity. These films combine the environmentally friendly and biodegradable characteristics of biopolymers with the ability to conduct electricity, making them suitable for a variety of applications. Conductive biopolymer films can be categorized according to different criteria, including the specific biopolymer type, the conductive additive employed, the preparation technique, and the intended uses, as depicted in Figure 14.3.

- **Conductive Polymer**

Blended biopolymer films: Films incorporating conductive polymers like polyaniline, polypyrrole, or polythiophene.

- **Carbon-Based**

Films containing carbon nanotubes (CNTs), graphene, or carbon black for conductivity.

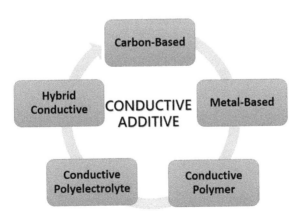

FIGURE 14.3
Conductive additives.

- **Metal-Based**

Films with added metallic conductive components like silver, gold, or copper nanoparticles.

- **Conductive Polyelectrolyte-Enhanced Biopolymer Films**

Films containing conductive polyelectrolytes, such as poly(3,4-ethylene-dioxythiophene) (PEDOT): poly(styrenesulfonate) (PSS), for improved conductivity.

- **Hybrid Conductive**

Films incorporating multiple types of conductive additives for synergistic effects.

14.3.1 Type of Preparation and Methods

Figure 14.4 shows the type of preparation and methods such as spin coating, solution casting, tape casting, spray coating, and sputtering.

- *Spin Coating*

Spin coating is one of the most practical and common techniques for depositing homogeneous thin coatings on a flat substrate. Commonly, an amount of solution is applied to a substrate, which is subsequently rotated at a high speed, such that centrifugal force distributes the fluid over the whole substrate surface. Transparent cellulose films were coated with an indium tin oxide (ITO) nanoparticle solution using a straightforward spin-coating technique,

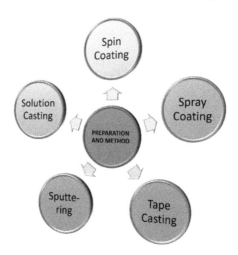

FIGURE 14.4
Types of preparation and methods.

producing flexible and transparent films layered with cellulose and ITO. ITO nanoparticles were evenly dispersed across a flat surface and exhibited strong adhesion at the interface between the cellulose and the ITO layer. The formation of cellulose gel can be observed on the cellulose film's surface, which can function as an adhesive for the bonding of ITO nanoparticles. This adhesion process occurs when the cellulose film is wetted with an appropriate solvent (Khondoker et al., 2012). The process of spin coating involves the removal of the solvent during high-speed spinning in order to create a solid film from a dissolved or dispersed component. The capability of silver-cellulose nanocrystals to develop a stable dispersion in chloroform facilitates the creation of a consistent cellulose film on the surface of a substrate by the spin-coating method, leading to the formation of regular films (Fortunati et al., 2014). Additionally, Qin et al. presented a spin-coating technique that combines metal sputtering to produce composite films of silk fibroin, platinum, and silver (SFPtAg). The spinning speed was set to 3000 rpm, and the spinning time was set to 60 seconds. Once the water had evaporated from the wafers, the SFPtAg composite films were carefully peeled off the substrate (Qin et al., 2019).

- *Solution Casting*

Typically, biodegradable and edible films are manufactured using the traditional casting method, in which a polymeric solution or suspension is poured onto a plate surface. This process does not enable the production of large-dimension films and demands extensive drying times in ovens with air circulating at 40–60 °C (Oliveira de Moraes et al., 2015). A thin film of cellulose nanofibres from sisal pulp and polyaniline (PANI) was created by casting onto the glass substrate, resulting in uniform films after 6 hours of drying at 50 °C (Serhan et al., 2019). Ampaiwong et al. investigated the tensile characteristics of carboxymethylcellulose/graphene oxide (CMC/GO) and carboxymethylcellulose/reduced graphene oxide (CMC/rGO) nanocomposite films produced by the solution casting method (Ampaiwong et al., 2019).

- *Tape Casting*

Tape casting is an effective technique for preparing ceramic layers on supports in electronics manufacturing. This method can also be used to make films that decompose naturally. The process of spreading the film-forming suspension can be done either on large supports or on continuous carrier tapes and enables adjustment of the film thickness.

- *Spray-Coating*

Miao et al. created a conductive film by applying a spray-coating technique, incorporating carbon nanotubes (CNTs), pristine graphene (PG), and PEDOT onto a composite film made of starch and chitosan. This film exhibited transmittance of 83.5% at 550 nm, a minimal sheet resistance of 46 sq^{-1}, and exceptional mechanical properties. This conductive film functioned as a flexible transparent electrode (Miao et al., 2018).

- *Sputtering*

To develop the CZA conductive film, the cellulose (C), zinc oxide (Z), and aluminium doped zinc oxide (A) conductive films were prepared by sputtering Aluminum doped zinc oxide (AZO) on an as-prepared CZ film with dimensions of 4 cm × 4 cm (Serhan et al., 2019).

There are two main approaches for combining starch with CNTs, including the mixing and casting method, which is assisted by a solution procedure and functionalization of the starch. Starch can also develop conductive films with high transparency. In the physical mixing method for preparing starch and CNTs, great attention should be paid to selecting the sequence of ingredient additions so as to avoid the creation of H-bonds between starch and CNTs (Chen et al., 2020).To create a conducting film that worked as a transparent flexible electrode, Miao et al. demonstrated spray-coated CNTs, PG, and PEDOT on a composite film of potato starch (Miao et al., 2018). The casting method is the method most often used for manufacturing starch films. To strengthen the films, plasticizers such as glycerol, sorbitol, and ethylene glycol must be added. The addition of plasticizer up to 60% concentration resulted in a decrease in the moisture content and water absorption of starch-plasticized films (Jawaid & Swain, 2017; Hazrol et al., 2021). In order to create conductive inks, Cataldi et al. employed a method of dispersing commercially available thermoplastic starch-based polymers and graphene nanosheets in an organic solvent. Subsequently, the inks were applied via spraying onto a substrate composed solely of cellulose, followed by a hot-pressing process to yield a pliable conductive film (Cataldi et al., 2015). Potato starch films were plasticized using glycerol, and graphene was then deposited onto the films by using a casting method to produce potato starch–graphene composite films.

14.3.2 Applications of Conductive Biopolymer Films

Conductive biopolymer films have gained significant interest due to their potential applications in various fields where both biocompatibility and electrical conductivity are desired. These films are typically created by incorporating conductive materials, such as carbon nanotubes, graphene, or metallic nanoparticles, into biopolymers like cellulose, chitosan, or proteins. Here are some notable applications of conductive biopolymer films, as shown in Figure 14.5:

Flexible Electronics: Conductive biopolymer films can be used as flexible substrates in electronic devices, such as flexible displays, sensors, and wearable electronics. Their flexibility and lightweight nature

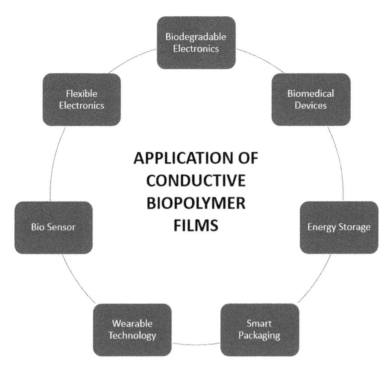

FIGURE 14.5
Applications of conductive biopolymer films.

make them ideal for creating bendable and stretchable electronic components.

Biomedical Devices: Conductive biopolymer films can be used in medical devices and implants. They can be incorporated into sensors that monitor physiological parameters and transmit data wirelessly, or they can be used in neural interfaces for bioelectronic applications.

Energy Storage: Conductive biopolymer films can be used in energy storage devices, such as supercapacitors and batteries. They can serve as electrodes or separator materials, contributing to the development of more environmentally friendly energy storage solutions.

Smart Packaging: Conductive biopolymer films can be integrated into packaging materials to create smart packaging that can monitor the freshness of perishable goods, detect temperature changes, or even display information about the contents.

Biodegradable Electronics: The combination of conductivity and biodegradability makes conductive biopolymer films suitable for applications where temporary electronics are needed, such as environmental monitoring systems that break down after use.

Bio Sensors: Conductive biopolymer films can be used to create bio-compatible sensors that can be integrated into the human body for monitoring purposes, such as glucose monitoring in diabetic patients.

Wearable Technology: Conductive biopolymer films can be integrated into wearable devices, such as smart clothing or accessories, to enable various functionalities like gesture recognition or health monitoring.

Conductive biopolymer films are a subset of materials that combine the electrical properties of conductive materials with the biocompatibility and sustainability of biopolymers. Biopolymers are polymers that are directly obtained from biomass, which is a natural, abundant, and underutilized source of renewable feedstocks. These feedstocks primarily consist of cellulose, protein, starch, and lignin. They are not only abundant and renewable but also biodegradable and biocompatible, resulting in numerous economic and environmental benefits (Zhang et al., 2017; Dang & Yoksan, 2021). These materials shield the body's components from mechanical deformation and environmental destruction. In addition, adhesion for the device can be provided by van der Waals forces, covalent bonds, or hydrogen bonds. Polymers are regularly used as substrates because they are inexpensive to produce and highly deformable (Piccone, 2022).

The most significant element that influences the structure of the conductive pathways is probably mechanical properties. Additionally, the filler network structure may suffer severe damage. The polymer matrix forms an electro-conductive physical network that allows the electrons to move throughout the composite more easily. To increase the strength of films, the majority of studies focus on the incorporation of modification techniques involving particles, fibres, and doping ions into the matrix. Polymer composites that incorporate conductive fillers, including graphene flakes, carbon black, metal nanowires, carbon nanotubes, and metal nanoparticles, have the ability to absorb mechanical deformation when fabricated as thin films. As a result, these composites are regarded as inherently elastic materials (Trung & Lee, 2017). Moreover, the application of conducting polymer films and coatings is widespread due to their distinctive characteristics, such as minimal density, excellent strength, ease of manufacturing, design flexibility, stability, and cost-effectiveness (Wang et al., 2009). However, these conventional films commonly demonstrate suboptimal water resistance and limited mechanical performance due to their inherent hydrophilic nature and insufficient interfacial adhesion. This constraint significantly restricts their capacity for extensive industrial manufacturing and pragmatic implementation. In order to address these concerns, there exists an extremely challenging yet crucial requirement to formulate a plan for the production of high-performance materials that possess exceptional conductivity characteristics as well as mechanical strength.

14.4 Types of Conductive Biopolymer Film

14.4.1 Starch-Based Films

Starch, classified as a polysaccharide, possesses several desirable characteristics such as renewability, affordability, biodegradability, and widespread accessibility. It is generated from natural polymers that are widely obtained from the roots, stems, and seeds of rice, corn, wheat, cassava, potatoes, and other crops (Xiang et al., 2022). Starch-based films are transparent, odourless, tasteless, and colourless. The starch can be classified as amylose or amylopectin based on its molecular chain structure. Furthermore, starch has excellent environmental degradability and biocompatibility. These characteristics make it a promising raw material for the production of environmentally friendly flexible electrodes. Among the documented sustainable materials used in biofilm conductors, starch has the advantages of acceptable water solubility, good film-forming properties, and a relatively low price compared with cellulose, lignin, chitosan, and proteins. Therefore, extensive research has been devoted to developing a film or gel using starch. Starch is currently the most extensively used biodegradable substance. Fifty per cent of the starch produced in Europe is utilized for non-food purposes, such as adhesives and paper binders, textiles, chemical manufacture, fermentation feedstock, and other industrial products (Ogunsona et al., 2018). Starch is easily converted into a film by heating and drying in the presence of water (Xiang et al., 2022). Starch films that are developed using natural starch are typically brittle and have a low degree of transparency.

Starch has been combined with plasticizers like water, glycerol, sorbitol, sugars, and various organic compounds to create thermoplastic starch (TPS) or plasticized starch (PS) through solution casting or melt-extrusion processes. Plasticization is the process through which the mechanical and thermal characteristics of a polymer change in the presence of plasticizer molecules (Singh & Genovese, 2021). The addition of plasticizers helped to improve the flexibility and reduce brittleness (Syafiq et al., 2022). However, it has poor mechanical qualities (fragility/brittleness), low thermal stability, high water vapour permeability, and poor resistance to external influences (humidity, ripping, plucking, etc.), and is incompatible with hydrophobic polymers. Hence, in order to augment the physical, mechanical, and barrier qualities, starch is combined with other natural and synthetic polymers or integrated with diverse nanomaterials. The utilization of chemical modifiers and the addition of nanofillers have been employed to increase the characteristics of TPS (Gamage et al., 2022).

Conductive biopolymer films are a class of materials that combine the properties of biopolymers, which are natural polymers derived from living organisms, with conductivity. These films find applications in various fields, including electronics, sensors, biomedical devices, and more. The type of conductive biopolymer films can be based on different factors such as the type of biopolymer used, the method of synthesis, and the intended application. The following are the types of starch-based films:

- *Corn Starch*

Corn starch film, often referred to as starch-based film or bioplastic film, is a type of biodegradable and environmentally friendly material derived from corn starch. A wearable sensor application for personal health monitoring was developed by using maize starch and calcium chloride to get a flexible, eco-friendly, transparent, and ionically conductive film (Liu et al., 2020). Films derived from corn starch exhibit intrinsic brittleness and lack the requisite strength for typical packaging purposes. Blending starch with carbon is an efficient method of achieving an effective combination of physical properties and improved processing characteristics. In particular, Peidayesh et al. fabricated TPS from corn starch filled with carbon black (CB) by compression moulding. They looked into the variation in electrical conductivity of elastomer-based composites as they were mechanically deformed in tensile or compressive modes. The TPS and CB composite's ultimate tensile strength is enhanced by the addition of CB as compared with pure TPS. When the CB content is increased from 0 to 10.5% by weight, the tensile strength increases from 0.30 to 3.18 MPa, while the Young's modulus increases from 0.81 to 32.77 MPa. Adding CB to the TPS matrix reduces the elongation at break. The electrical conductivity of TPS with embedded CB was increased as a result of the addition of CB between 0 and 5.5 wt.%. This showed that a conducting network had not yet formed in the TPS matrix, and the CB particle agglomerates were still separated. In the percolation concentration area between 5.5 and 10.5 wt.% CB, a further increment in CB concentration led to a composite conductivity increase from 3.1×10^{-7} to $4.2 \times 10{-3}$ S cm^{-1}. The electrical conductivity also increased as the percentage of CB in the composites was increased (Peidayesh et al., 2021).

The research reported by Chaleat et al. focused on the study of polymer blend electrolytes derived from corn starch and chitosan, with the addition of ammonium iodide (NH_4I) as a dopant. The polymer consisting of 80% starch and 20% chitosan, produced by the solution cast technique, was determined to be the most amorphous and appropriate host material. Fourier transform infrared spectroscopy (FTIR) indicated interactions between starch, chitosan, and NH_4I. The polymer substrate achieved a maximum room temperature conductivity of $(3.04 \pm 0.32) \times 10^{-4}$ S cm^{-1} when doped with 40% NH_4I (Chaléat et al., 2014).

Previous research was focused on creating and characterizing corn TPS composites with varying amounts of MWCNTs (multi-walled carbon nanotubes). The composites were made through melt extrusion followed by injection moulding, using glycerol and stearic acid as plasticizers for the corn starch. The addition of 1% MWCNTs increased TPS crystallinity by 20%, raising tensile strength and elastic modulus significantly. However, the effect on

TABLE 14.1

Properties of Corn Starch Biofilm

Type of Starch	Conductive Additive	Method	Conductivity (S cm^{-1})	Tensile Strength (MPa)	References
Corn	10.5% Carbon black	Compression moulding	4.2×10^{-3}	3.18	(Peidayesh et al., 2021)
	Ammonium iodide20% chitosan	Solution casting	3.04×10^{-4}	–	(Chaléat et al., 2014)
	1% MWCNT	Solution casting	2×10^{-4}	0.29	(Yurdakul et al., 2013)
	Polyvinyl alcohol25% graphene	Evaporative casting	2.24×10^{-2}		(Bin-Dahman et al., 2018)

mechanical properties was less pronounced at lower MWCNT loadings (0.1% and 0.5%), though it improved toughness. MWCNT-containing composites had much higher electrical conductivity than pure TPS. Scanning electron microscopy (SEM) analysis revealed that MWCNTs influenced the TPS fracture surface. Scanning transmission electron microscopy (STEM) showed MWCNT clusters near TPS pores, enhancing conductivity. High-resolution transmission electron microscopy (HRTEM) images confirmed MWCNT distribution between stearic acid and the starch matrix. Future research will investigate the effects of MWCNT on TPS crystallization kinetics.

A previous study focused on making and studying corn TPS composites with varying MWCNT levels. The composites were created using melt extrusion followed by injection moulding, using glycerol and stearic acid as plasticizers for corn starch. Incorporating 1% MWCNTs raised TPS crystallinity by 20% according to X-ray diffraction (XRD) measurements. TPS composites with 1% MWCNTs showed significantly improved tensile strength and elastic modulus compared with pure TPS. However, at lower loading rates (0.1% and 0.5%), MWCNTs had less impact on mechanical properties but improved toughness. Moreover, TPS composites with MWCNTs exhibited electrical conductivity almost 100 times higher than pure TPS. The properties of corn starch biofilm are presented in Table 14.1.

- *Potato Starch*

Potato starch is one of the most commonly used commercial starches for sizing paper and textiles, as well as stiffening washed fabrics (Jiang et al., 2016). Insufficient for use in some packaging materials, starch has some drawbacks when compared with conventional synthetic polymers, including a strong hydrophilic nature and low mechanical characteristics. Potato starch keeps its granular structure and lacks good conductivity. Hence, starch was reinforced by incorporating components like nanoparticles, clays, crosslinking, cellulose, and graphene. It has previously been reported that adding glycerol to starch-based materials improves their film conductivity. The conductivity of potato starch films plasticized with glycerol increased with glycerol concentration and temperature (Valencia et al., 2014). In a prior study, glycerol plasticized PS films were combined with graphene to develop potato starch/graphene composite films. Tensile strength was enhanced by incorporating graphene layers within the starch network. This had a significant impact on its mechanical properties at the interface (Gürler & Torğut, 2021). Danilo et al. demonstrated a polymer film using potato starch and iron filler as active layers, in which the starch improved the uniformity of the film. When combined with iron particles and HCl, thermoplastic starch (TPS) exhibited the highest electrical conductivity ($7.42 \times 10^{-3} s/m$) (Battistelli et al., 2020). Jie et al. prepared graphene quantum dots (GQDs) from citric acid and incorporated them into TPS from potato starch to prepare GQDs/TPS composite films using a melt-extrusion technique with glycerol and distilled water as plasticizers. Due to the brown colour of the GQDs and increased scattering effects of the loaded films, a high GQD loading reduced the transparency of the film (Jie Chen et al., 2019). Ma et al. prepared an electrically conductive composite by utilizing graphene oxide (GO) as a filler and potato starch. The abundant oxygen-containing groups of GO are capable of forming hydrogen bond interactions with starch, which could form a stronger interaction between these materials. Important roles are demonstrated by these interactions, and the uniform dispersion of GO in the PS matrix, in enhancing the mechanical and moisture barrier characteristics (Ma et al., 2013). In previous research, composite films were created using a casting method, blending graphene with glycerol plasticized PS. Analytical methods like attenuated total reflectance (ATR)-FTIR and XRD were employed to investigate the characteristics of the films, alongside a series of assessments to determine their capacity for swelling, transparency, solubility, tensile strength, and electrical conductivity. When the graphene content in the films increased from 0% to 1%, tensile strength improved from 1.035 to 1.681 MPa, but was reduced to 1.175 MPa at 2% graphene. Electrical conductivity also increased with higher graphene concentration (Gürler & Torğut, 2021). Tensile strength was improved by incorporating graphene layers within the starch network. This had a significant impact on the mechanical properties of the material. The surface interaction between starch molecules may be responsible for this augmentation or enhancement. A PS matrix and the grafted starch in graphene nanosheet (GN)-starch may interact through hydrogen bonds. Mechanical properties were enhanced by these interactions and the uniform distribution of GN-starch sheets in the PS matrix.

According to previous study, by adding rGO to the potato starch matrix, the tensile strength was increased from 20 to 42 MPa and the elongation at break from 67% to 80%. rGO had a reinforcing impact on the PS matrix.

TABLE 14.2

Properties of Corn Starch Biofilm

Type of Starch	Conductive Additive	Method	Conductivity (S cm^{-1})	Tensile Strength (MPa)	References
Potato	1% Graphene	Solution casting	3.9×10^{-4}	1.681	(Gürler & Torğut, 2021)
	Iron	Solution casting	1.51×10^{-3}	0.70	(Battistelli et al., 2020)
	0.5% MWCNTs	Solution casting	1.10×10^{-4}	9.8	(Domene-López et al., 2020)
	2% MWCNTs	Solution casting	3.9×10^{-4}	1.175	(Gürler & Torğut, 2021)

The homogeneous dispersion of rGO inside the PS matrix, which led to significant interfacial contacts between the rGO and the polymer matrix, may be responsible for this improvement in mechanical performance. It can be stated that the application of tensile stress to the nanocomposite leads to a reduction in loading stresses on the starch matrix. This phenomenon occurs due to the challenging separation of fillers from the matrix, which in turn hinders the transmission of the applied force. Additionally, the hydrogen bonding interactions between the hydroxyl groups of starch and the oxygen functional groups of rGO may enhance the tensile properties (Sandhya et al., 2019). The maximum tensile strength and Young's modulus were also significantly increased by using MWCNTs as reinforcing agents in starch matrices. Adding 0.25 wt.% of MWCNTs to the plasticized potato starch increased the tensile strength and Young's modulus to 9.8 MPa and 0.376 GPa, respectively. At 5 wt.% MWCNT content, the elongation at break increased almost 70%, probably due to the good MWCNT dispersion in the starch matrix. MWCNTs also generate a conduction channel for electrons to flow across the film. The maximum electrical conductivity was found in a sample containing plasticized potato starch and 0.5 wt.% MWCNTs, measuring 1.10×10^{-4} S m^{-1} (Domene-López et al., 2020). The properties of potato starch biofilm are presented in Table 14.2.

- *Cassava Starch*

Cassava is one of the major sources of industrial starch, making it an abundant and affordable source of starch. Cassava starch (CS) is derived from natural polymers, and this material will have enhanced mechanical properties, increased thermal stability, and biodegradability. A previous study reported that the addition of cassava peel starch to $CuSO_4$ resulted in the formation of more pores, hence decreasing the density of the film (Humaidi, 2020). In another report, the electrochemical impedance spectroscopy technique was employed to reveal the conduction mechanism and electrical properties of a composite conductive polymer synthesized from plasticized CS and PEDOT. It showed that the polymer composite of starch and PEDOT has high conductivity and charge-storing capability, making this material of interest for potential charge storage technological applications such as in batteries, accumulators, or electrochemical actuators (Alvaro et al., 2018).

CS-CuSO$_4$ plastic film has been successfully developed by the melt intercalation method, mixing the raw materials at a temperature of 80–90 °C. The conductivity value is determined by the amount of CuSO$_4$ added; the more CuSO$_4$ added, the stronger the conductivity value. The maximum Young's modulus of the material was found in the variation of 95% cassava and 5% CuSO$_4$ by mass, with a value of 0.005 GPa, and the minimum value in the variation of 75% cassava and 25% CuSO$_4$, with a value of 0.001 GPa. This means that a decrease in the amount of starch and the addition of CuSO$_4$ can reduce the value of the Young's modulus of the material. The value of tensile strength tends to decrease with a reduction in starch content and the addition of CuSO$_4$. The increase in conductivity value is consistent with the increasing percentage of filler composition, with values ranging from 7×10^{-5} S m^{-1} to 7.3×10^{-4} S m^{-1} (Humaidi, 2020). In a previous study, the characteristics of CS biocomposites were modified by mixing GO at varied concentrations via the starch intercalation method. This approach was found to enhance the starch structure, leading to the purified cassava powder (CP) biocomposites' characteristics being improved by this method because it improved the starch structure. The greatest tensile strength value was for 10% GO, measured with a tensile strength of 1.6 MPa, which increased by up to 56% from 1% GO. The improved tensile strength can be attributed to the GO/CP matrix's great interfacial interaction with cassava-based starch. The Young's modulus was increased by a 10% GO concentration to a value of 189 MPa. This value is about three times higher than 59 MPa for CS base without GO. The elongation at break was steadily reduced from 4.2 to 2.8 mm as 10% GO was added to the starch matrix. Incorporating GO into a polymer filler reduces hydrogen bonds inside the polymer, which makes the polymer chain more breakable (Zaki et al., 2021).

In previous research, CS was used to create a conductive and biodegradable material on a zinc anode (ZnC) in a zinc-air fuel cell (ZAFC) using an electrochemical method. The effectiveness of this process was tested by measuring conductivity using a four-point probe instrument. The results indicated that the conductivity of the anode improved after depositing CS. The study also explored the impact of different concentrations of cassava on conductivity, revealing a 30% increase in conductivity values compared with the original conductivity of pure zinc. For instance, when 4 grams of CS were used, the conductivity rose from 0.079 S cm^{-1} to 0.105 S cm^{-1}.

14.4.2 Cellulose-Based Film

Cellulose is a renewable and biodegradable substance with large natural reserves. Cellulose, with its unique supramolecular structure and material characteristics, can assist in awakening the potential for conducting polymers in an interplay between the two materials. Cellulose has material qualities that include the ability to conduct electricity. Due to the insulating properties of natural cellulose, conductive materials should be incorporated in order to develop a conductive cellulose substrate. Cellulose plays a complementary role to the brittle conjugated polymers due to its ability to experience additional derivatization, its rigid and oriented molecular structure, and its inherent strength, stability, and film-forming properties. Cellulose also imparts the properties of a stable and durable carrier component (Rußler et al., 2011). Cellulose material has undergone vigorous development in order to produce a flexible substrate that is competitive and reduces the production and consumption of synthesized materials. As a transparent and flexible substrate, it has demonstrated considerable potential for use in the development of future electronic goods. Low-priced chemically modified cellulose derivatives are flexible, making the film dense and transparent, and are resistant to mechanical deformation and chemical degradation (Wang & Huang, 2021). It makes a great candidate for conductive material attachment.

Shiqi Wang and his coworkers studied the possibility of using cellulose nanofibres (CNFs) from sisal pulp as dispersants to scatter and stabilize cellulose nanocrystals@polyaniline (CNCs@PANI), resulting in the development of high-strength conductive films. It has the potential to be used in conductive ink, sensors, flexible supercapacitors, biomedicine, and other applications (Serhan et al., 2019). In a previous study, we produced a cellulose from bamboo, dissolving pulp fibres and developing a conductive film that is transparent, flexible, and has a sandwich structure of cellulose, ZnO, and AZO. The high-performance CZA film was also used to create a flexible electroluminescent device. The ZnO can firmly bond the fundamentally incompatible cellulose and AZO, hence enhancing the film's stability and durability (Serhan et al., 2019). The combination of natural polymers such as cellulose and graphene improved the electrical conductivity, mechanical strength, and thermal stability of composite films. Zhan et al. demonstrated the incorporation of cellulose nanofibrils derived from wood into nanoscale layers of PG, resulting in the formation of a layered structure that is held together by robust hydrogen bonding (Zhan et al., 2019). Cellulose, a naturally existing polymer with excellent mechanical properties and film-forming abilities, could be used to compensate for the mechanical weakness of these conjugated polymer structures. Cellulose is typically employed as a flexible transparent substrate because the network structure created by the long fibres has a higher mechanical durability and strength.

14.4.3 Protein-Based Films

- *Soy Protein Isolate (SPI)*

SPI, a natural polymer that is sustainable, degradable, and biocompatible, is being considered as an alternative substrate for green flexible electronics. SPI has also been employed as reductant, template, and capping agent in the synthetic process. A novel form of nanocomposite hybrid film was created by combining SPI, which possesses exceptional mechanical capabilities, with high–aspect ratio gold nanosheets (NSs). The applications for these films, including sensors and electrical, micromechanical, and biological devices, are severely restricted. Additionally, utilizing a filtering technique, a strong gold/SPI hybrid nanocomposite film was finally developed from gold NSs reduced by SPI and the extra chemically modified SPI. By changing the mass percentage of gold NSs in the film, the hybrid film's conductivity can be adjusted. Notably, the gold/SPI hybrid nanocomposite film demonstrates an exceptionally robust and sensitive reaction to humidity, offering promise in the innovative design of biosensors, e-skin, and actuators (Ling et al., 2016).

SPI is also being explored as a potential material for electronic devices due to its conductivity properties. Previously, researchers processed SPI through extrusion to create pellets, which were then used to produce transparent and uniform films using hot pressing. These films demonstrated semiconductor behaviour, evident from their pinched hysteretic intensity--voltage curve. In contrast to traditional rigid electronic devices, SPI films were pliable and hydrophilic. This characteristic is valuable for applications involving contact with the human body, where adaptable and soft electronic devices are preferable. Overall, these easily manageable films, derived from natural materials, offer a promising avenue for biocompatible organic electronics due to their flexibility and compatibility with the human body. This impacted the distribution of charges and electron transport. The electrical conductivity of the SPI films was measured at 9.889×10^{-4} S m^{-1}, which is comparable to electrical semiconductors like silicon and higher than that of other natural polymers (Guerrero et al., 2021).

SPI, which has been extensively studied, has a significantly greater mechanical property than polysaccharide and lipid biomaterials due to its high protein content and excellent film-forming capability (Kang et al., 2016). SPI is also being explored as a promising option for creating environmentally friendly flexible electronics, especially for wearable devices. Wei et al. made conductive film–based strain sensors that can monitor human motion effectively and possess good fatigue resistance, strong mechanical properties, eco-friendliness, and motion-sensing capabilities. To address this, these researchers developed SPI-based nanocomposite films by adding surface-hydroxylated high–dielectric constant inorganic nanoparticles (BaTiO$_3$) and a biodegradable SPI base. One specific film, called SPI-HBT0.5-GL0.5, containing 0.5 wt.% of BaTiO$_3$ and glycerin, exhibited a range of useful

TABLE 14.3

Properties of SPI Starch Biofilm

Type of Protein	Conductive Additive	Method	Conductivity (S cm^{-1})	Tensile Strength (MPa)	References
SPI	20% Glycerol	Hot pressing	9.98×10^{-6}	9.2	(Guerrero et al., 2021)
	>65% Gold	Vacuum filtration	10^2	10	
	Glycerol 0.5% high-dielectric constant inorganic filler BaTiO$_3$ (HBT) 0.5%	Solution casting	2.4×10^{-3}	21.63	(Wei et al., 2021)

characteristics such as toughness, tensile strength, conductivity, translucence, recyclability, and thermal stability (Wei et al., 2021). The properties of SPI biofilm are presented in Table 14.3.

- *Silk Fibroin*

Silk primarily consists of two major protein components, known as silk fibroin (SF) and sericin. SF constitutes the primary constituent of the fibre, with sericin serving as a cohesive agent responsible for the binding of the fibroin fibres. Silk is a biomaterial with notable mechanical strength and a diverse array of functional and mechanical qualities, making it suitable for various biomedical applications. The brittleness and weakness of regenerated silk fibroin (RSF) fibres typically restrict their extensive utilization as a structural material. A study created a conducting film using a combination of SF and graphene. This author and his group then tested the film to see how it impacted the development of neurons from induced pluripotent stem cells (iPSCs). The film was found to promote the iPSCs' transformation into neurons, and this effect became more significant as the amount of graphene in the film increased. The best outcome was achieved with a 4% graphene content. Overall, the study indicates that a 4% graphene and SF film could serve as a promising biomaterial scaffold for neural regeneration applications. Keratin, the primary structural protein of keratinaceous tissues, is part of a group of structural proteins that produce mechanical strength and toughness due to their fibrillary structure and substantial disulfide crosslinking. SF, a natural silk fibre obtained from the cocoon of the silkworm *Bombyx mori*, has enormous potential to become a sustainable, biocompatible, and biodegradable material platform for the manufacture of bio-derived materials for a wide range of technical and biomedical applications. Natural *Bombyx mori* cocoons are mostly composed of two proteins: SF (70–80 wt.% of a cocoon) and sericin (25–30 wt.% of a cocoon). Combining SF composite films with single-walled carbon nanotubes, a novel conductive film was developed in a previous study. SF also can be developed into transparent thin films that could serve as good optoelectronic substrates. To expand the functions of SF materials in electronics, an SF solution was doped with other materials, such as ultra-long silver nanowires for coating and platinum to develop the film, and the conductivity was improved via the sputtering method. The film also showed potential for practical uses due to its almost linear temperature dependence (Qin et al., 2019). Li et al. reported the fabrication of strong SF with graphene-based nanomaterials (GBN) composite films through a gel–film transformation method, in which SF-coated GBN sheets were stacked together tightly. A composite film containing 15 wt.% SF showed a high tensile strength of 300 MPa, elongation at break 1.5%, and a Young's modulus of 26 GPa. The interfacial interactions among building units possessing distinct architectures play a crucial role in augmenting mechanical properties, facilitating nonlinear deformation, and transforming inherently brittle materials into remarkable substances capable of undergoing inelastic deformation. These interactions enable the redistribution of stresses around defects and the efficient dissipation of energy (Li et al., 2019). In a previous study, Hu et al. (2013) provided evidence that an SF–GO multilayer film, produced by vacuum-assisted filtration, exhibited a maximum tensile strength of 130 MPa. Additionally, the film displayed an elongation at break of 2.8% and a Young's modulus of 13 GPa. These values were found to be approximately two to three times greater than those observed in the pure GO paper. Subsequently, the film was sandwiched between two layers of thin aluminium foil in order to facilitate an electrochemical reduction process. Al ions reduced GO in the SF–GO layered film during this procedure; the degree of reduction was adjustable by changing the reaction duration and the position of the aluminium foil (Hu et al., 2013). The addition of GO inhibited SF crystallization, resulting in lower crystallinity, reduced crystallite sizes, and the development of new interphase zones in the synthetic silk.

Nilogal et al. (2021) discussed the creation of bionanocomposites (BNCs) using SF and rGO. The study confirmed the reduction of GO to rGO within SF by observing a plasmon resonance band in UV-visible spectra. XRD analysis showed a decrease in the GO peak intensity with longer reaction times, indicating the transformation to rGO in the SF matrix. SEM revealed the presence of rGO in the SF matrix, particularly with higher GO amounts, causing broken graphene sheets that enhanced surface roughness and physical contact between SF and rGO. HRTEM depicted rGO clusters with fewer layers, fewer wrinkles, and folding. Raman spectroscopy displayed a higher intensity ratio of the D to G band (ID/IG), confirming rGO formation in the BNC. The research also assessed the impact of rGO on electrical conductivity, demonstrating an increase in DC conductivity from 1.28×10^{-9} to 82.4×10^{-9} S cm^{-1} with increasing GO content in SF. Consequently, the study highlights the transition from

TABLE 14.4

Properties of Silk Fibroin Starch Biofilm

Type of Protein	Conductive Additive	Method	Conductivity (S cm^{-1})	Tensile Strength (MPa)	References
Silk Fibroin	2% Graphene	Solution casting	Conductivity increase	4.37	(Niu et al., 2018)
	0.5% Graphene oxide	Vacuum filtration	13.5	153	(Hu et al., 2013)
	0.2% Graphene oxide	12.5	62.1 × 10^{-9}	26.2	(Nilogal et al., 2021)

insulation to improved conductivity in the SF biopolymer due to the incorporation of rGO. The properties of silk fibroin biofilm are presented in Table 14.4.

- *Keratin*

Keratin is a fibrous protein that serves as a primary structural component in several biological structures, predominantly observed in hair, fingernails, scales, feathers, and wools. The compound exhibits favourable chemical stability and demonstrates insolubility in both aqueous and organic solvents (Xue et al., 2019). Keratin also has proved to be a suitable material in green organic field-effect transistors (OFETs) due to its peptide bond–induced degradation. Singh et al. reported a new efficient pathway for the development of biodegradable organic thin-film transistors (OTFTs) through the inclusion of keratin protein thin film from chicken feathers as a gate dielectric thin film in these transistors. Keratin protein has a high concentration of sheet structure, which gives it good film-forming properties on Si substrate, making it ideal for the fabrication of organic electronic devices (Singh et al., 2017). Due to the presence of hydrophilic functional groups (carbonyl, hydroxyl, and epoxy) and a proton conductive structure, GO is a graphene derivative that has attracted significant interest as a sensing material for humidity sensors (Huang et al., 2020). Cataldi et al. developed protein-based electronic materials by utilizing the keratin from waste wool clips, combined with graphene. The fabrication, characterization, and assembly of resistors, planar capacitors, and inductors generated analogue electrical circuits, such as high-pass filters or resonators. A water-based ink combining keratin and graphene was utilized to functionalize cellulose in order to produce flexible electrodes with outstanding sheet resistance, and the electrical conductivity after folding/unfolding cycles was determined (Cataldi et al., 2019). Ferraro and coworkers reported that the addition of polyglycerol to the keratin-graphene nanocomposite increased the elongation at break of the film. When conditioned in a low-humidity environment, the highest-performing electrode had a high Young's modulus but poor ductility. The samples were significantly softened in a high-humidity environment (100% relative humidity), which decreased the Young's modulus and increased the extensibility (Ferraro et al., 2016). Additionally, the functionalized GO enhances the mechanical properties of the keratin biopolymer and has good compatibility with it. Wenjing Yuan et al. developed graphene-coated human hair strain sensors that exhibited great reliability and durability (Serhan et al., 2019).

14.5 Conclusions

In conclusion, conducting biopolymer films represent a fascinating and promising area of research at the intersection of materials science, biotechnology, and electronics. These films combine the advantageous properties of biopolymers, such as biocompatibility and sustainability, with the unique conductivity imparted by various conducting additives or modifications. The development of conducting biopolymer films holds great potential for numerous applications across diverse fields. In electronics, these films could lead to flexible and lightweight organic electronic devices, wearable sensors, and bioelectronic interfaces that seamlessly integrate with the human body. The biomedical sector could benefit from bioactive films for controlled drug release, tissue engineering scaffolds, and biosensing platforms that offer improved compatibility and reduced risk of adverse reactions. Additionally, conducting biopolymer films may find utility in energy storage, as components in supercapacitors and energy harvesting devices, due to their properties and abundant source materials. However, challenges still remain in terms of optimizing the conductivity, mechanical strength, and long-term stability of these films. Further research is needed to fine-tune the fabrication processes, explore novel conducting additives, and enhance the overall performance of these materials. Additionally, scalable manufacturing methods must be developed to enable large-scale production and integration into commercial products. As the world continues to emphasize sustainable and environmentally friendly solutions, conducting biopolymer films offer an exciting avenue for addressing both technological needs and ecological concerns. With ongoing advancements and interdisciplinary collaboration, these films have the potential to revolutionize various industries and pave the way for a more interconnected and sustainable future.

Acknowledgements

N.F. Rahmat would like to thank the Malaysian government and Ministry of Higher Education (MOHE) for sponsoring their doctorate study.

References

Abhilash, M., & Thomas, D. (2017). Biopolymers for biocomposites and chemical sensor applications. In *Biopolymer Composites in Electronics* (Issue October). https://doi.org/10.1016/B978-0-12-809261-3.00015-2

Alvaro, A. A., Reinaldo, J. P., & Manuel, S. P. (2018). Impedanciometric study of conducting polymer composite films from cassava starch/ poly (3,4-ethylene-dioxythiophene). *Advance Journal of Food Science and Technology, 15*(SPL), 28–32. https://doi.org/10.19026/ajfst.14.5869

Ampaiwong, J., Rattanawaleedirojn, P., Saengkiettiyut, K., Rodthongkum, N., Potiyaraj, P., & Soatthiyanon, N. (2019). Reduced graphene oxide/carboxymethyl cellulose nanocomposites: Novel conductive films. *Journal of Nanoscience and Nanotechnology, 19*(6), 3544–3550. https://doi.org/10.1166/jnn.2019.16120

Battistelli, D., Ferreira, D. P., Costa, S., Santulli, C., & Fangueiro, R. (2020). Conductive thermoplastic starch (Tps) composite filled with waste iron filings. *Emerging Science Journal, 4*(3), 136–147. https://doi.org/10.28991/esj-2020-01218

Bin-Dahman, O. A., Rahaman, M., Khastgir, D., & Al-Harthi, M. A. (2018). Electrical and dielectric properties of poly(vinyl alcohol)/starch/graphene nanocomposites. *Canadian Journal of Chemical Engineering, 96*(4), 903–911. https://doi.org/10.1002/cjce.22999

Cataldi, P., Bayer, I. S., Bonaccorso, F., Pellegrini, V., Athanassiou, A., & Cingolani, R. (2015). Foldable conductive cellulose fiber networks modified by graphene nanoplatelet-bio-based composites. *Advanced Electronic Materials, 1*(12). https://doi.org/10.1002/aelm.201500224

Cataldi, P., Condurache, O., Spirito, D., Krahne, R., Bayer, I. S., Athanassiou, A., & Perotto, G. (2019). Keratin-graphene nanocomposite: Transformation of waste wool in electronic devices [research-article]. *ACS Sustainable Chemistry and Engineering, 7*(14), 12544–12551. https://doi.org/10.1021/acssuschemeng.9b02415

Chaléat, C., Halley, P. J., & Truss, R. W. (2014). Mechanical properties of starch-based plastics. *Starch Polymers: From Genetic Engineering to Green Applications*, 187–209. https://doi.org/10.1016/B978-0-444-53730-0.00023-3

Chen, J., Long, Z., Wang, S., Meng, Y., Zhang, G., & Nie, S. (2019). Biodegradable blends of graphene quantum dots and thermoplastic starch with solid-state photoluminescent and conductive properties. *International Journal of Biological Macromolecules, 139*, 367–376. https://doi.org/10.1016/j.ijbiomac.2019.07.211

Chen, Y., Guo, Z., Das, R., & Jiang, Q. (2020). Starch-based carbon nanotubes and graphene: Preparation, properties and applications. *ES Food & Agroforestry*, 13–21. https://doi.org/10.30919/esfaf1111

Dang, K. M., & Yoksan, R. (2021). Thermoplastic starch blown films with improved mechanical and barrier properties. *International Journal of Biological Macromolecules, 188*(August), 290–299. https://doi.org/10.1016/j.ijbiomac.2021.08.027

Domene-López, D., Delgado-Marín, J. J., García-Quesada, J. C., Martín-Gullón, I., & Montalbán, M. G. (2020). Electroconductive starch/multi-walled carbon nanotube films plasticized by 1-ethyl-3-methylimidazolium acetate. *Carbohydrate Polymers, 229*(October), 115545. https://doi.org/10.1016/j.carbpol.2019.115545

Ferraro, V., Anton, M., & Santé-Lhoutellier, V. (2016). The 'sisters' α-helices of collagen, elastin and keratin recovered from animal by-products: Functionality, bioactivity and trends of application. *Trends in Food Science and Technology, 51*, 65–75. https://doi.org/10.1016/j.tifs.2016.03.006

Fortunati, E., Mattioli, S., Armentano, I., & Kenny, J. M. (2014). Spin coated cellulose nanocrystal/silver nanoparticle films. *Carbohydrate Polymers, 113*, 394–402. https://doi.org/10.1016/j.carbpol.2014.07.010

Gamage, A., Thiviya, P., Mani, S., Ponnusamy, P. G., Manamperi, A., Evon, P., Merah, O., & Madhujith, T. (2022). Environmental properties and applications of biodegradable. *Polymers, 14*.

Guerrero, P., Garrido, T., Garcia-Orue, I., Santos-Vizcaino, E., Igartua, M., Hernandez, R. M., & de la Caba, K. (2021). Characterization of bio-inspired electro-conductive soy protein films. *Polymers, 13*(3), 1–15. https://doi.org/10.3390/polym13030416

Gürler, N., & Torğut, G. (2021). Graphene-reinforced potato starch composite films: Improvement of mechanical, barrier and electrical properties. *Polymer Composites, 42*(1), 173–180. https://doi.org/10.1002/pc.25816

Hazrol, M. D., Sapuan, S. M., Zainudin, E. S., Zuhri, M. Y. M., & Wahab, N. I. A. (2021). Corn starch (Zea mays) biopolymer plastic reaction in combination with sorbitol and glycerol. *Polymers, 13*(2), 1–22. https://doi.org/10.3390/polym13 020242

Hu, K., Tolentino, L. S., Kulkarni, D. D., Ye, C., Kumar, S., & Tsukruk, V. V. (2013). Written-in conductive patterns on robust graphene oxide biopaper by electrochemical microstamping. *Angewandte Chemie - International Edition, 52*(51), 13784–13788. https://doi.org/10.1002/anie.201307830

Huang, X. M., Liu, L. Z., Zhou, S., & Zhao, J. J. (2020). Physical properties and device applications of graphene oxide. *Frontiers of Physics, 15*(3). https://doi.org/10.1007 /s11467-019-0937-9

Humaidi, S. (2020). Preparation and characterization of conductive plastics using cassava peel waste and addition of $CuSO_4$. *Journal of Technomaterials Physics, 2*(1), 34–41. https://doi.org/10.32734/jotp.v2i1.5272

Jawaid, M., & Swain, S. K. (2017). Bionanocomposites for packaging applications. *Bionanocomposites for Packaging Applications*, January, 1–290. https://doi.org/10 .1007/978-3-319-67319-6

Jiang, S., Liu, C., Wang, X., Xiong, L., & Sun, Q. (2016). Physicochemical properties of starch nanocomposite films enhanced by self-assembled potato starch nanoparticles. *LWT - Food Science and Technology, 69*, 251–257. https://doi.org/10 .1016/j.lwt.2016.01.053

Kang, H., Song, X., Wang, Z., Zhang, W., Zhang, S., & Li, J. (2016). High-performance and fully renewable soy protein isolate-based film from microcrystalline cellulose via bio-inspired poly(dopamine) surface modification. *ACS Sustainable Chemistry and Engineering, 4*(8), 4354–4360. https://doi.org/10.1021/acssus-chemeng.6b00917

Khondoker, M. A. H., Yang, S. Y., Mun, S. C., & Kim, J. (2012). Flexible and conductive ITO electrode made on cellulose film by spin-coating. *Synthetic Metals, 162*(21–22), 1972–1976. https://doi.org/10.1016/j.synthmet.2012.09.005

Li, K., Li, P., & Fan, Y. (2019). The assembly of silk fibroin and graphene-based nanomaterials with enhanced mechanical/conductive properties and their biomedical applications. *Journal of Materials Chemistry B, 7*(44), 6890–6913. https://doi .org/10.1039/c9tb01733j

Ling, S., Liang, H., Li, Z., Ma, L., Yao, J., Shao, Z., & Chen, X. (2016). Soy protein-directed one-pot synthesis of gold nanomaterials and their functional conductive devices. *Journal of Materials Chemistry B, 4*(21), 3643–3650. https://doi.org/10.1039/c6tb00616g

Liu, P., Ma, C., Li, Y., Wang, L., Wei, L., Yan, Y., & Xie, F. (2020). Facile preparation of eco-friendly, flexible starch-based materials with ionic conductivity and strain-responsiveness. *ACS Sustainable Chemistry and Engineering, 8*(51), 19117–19128. https://doi.org/10.1021/acssuschemeng.0c07473

Ma, T., Chang, P. R., Zheng, P., & Ma, X. (2013). The composites based on plasticized starch and graphene oxide/reduced graphene oxide. *Carbohydrate Polymers, 94*(1), 63–70. https://doi.org/10.1016/j.carbpol.2013.01.007

Miao, J., Liu, H., Li, Y., & Zhang, X. (2018). Biodegradable transparent substrate based on edible starch-chitosan embedded with nature-inspired three-dimensionally interconnected conductive nanocomposites for wearable green electronics. *ACS Applied Materials and Interfaces, 10*(27), 23037–23047. https://doi.org/10.1021/acsami.8b04291

Nilogal, P., Uppine, G. B., Rayaraddi, R., Sanjeevappa, H. K., Martis, L. J., Narayana, B., & Yallappa, S. (2021). Conductive in situ reduced graphene oxide-silk fibroin bionanocomposites. *ACS Omega, 6*(20), 12995–13007. https://doi.org/10.1021/acsomega.1c00013

Niu, Y., Chen, X., Yao, D., Peng, G., Liu, H., & Fan, Y. (2018). Enhancing neural differentiation of induced pluripotent stem cells by conductive graphene/silk fibroin films. *Journal of Biomedical Materials Research - Part A, 106*(11), 2973–2983. https://doi.org/10.1002/jbm.a.36486

Ogunsona, E., Ojogbo, E., & Mekonnen, T. (2018). Advanced material applications of starch and its derivatives. *European Polymer Journal, 108*(September), 570–581. https://doi.org/10.1016/j.eurpolymj.2018.09.039

Oliveira de Moraes, J., Scheibe, A. S., Augusto, B., Carciofi, M., & Laurindo, J. B. (2015). Conductive drying of starch-fiber films prepared by tape casting: Drying rates and film properties. *LWT - Food Science and Technology, 64*(1), 356–366. https://doi.org/10.1016/j.lwt.2015.05.038

Peidayesh, H., Mosnáčková, K., Špitalský, Z., Heydari, A., Šišková, A. O., & Chodák, I. (2021). Thermoplastic starch–based composite reinforced by conductive filler networks: Physical properties and electrical conductivity changes during cyclic deformation. *Polymers, 13*(21). https://doi.org/10.3390/polym13213819

Piccone, A. (2022). Flexible and stretchable bioelectronics inform cardiac health. *Scilight, 2022*(2), 021107. https://doi.org/10.1063/10.0009349

Qin, X., Peng, Y., Li, P., Cheng, K., Wei, Z., Liu, P., Cao, N., Huang, J., Rao, J., Chen, J., Wang, T., Li, X., & Liu, M. (2019). Silk fibroin and ultra-long silver nanowire based transparent, flexible and conductive composite film and its temperature-Dependent resistance. *International Journal of Optomechatronics, 13*(1), 41–50. https://doi.org/10.1080/15599612.2019.1639002

Rußler, A., Sakakibara, K., & Rosenau, T. (2011). Cellulose as matrix component of conducting films. *Cellulose, 18*(4), 937–944. https://doi.org/10.1007/s10570-011-9555-6

Sandhya, P. K., Sreekala, M. S., Padmanabhan, M., Jesitha, K., & Thomas, S. (2019). Effect of starch reduced graphene oxide on thermal and mechanical properties of phenol formaldehyde resin nanocomposites. *Composites Part B: Engineering, 167*, 83–92. https://doi.org/10.1016/j.compositesb.2018.12.009

Serhan, M., Sprowls, M., Jackemeyer, D., Long, M., Perez, I. D., Maret, W., Tao, N., & Forzani, E. (2019). Total iron measurement in human serum with a smartphone. *AIChE Annual Meeting, Conference Proceedings*, 2019-November. https://doi.org /10.1039/x0xx00000x

Singh, A. A., & Genovese, M. E. (2021). Green and sustainable packaging materials using thermoplastic starch. *Sustainable Food Packaging Technology*, 133–160. https://doi.org/10.1002/9783527820078.ch5

Singh, R., Lin, Y. T., Chuang, W. L., & Ko, F. H. (2017). A new biodegradable gate dielectric material based on keratin protein for organic thin film transistors. *Organic Electronics*, 44, 198–209. https://doi.org/10.1016/j.orgel.2017.02.024

Syafiq, R. M. O., Sapuan, S. M., Zuhri, M. Y. M., Othman, S. H., & Ilyas, R. A. (2022). Effect of plasticizers on the properties of sugar palm nanocellulose/cinnamon essential oil reinforced starch bionanocomposite films. *Nanotechnology Reviews*, 11(1), 423–437. https://doi.org/10.1515/ntrev-2022-0028

Trung, T. Q., & Lee, N. E. (2017). Recent progress on stretchable electronic devices with intrinsically stretchable components. *Advanced Materials*, 29(3), 1–29. https://doi.org/10.1002/adma.201603167

Valencia, G. A., Henao, A. C. A., & Zapata, R. A. V. (2014). Influence of glycerol content on the electrical properties of potato starch films. *Starch/Staerke*, 66(3–4), 260–266. https://doi.org/10.1002/star.201300038

Wang, X. S., Tang, H. P., Li, X. D., & Hua, X. (2009). Investigations on the mechanical properties of conducting polymer coating-substrate structures and their influencing factors. *International Journal of Molecular Sciences*, 10(12), 5257–5284. https://doi.org/10.3390/ijms10125257

Wang, Y., & Huang, J. T. (2021). Transparent, conductive and superhydrophobic cellulose films for flexible electrode application. *RSC Advances*, 11(58), 36607–36616. https://doi.org/10.1039/d1ra06865b

Wei, Y., Jiang, S., Li, X., Li, J., Dong, Y., Shi, S. Q., Li, J., & Fang, Z. (2021). 'Green' Flexible Electronics: Biodegradable and Mechanically Strong Soy Protein-Based nanocomposite Films for Human Motion Monitoring. *ACS Applied Materials and Interfaces*, 13(31), 37617–37627. https://doi.org/10.1021/acsami.1c09209

Xiang, H., Li, Z., Liu, H., Chen, T., Zhou, H., & Huang, W. (2022). Green flexible electronics based on starch. *Npj Flexible Electronics*, 6(1). https://doi.org/10.1038/ s41528-022-00147-x

Xue, Y., Lofland, S., & Hu, X. (2019). Thermal conductivity of protein-based materials: A review. *Polymers*, 11(3). https://doi.org/10.3390/polym11030456

Yurdakul, H., Durukan, O., Seyhan, A. T., Celebi, H., Oksuzoglu, M., & Turan, S. (2013). Microstructural characterization of corn starch-based porous thermoplastic composites filled with multiwalled carbon nanotubes. *Journal of Applied Polymer Science*, 127(1), 812–820. https://doi.org/10.1002/app.37794

Zaki, N. N. M., Yunin, M. Y. A. M., Adenam, N. M., Noor, A. M., Wong, K. N. S. W. S., Yusoff, N., & Adli, H. K. (2021). Physicochemical analysis of graphene oxide-reinforced cassava starch biocomposites. *Biointerface Research in Applied Chemistry*, 11(5), 13232–13243. https://doi.org/10.33263/BRIAC115 .1323213243

Zhan, Y., Xiong, C., Yang, J., Shi, Z., & Yang, Q. (2019). Flexible cellulose nanofibril/ pristine graphene nanocomposite films with high electrical conductivity. *Composites – Part A: Applied Science and Manufacturing, 119*(October 2018), 119–126. https://doi.org/10.1016/j.compositesa.2019.01.029

Zhang, B., Xie, F., Shamshina, J. L., Rogers, R. D., McNally, T., Wang, D. K., Halley, P. J., Truss, R. W., Zhao, S., & Chen, L. (2017). Facile preparation of starch-based electroconductive films with ionic liquid. *ACS Sustainable Chemistry and Engineering, 5*(6), 5457–5467. https://doi.org/10.1021/acssuschemeng.7b00788

Index

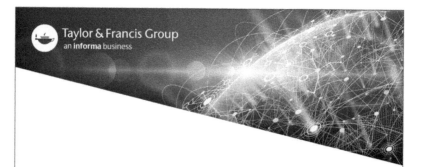